"An indispensable guide to building things, fixing things, and being a better, bolder version of yourself. If you've ever wondered how to change a tire, fix a running toilet, construct a birdhouse, or just feel a little braver in your own skin, this is the book for you. It's accessible, inspirational, and actionable—and will help you understand why so many girls have benefited so much from Emily Pilloton and her mission to get us all fearing less and building more."

—MELINDA GATES, **PHILANTHROPIST AND AUTHOR OF *THE MOMENT OF LIFT***

"When I started this book, my house had a toilet that hadn't flushed well for months. Halfway through *Girls Garage*, a delightful guide to all things construction—I realized—holy cow, what am I waiting for? I can fix my own toilet! *Girls Garage* is more than a guide book, it is a clarion call for girls (of all ages) to take their destinies into their own hands. Learn to wield a tool, this book says, and you will build not only cool bookshelves, planters, and doghouses, but your own confidence. Encouraged by Pilloton's enthusiastic certainty in my greatness, as well as the many inspirational bios of other female builders, I fixed that toilet in no time. This is a book I wish I'd had as a kid, an ebullient manifesto to the DIY spirit in every girl."

—CAROLINE PAUL, **AUTHOR OF *THE GUTSY GIRL***

"*Girls Garage* is a do-it-yourself book for young girls everywhere. It empowers the next generation of girls to be self-reliant by providing a practical guide on how to use common tools and demystifying common repairs. Along the way, girls hear the encouraging words of strong female role models who all sing in unison: 'You can do this and you are not alone!'"

—LUZ RIVAS, **CALIFORNIA STATE ASSEMBLYWOMAN, 39TH DISTRICT, AND FOUNDER OF DIY GIRLS**

GIRLS
GARAGE

GIRLS GARAGE

How to Use Any Tool, Tackle Any Project, and Build the World You Want to See

By EMILY PILLOTON

Illustrated by KATE BINGAMAN-BURT

chronicle books · san francisco

This book is intended for girls 14 and up, with skilled adult supervision. Even if you're an adult, you should work with a buddy while operating tools and working on building projects. While this book provides some instruction on how to use different types of tools, you should always read your own tool manuals for each specific tool you are using and follow all the safety precautions indicated. The authors and Chronicle Books disclaim any and all liability resulting from injuries or damage caused during the production or use of the projects discussed in this book.

Library of Congress Cataloging-in-Publication Data available.

ISBN 978-1-4521-6627-8

Photo credits:
Title page/Girls Garage: *photo by Emily Pilloton*
Introduction/baby Emily: *photo by Anna Pilloton*
Introduction/welding: *photo by Emily Pilloton*
Introduction/farmers' market: *photo by Brad Feinknopf Photography*
Bibi Amina: *photo by Safiullah Baig*
Patrice Banks: *photo by Girls Auto Clinic*
Tiarra Bell: *photo by Tamira Bell*
Kari Byron: *photo by Kari Byron*
Erica Chu: *photo by Emily Pilloton*
Quetzalli Feria Galicia: *photo by Emily Pilloton*
Tami Gamble: *photo by Texasblewprints*
Jeanne Gang: *photo by Sally Ryan*
Miriam E. Gee: *photo by Sean Wittmeyer*
Evelyn Gomez: *photo by Eric Quintero*
Kay Morrison: *photo by Emily Pilloton*
Allison Oropallo: *photo by Gretchen Gottwald*
Simone Parisi: *photo by Emily Pilloton*
Liisa Pine: *photo by Claire Porter*
Kia Weatherspoon: *photo by A Little Bit of Whimsy Photography*

Manufactured in China.

Illustrations by Kate Bingaman-Burt.
Design by Jennifer Tolo Pierce.
Typeset in Rational.
The custom Girls Garage stencil font was made by Tom Tian of Firebelly Design.
The color story, icons, and textures were made for Girls Garage by Firebelly Design.

10 9 8 7 6 5 4 3 2 1

Chronicle Books LLC
680 Second Street
San Francisco, California 94107

Chronicle Books—we see things differently. Become part of our community at www.chroniclekids.com.

For the women who built the world for me:
Anna, Margaret, and Vivette.
And to the girls who will build the future.
—E. P.

CONTENTS

INTRODUCTION: FEAR LESS. BUILD MORE.

To my sisters-in-building,

This book is about building, tools, and how to make awesome things with your own two hands. But it's also about identifying as female, overcoming fear, learning skills you haven't tried before, and feeling the power that comes from making an imprint on the world.

I love building so much! I think about it all day, do it almost every day, and go through withdrawal when I don't do it. I love the idea that in just a few hours, you can build something from scratch, point to it when you're done, and say, "I built that." Building something means you've changed the world a little bit by bringing something new into it using your hands, head, and heart.

Whether you're young, older and wiser, adventurous or shy, good at math or not, from a neighborhood you love or want to escape, the fact that you've picked up this book means you're already a brave builder girl. Welcome to the club. I want you to love building as much as I do, or at least experience the magic of that feeling when you get to say, "I built that." I want you to take this book, try something new, and walk away feeling stronger and more powerful than yesterday.

I'm the founder and director of a nonprofit program called Girls Garage which I started in 2013 to give young girls the tools to build the world they want to see. I studied architecture, worked in architecture firms and as a furniture designer, and after years of making other people's ideas a reality, and deeply missing working with my hands, I quit. I wanted to build real projects with and for communities—and especially youth—because I have always believed that design and architecture can change the world (even if just the world of one person).

Since 2008, I've taught architecture and construction in public high schools in rural North Carolina and the San Francisco Bay Area. I've designed and built some amazing things with my middle and high school students, such as a 2,000-square-foot farmers' market, micro homes for the homeless, and a school library. And, closest to my heart, I opened Girls Garage, the first ever design and building work space for girls in the country, because I noticed that on construction sites or in the classroom, in a mixed-gender environment, my young female students often acted differently than when it was just a few of us women. (I am guilty of this as well.) We sometimes censored our comments, or gave up responsibilities, even though we knew how to use the miter saw as well as anyone. Or sometimes we would be flat-out told we didn't know what we were doing, tools taken from our hands. There were social dynamics at work that would often limit our ability to soar, and I wanted to tear down those barriers. Above all, I wanted to build a community of fearless builder girls.

Girls Garage is a bright, beautiful 3,600-square-foot workshop in Berkeley, California, where pre-teen and teen girls of diverse backgrounds come together to do audacious, brave things as young builders. In one week, I might get to weld with a nine-year-old, build a toolbox with that girl's older sister, deliver handmade furniture to the local women's center, and host a group of female engineers for a team-building exercise. We weld and cut wood and frame walls, pick locks, screen-print, make our own laser-etched skateboards, and more. The girls who walk through our door after school and over the summer, year after year, have their names on the

wall (literally) and know this is a safe space where their craziest ideas can be brought to life.

Girls Garage is particularly special in four ways.

1/ **It is a community and family that thrives through face-to-face experiences with highly skilled instructors and mentors.** All our instructors are female (a few shared their stories for this book!) and have incredible resumes, with decades of experience in architecture, carpentry, metalwork, education, and leadership. Our girls can look at our instructors and see themselves, and vice versa. We build these relationships as pathways, so girls always have something new to learn as they grow up at Girls Garage.

2/ **The experiences we have at Girls Garage are always rooted in togetherness and giving; we hold the other end of a 2×4 for our teammate, and we build projects that live in the community.** We rely on multiple voices and ideas to make all of our work better. I think many of our girls would say it matters to have me, or any of our instructors or a fellow classmate, standing next to them, helping them discover their power as a builder. I hope this book comes to feel like a stand-in for one of us next to you, a trusted friend to hold up the other end of a 2×4 or remind you that you can do it! Working together also extends beyond our walls: we build projects (furniture, sandboxes, greenhouses, and more) for our neighbors, other nonprofits, and schools. Our greatest service is to use our skills to make tangible change in our community.

3/ **All of our stories and voices matter and are honored—our races, our families, our backgrounds, our limitations, and our fears; in fact, these are what make building so meaningful at Girls Garage.** For example, when we build furniture for the women's shelter nearby, we invite and honor the experiences of girls who have experienced homelessness or domestic violence in their own lives. We hold space for hard discussions, where girls can be vulnerable and voice their questions or challenges with identity, mental health, race, family, and relationships. We hold these stories for each other as we build. Our stories are not only important, they make our work more personal and meaningful.

4/ **Girls Garage is also a physical place in the world, with an address and walls.** I have always loved architecture because it says, "This space is ours to work in together as a community." Especially for girls and women, having our own spaces can create a sense of independence and identity. In her coming-of-age book *The House on Mango Street,* Sandra Cisneros describes this female desire for space:

Not a flat. Not an apartment in back. Not a man's house. Not a daddy's. A house all my own. With my porch and my pillow, and my pretty purple petunias. My books and my stories. My two shoes waiting beside the bed. Nobody to shake a stick at. Nobody's garbage to pick up after. Only a house as quiet as snow, a space for myself to go, clean as paper before the poem.

Half a century earlier, in *A Room of One's Own,* Virginia Woolf wrote about space as a catalyst for women to make change:

Women have sat indoors all these millions of years, so that by this time the very walls are permeated by their creative force, which has, indeed, so overcharged the capacity of bricks and mortar that it must needs harness itself to pens and brushes and business and politics.

Space matters. And Girls Garage is *our* space.

Sometimes people ask me, "Why is it Girls Garage? Shouldn't there also be a Boys Garage?" I tell them, of course, that building is a powerful experience for all people, especially at a young age. But access to certain spaces has been historically limited for women and, in my opinion, "Boys Garage" has been synonymous with "every garage in America." We girls and women need to create these spaces for ourselves.

All this matters even more right now, and will matter in the near and distant future, as we grapple with a new wave of feminism in which the words "me too" have grown louder as a rallying cry. It is both a scary and an exciting time to be female-identifying. As women and minority women, we're grabbing at anything we can to feel safe, to feel equal, and at the same time, holding one another's hands as never before and saying, "We matter." Girls are underrepresented in science, technology, computer science, engineering, math, architecture, the building trades, and related fields. Some chalk this up to "interest," claiming, "Girls just aren't into that." But we know that "interest" is influenced by factors that are more complex and that what we see as "lack of interest" is actually code for "lack of warm welcome," or outright oppression and discrimination. If women and girls are not welcome in some of these fields, either overtly or covertly, why are we shocked when we see these statistics? It's also worth noting that Girls Garage has had a wait list since the day we opened our doors; I categorically oppose the notion that girls aren't interested.

Having worked with young girls who *do* opt in, in huge numbers, who wield welders and power tools and are always asking for more, I can tell you without question that girls are hungry. Girls want to know how the world works, to negotiate technology and tools, and to build their own future. I hope this book gives every girl a few more tools to make that happen.

I acknowledge and am sensitive to the use of the word "girl." "Girl" is binary—it implies you are either a boy

or a girl. It can also come with a negative connotation, like when someone says you "throw like a girl." I hope this book resonates with anyone who identifies as a girl or with the more female side of the gender spectrum as well as non-binary and gender non-conforming youth. And let "girl" be read as a categorically positive descriptor, because girls are awesome and hold so much power. I also hope that brothers and fathers and friends of all kinds will truly hear the messages in this book and support the girls who use it.

I should also tell you I'm a biracial woman—my mother is Chinese and my father is French—who grew up in a predominantly white community. This is important because I think lots of young girls feel in some way like they don't belong. I spent much of my childhood trying to fit in, to have a cuter little button nose, to have slightly less slanty eyes.

Building and making was the first activity I discovered that helped me make sense of the world and how I fit into it. Building taught me that it was okay to be "both": Chinese and French, a nerd and an artist, afraid and brave. The first toy I remember obsessing over was the Quadro, a German building set that included PVC pipes, elbow joints, and small plastic fasteners you could assemble into inhabitable houses and fort structures. My father took me to open houses nearly every Sunday, and we critiqued and discussed the homes' floor plans and materials. And then we'd go home and draw blueprints of our own home, or my room, which I'd redesign nearly every month. I loved spending this creative time with my dad and reinventing my own space over and over.

I can point to five specific moments as a young(ish) person that cemented my love of building.

1/ **When I was about eight, I took apart my grandmother's old-school rotary-dial phone because I wanted to figure out what made it ring.** When my grandmother walked in and saw the wreckage all over her carpet, she looked at me and said, "Oh, that's okay, I don't get any phone calls anyway!" In that moment, I had figured out one small detail about how the world worked, and my grandmother gave me permission to keep exploring and following my curiosities.

2/ **In my junior year of high school, I started a Habitat for Humanity campus chapter with the support of my environmental science and economics teachers as faculty sponsors.** As a student group, we traveled around the San Francisco Bay Area, volunteering on house builds. This taught me the importance of starting before you're ready (what did I know about building houses?!) and committing to incremental progress and learning.

3/ **The following summer, I fundraised and registered myself for a summer trip with other teenagers from around the country to travel to Belize and build a public park in a small village.** During one month, we cut down a field of weeds with machetes, hand-mixed hundreds of bags of concrete, and built a stage, a gazebo, and benches to transform the park. This trip taught me two things: my body was physically stronger than I'd ever thought possible, and building could be a form of service within and for a community.

4/ **In graduate school in Chicago, I walked into the fabrication shop and said, "Teach me how to weld."** I learned how to braze and MIG weld in a matter of days—and haven't stopped since. Fusing metal felt superhuman and opened up a whole new world of possible ideas to

build. Welding taught me (literally and metaphorically) how to bring the fire to my work.

5/ **And lastly, in 2010, I taught my first class of high school students in rural North Carolina.** Over one year, we designed, engineered, and constructed a 2,200-square-foot farmers' market pavilion for our small town of Windsor (population: 3,500 people). We cut the ribbon at the opening ceremony, and my students stood in awe inside the building they had dreamed up. We'd proven everyone wrong and built something that changed the face of the town, and it still stands today as a hub for gathering and community. This project showed me that young people are the ultimate dreamers and collaborators, and that the most audacious building projects are always possible.

Each moment had a profound impact on how I think and go about building today. It is through building that I grew up and came to love and understand my race, gender, and story. Above everything else, building is a way for me (and you) to have a voice, to exercise power, to be a free and independent woman, and to play an active role in the physical world.

This book is my attempt to help you bring out your brave builder girl, too, so that years from now you can look back on your own life as a builder and pass it on. I should tell you that the use of the word "brave" in this book is an intentional replacement of the word "fearless." Because no one is fearless. We're all afraid of something: the dark, using a miter saw for the first time, or open-water swimming (I'm terrified!). But fear is an invitation to be brave and to grow! Our goal shouldn't be to live without fear, but to acknowledge that fear is unavoidable and to act bravely in spite of those fears. Bravery is something you can practice, something you can choose.

So let's acknowledge that we're all often scared or don't know where to start, and let's get started anyway. Read on for some tips to bring that brave builder girl out of the woodwork.

The best way to start is to start (before you're ready). Lots of people ask me, "Where do I start?" I have no good answer other than "You just do." Your goal is not to hit it out of the park on your first try. Your goal is to start before you're ready, wherever you are, and make small improvements every time you do it again.

Acknowledge your fear, and then tell it to go jump off a cliff. Sometimes when I'm teaching welding to a first-timer, instead of pretending not to be scared, we just say out loud, "I'm scared." And then we tell our fear, "Get outta here!" Fear is normal and healthy, and so is your innate bravery to act in spite of it.

Remember that as a girl, you are MADE FOR THIS. Girls and women are purposeful, self-reflective, and good at negotiating complex problems. Building requires all these skills. Do not let anyone tell you that building is just for boys; as women, we are already wired to build beautifully.

Break it down. I love solving crossword puzzles. I stare at all the empty boxes and then pick one clue I know I can tackle and fill in the word. That word unlocks another word that is connected to it, and so on, until the whole puzzle is complete. Building can seem complex, like a giant impossible task (go build a playhouse!), but it can also be broken into smaller pieces that all add up to the whole (just start by cutting your wood).

Use the correct (specific) language. At Girls Garage, "thing" is a bad word; if you say it, you have to do ten push-ups. Language is power, and intentional language tells the world you know what you want. Learn the specific names for the correct objects and tools. Walk into a hardware store and say exactly what you need.

Put some muscle into it. I mean this quite literally. Building is a physical activity. You don't have to have giant biceps, but you may be surprised just how much strength you possess for swinging a hammer or driving a screw or lifting a piece of plywood. Your own body is your first and strongest tool.

Practice the hard parts. I have always been awful at cutting angled wood pieces, and joining them in any geometrically precise way. You'll likely try something and realize it's not your most natural skill. This is the skill you should practice the most!

You are always your harshest critic. Remember that everyone else is just as befuddled and self-conscious as you are. Realizing that we're all imperfect and vulnerable can be a great comfort under pressure! Don't be so hard on yourself, and remember that the people you love are always on your side.

Your identity as a builder matters. Who you are is *not* separate from what you build or why. The fact that you are a young girl of color, or that you're a superstar athlete, or that your parents are divorced, or that you've experienced violence or abuse, or that you're an animal lover, or anything else personal to you, matters. If there are things that are personal to you, use those things as inspiration. For example, if you are an avid supporter of rescued animals, call up the local animal shelter and build them a doghouse. One of my teen girls was nervous about learning the basics of carpentry because her father worked in construction and did not want her to follow in his footsteps. But she was proud of her family and wanted to learn; making her father proud became a motivator for her. Bring your heart and your story to what you create, and scream who you are through your tools.

Go big or go home. If you want to build a house, figure out how you might build an actual house. Don't settle on a cardboard model. Every single project I've ever thought was out of my league was actually just a few leaps away. Of course, you may need small steps to get to that big one, but always keep your eye on the biggest, baddest, most audacious goal.

Find your mentors and ask for help. It's hard to achieve the seemingly impossible alone. The good news is that most people genuinely want to help. Mentors committed to supporting you as a person and as a builder are everywhere! Send emails to people you admire. Ask to be an intern (like Tiarra Bell, page 271). Go to events and meet people. Ask your teachers to share their experiences. Open yourself up to learning from others at every step.

Support your fellow brave builder girls. Invite others in. It's been way too many years (centuries! millennia!) that women have not been invited into certain spaces. Now it's our turn. Be a sister in spirit to your fellow builders, and share what you know generously.

Finish what you start. It can be so easy to walk away from a project that just isn't working. But I encourage you to finish it, even if it ends up being a huge disaster. The act of finishing is always an important lesson and also reinforces the idea that imperfection isn't the same as failure—it is merely a moment to inform your next step.

Whew! That's a lot. So—what's in this book, and how do you use it to accomplish all these things?

SAFETY AND GEAR

Safety First! is, well, first, because it is the most important thing, always! You'll learn about the proper gear you need to build both bravely and safely. **Please read the chapter Safety and Gear (page 21) before building anything or picking up any tool!** Each project also includes a safety checklist. As a rule, always build with a skilled adult, read the manufacturer's manual for all of your own tools, and wear safety glasses and the proper protection for your body and head at all times. **Throughout the book you will also see Safety Alerts and safety-related text in red.**

THE TOOLBOX

In the biggest section of the book, there is an exhaustive collection of more than 175 tools. They are organized according to how you might use them in an actual building project, and in order within each category based on how common their use is. This section will become a great resource as you pause to find the right tool for the job or remind yourself how to use it. Tools are the foundation of any great building project! Tools have a character, with a story worth telling, which you'll find as you get to know one another. There are, of course, many more tools in the world than in this book, but this Toolbox should give you a great start.

I love tools because they increase our physical power in the world. This is not metaphorical but quantifiable! Our ancestors used stones to crush seeds and food against a larger rock. But when they added a wood or a bone handle to that stone, that extended the length of their swing—BAM! The force of their blows multiplied.

Tools also help us understand and alter the world. We can use tools to take anything apart to see how it works (like I did with my grandmother's phone), and also to unite parts into something new.

Tools bring us together. With our magnified power, we need more than one person to harness it. No barn can be raised by one person. We need each other to make tools work, which makes building an important social act. Building is bonding!

There is a tool for seemingly everything! I love that as humans we are wired to think, "I can make a tool for that!" As you'll learn, there are tools and hardware for even slight variations of the same task. You'll discover so many kinds of hand tools, screws, nails, and bolts, you'll wonder if they're all necessary. They are! And you'll learn which one you need at which time and for what job. Matching tools and tasks is like making puzzle pieces fall into place—and it's immensely satisfying.

ESSENTIAL SKILLS

This is a collection of useful strategies and skills that can help you take the leap into building—like how to use geometry to make sure something is "square," or fix your running toilet, or tie four really helpful knots. These aren't projects (they don't have a tangible end product), but they are great starters for fixing and solving problems in creative ways.

BUILDING PROJECTS

Each of these eleven projects is achievable in a short amount of time without too much cost. They range from carving your own wooden spoon to making your own toolbox, etching your own steel ruler to framing a doghouse structure. As a collection, they touch on a range of essential skills and call upon many tools from the toolbox. I also hope they inspire your own project ideas!

PROFILES OF FEMALE BUILDERS

Fifteen women and their stories are the soul of this book, and frankly, I could have included dozens more. You'll see the profiles pop up throughout like cheerleaders encouraging you to keep going! I hope reading about these builder women's lives and journeys will give you a sense of community and remind you that you are not alone. From my own colleagues and Girls Garage girls to an architect in Chicago, a carpenter in Pakistan, and a World War II welder, these women are paving the way and inviting us to join them. I asked each of these women to contribute to this book, and they all enthusiastically said yes because of their belief in YOU.

Now it's time to get started! Be sure to read the safety section first—and then dive in!

GO FORTH AND
BUILD BRAVELY,

SAFETY AND GEAR

I've seen a number of memorable safety signs on construction sites and in woodshops. My favorites are SAFETY RULES ARE YOUR BEST TOOLS! and SAFETY GLASSES: ALL IN FAVOR, SAY "EYE"! Despite the cheesy slogans, safety is absolutely your first consideration in any project.

While building can be a fun and challenging activity for all ages, this book is intended for builders ages 14 and up. It's really important for young people to have a skilled adult with them when using ANY AND ALL tools, and even for grown-ups, it's best to have a buddy. In particular, don't operate any power tools without the supervision and hands-on support of a skilled adult! Some tools should only be operated by adults, as I have indicated in their descriptions. It's also a great practice to wear safety glasses at all times; just put them on and leave them on.

Also, every specific tool is slightly different depending on the manufacturer. You should always read the manufacturer's manual to understand the specific instructions for the tool in front of you.

As for safety gear, I think of it as putting on my armor for battle—my uniform for the awesome work of building. Here is a must-have list of safety gear. Use it as a quick-reference checklist for all your building adventures.

- A skilled and experienced adult builder buddy
- Safety glasses (wear at all times!)
- Ear protection, if using power tools
- Dust mask or respirator, if using saws, sanders, or paints and stains
- Closed-toe shoes or work boots
- Long pants
- Hair tied back
- No loose clothing, hoodie strings, or jewelry
- Short sleeves, or sleeves rolled up to your elbows

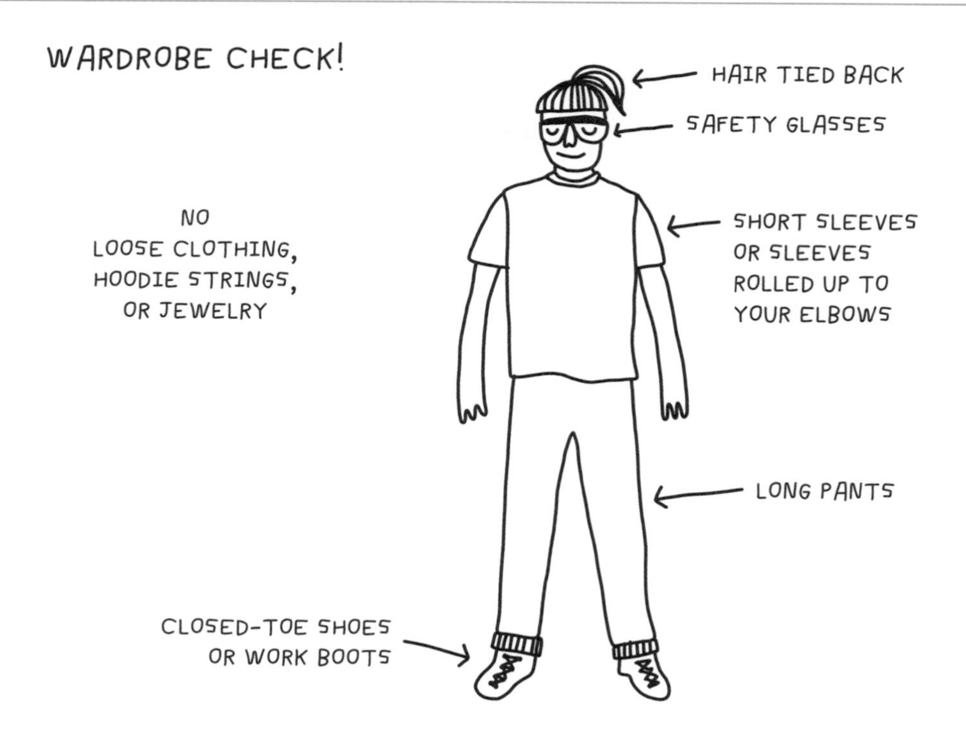

Safety glasses

Buy a pair you love and **put them on before you pick up any tools, then leave them on.** The best safety glasses are the ones you actually wear. You can buy safety glasses in as many styles and colors as sunglasses, so pick a pair that fits your face comfortably.

Safety glasses for building-related work are made from high-impact-resistant plastics that are stronger than your regular glasses. If you already wear glasses, you can buy a pair that fits safely over your existing frames. My favorites are a lightweight pair with clear lenses, a rubber nose bridge, and brightly colored arms so I can find them easily. I also like safety glasses with UV protection (which most have), so I can wear them as an extra layer of protection while working outside, or underneath a welding mask.

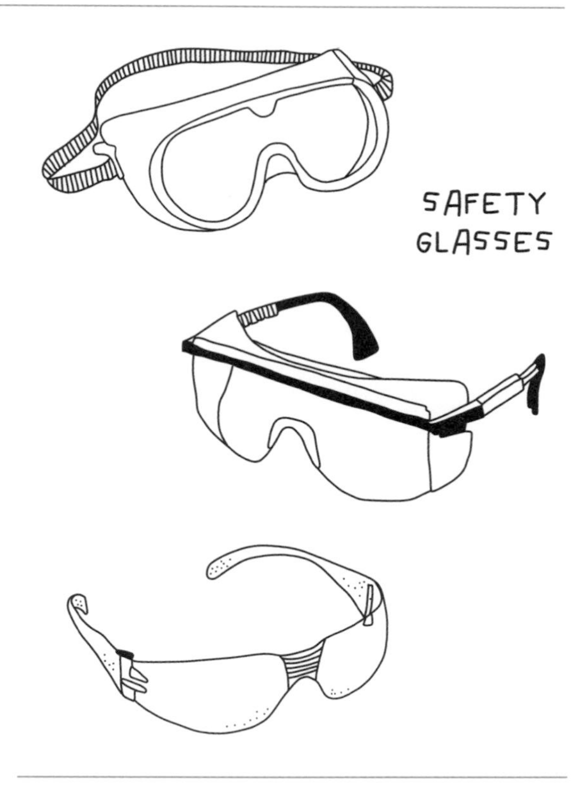

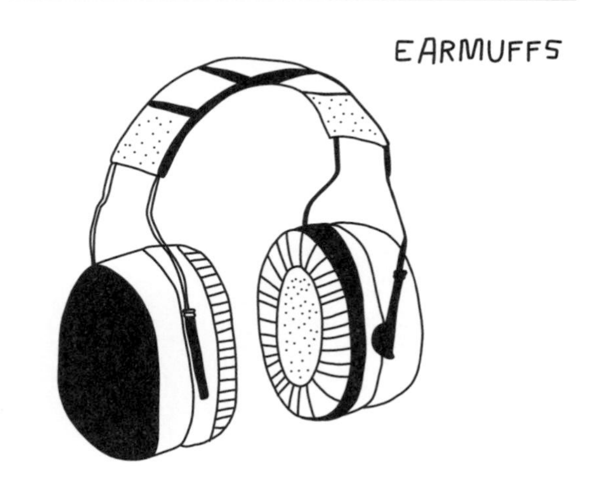

Earmuffs and earplugs

Power tools are loud! It is important to protect your ears from harmful decibels—especially over the course of a project (or a lifetime of working on many projects). Any sound over 85 decibels can be damaging to your ears. For reference, a normal conversation registers at about 60 decibels. A jigsaw, circular saw, and belt sander can all easily reach levels of more than 100 decibels. Just like safety glasses, choose ear protection that is comfortable to you. I prefer earmuffs, but you can also buy earplugs that come on a handy headband or string so you can rest them around your neck when not needed. Headphones or ear buds used for listening to music are not the same as ear protection and should never be used in a shop. **Do not listen to music while using tools or working on building projects;** it's important that you can always hear clearly in case there are changes in the sounds of your tools' operation, or someone is trying to get your attention for safety purposes.

Tool belt

You may not think of a tool belt as a safety item, but when there's a place for every tool and every tool is in its place (on your hip!), your work area will be that much safer. When all your must-have tools are on your body, you don't have to maneuver around the shop as much. An apron-style tool belt is lightweight enough to wear comfortably and holds all the basics. Most tool belts have about four big square pockets for tools or stashes of nails and screws, a hammer hook, and a center pocket for a tape measure.

Hard hat

Wear a hard hat anytime you're working on something overhead, or when there is action going on above you that might lead to something falling on you. Hard hats are made of high-impact-resistant plastics and have an adjustable ratchet strap to fit your head snugly. They also come in a ton of fun colors and patterns. Or cover them in stickers for extra customization!

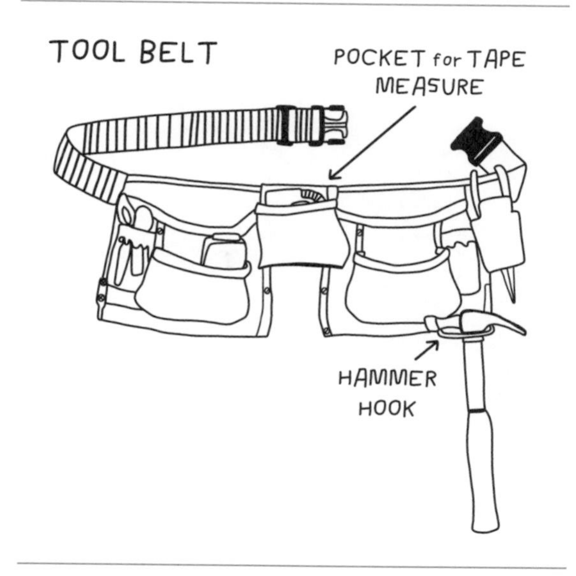

WORK BOOTS

HEAVY-DUTY LEATHER GLOVES

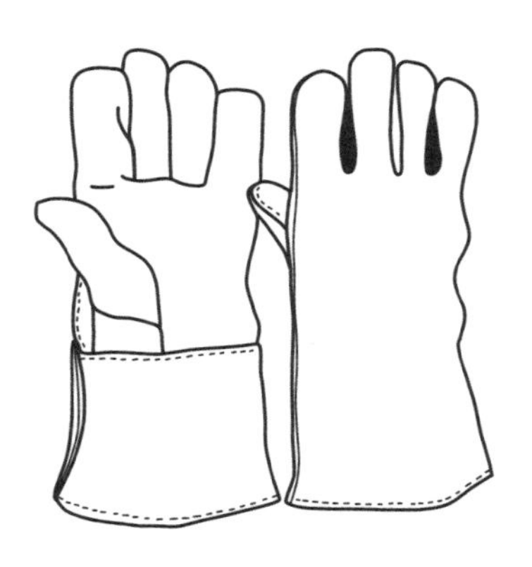

RUBBER-DIPPED WORK GLOVES

Work boots

Protect those little piggies! A sturdy pair of work boots are an absolute must-wear item, and will save you from so many toe and foot injuries, stubbed toenails, and more. The standard for construction sites is a steel-toe boot with a built-in steel plate. I wear slip-on leather work boots that come up high enough to protect me from dangers like small nails or welding sparks falling in. A good pair of work boots lasts for ages and only looks better over the years. (You'll be able to point to certain stains and scratches and say, "That's from when I built my bookshelf!")

Gloves

You actually should *not* wear gloves while using hand or power tools, because your bare hands will always give you the best grip. But when handling sharp or rough materials, gloves offer important protection. A pair of heavy-duty leather gloves and a lightweight pair of coated gloves (usually a latex or rubber coating) are great to have handy. Lightweight dipped or coated gloves are thin and flexible enough for your fingers to move freely and grip objects, so use them for tasks like carrying plywood. For jobs where you need more protection from sharp objects, like metal shards, use the heavy-duty leather work gloves. Also, there are special gloves made for welding. Make sure to wear them when welding and handling just-welded metal!

Face shield

A face shield makes you feel like an instant hard-core builder. Use a face shield when you need more face protection than safety glasses alone can provide but still need total visibility. Anytime you have projectile sawdust or wood shavings (like when using a lathe) or sparks (like from using an angle grinder), a face shield is a great choice. You can get them with clear or colored replaceable masks.

Respirator and dust mask

I had asthma as a child that was aggravated by certain junk in the air. Whether you're prone to respiratory irritation or not, wearing a respirator or dust mask is an important way of protecting your lungs from toxic or aggravating substances. You can use soft dust masks with an elastic band for sawdust-heavy work, or a respirator with replaceable cartridges that protect specifically against dust, particulates, or chemicals.

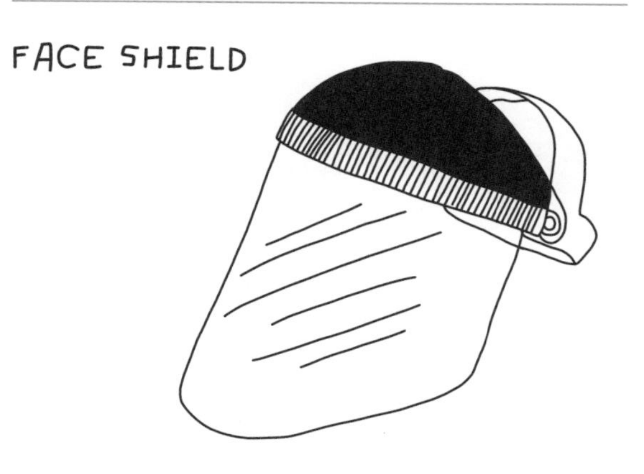

FIRST AID KIT

Hair ties

Dangling tendrils are a bad idea around power tools, spinning blades, and sanders. Always tie your hair back from your face before building. Aren't we glad scrunchies are back in fashion?

First aid kit

Make sure to have a fully stocked first aid kit available and within reach! If an injury occurs, bandages, gauze, tweezers for splinters, and antiseptic wipes are essential.

Other important safety precautions

I can't stress enough the importance of a clean work space! Keep your work area organized, and clean up debris and dust as you work. A dusty floor can be a slipping hazard. On your work table, keep all tools organized and be aware of your power cords, unplugging any tools that are not in use. Maintain ample space for you and your work buddy to move around safely.

Lastly, it's always a good rule to do one thing, with one tool, at a time. The more tasks you're trying to juggle, the more likely you are to forget something and end up hurting yourself. Take your time and move smoothly.

EVELYN GOMEZ

Engineer and STEM Educator

Sylmar, California

When I met Evelyn, she told me that as a Latina high school student (and valedictorian) growing up in Los Angeles, she wanted to go to the Massachusetts Institute of Technology simply because it seemed impossible, and she was drawn to the impossibly hard.

After earning her degree from MIT in aerospace, aeronautical, and astronautical engineering, she went on to earn a master's degree in education from Harvard and a master of science degree in aerospace, aeronautical, and astronautical engineering from UCLA. Her area of expertise in graduate school was space hardware design, rapid prototyping, and manufacturing.

Needless to say, Evelyn has quite the impressive resume. But never satisfied with what's expected, she brought her engineering skills to young girls from her neighborhood. As the former executive director and a current board member of DIY Girls, she has inspired and supported hundreds of young girls onto paths just like hers. She has taught engineering and electronics and tackled wicked problems with her girls, helping to set them on their own "impossible" paths.

"I first discovered making at a very young age. My dad is a jack-of-all-trades and would fix everything and anything around the house, doing electrical, automotive, and mechanical work. I grew up watching him use all kinds of tools and became very interested in using them myself. However, he never let me help, partly because I was a girl and partly because he thought I was too young. But he did instill in me an interest in learning to do things myself. My first mentors were my parents. They are both immigrants from Mexico who came to the United States to give their family a better life. They taught me the value of hard work from a young age and made many sacrifices so I could make a living using my brain, not just my hands.

"While deciding which college to attend, I visited many schools, but one really impressed me—MIT. It was filled with people experienced in making and building incredible devices. But because I didn't have the same experience, it was intimidating, and I was terrified of being discovered as a fraud and an imposter. I also knew, though, that by going to MIT, I would be pushed way beyond my comfort zone to become a better version of myself. I was accepted, and decided to put my fear aside and go!

"At MIT, I had access to machine shops, woodshops, and glass-blowing studios. My classes and research in aerospace engineering required

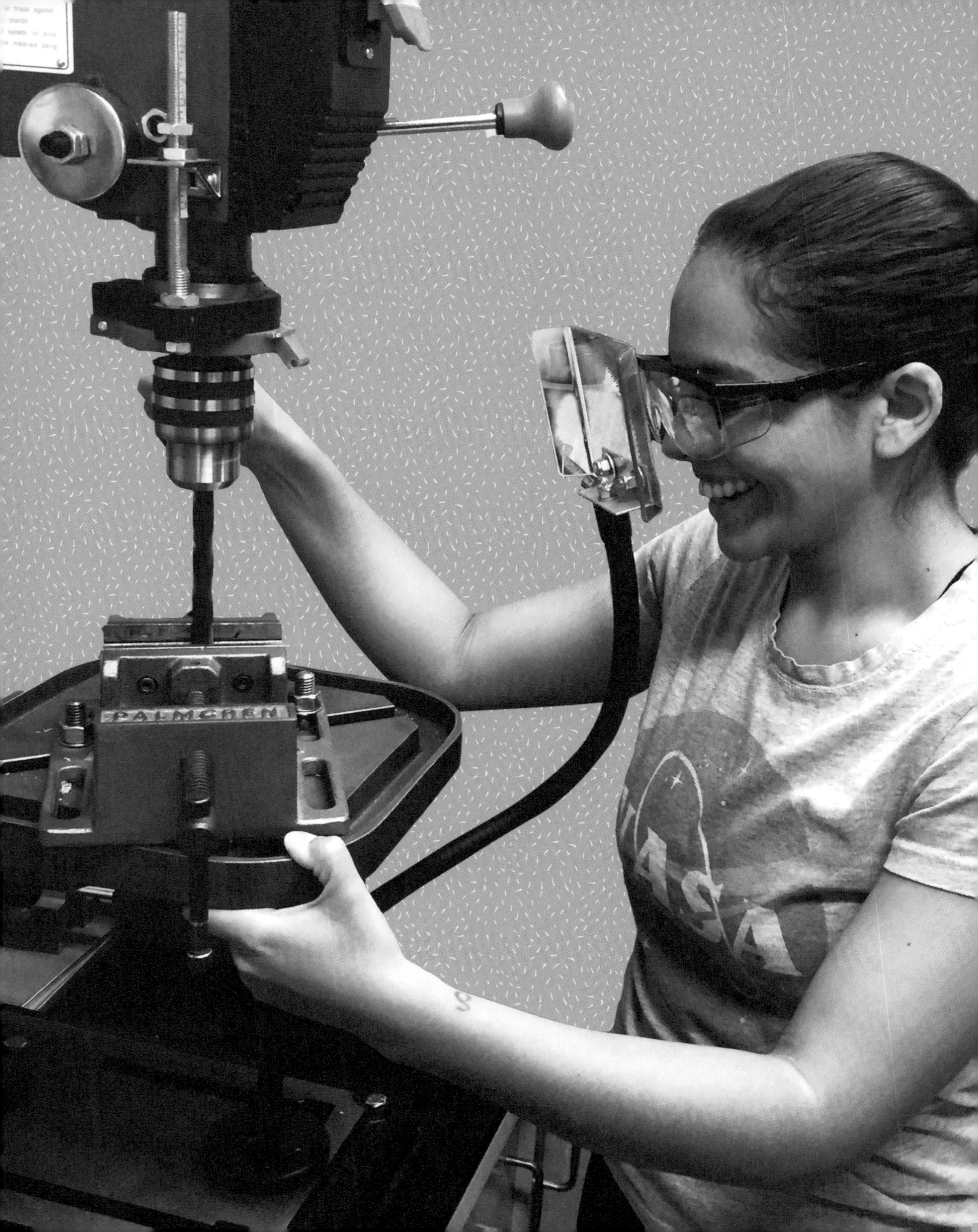
PALMGREN

that I learn my way around the lab. My very first project was designing and building a water bottle rocket, and it only got more complex and way more fun from there! To this day, one of the projects I'm most proud of is an autonomous aerial vehicle I built during my senior design class at MIT.

"At DIY Girls, I loved connecting young girls interested in STEM careers to experiences that would help them realize their dreams of becoming engineers and scientists. One of my favorite projects was building a solar-powered shelter for homeless people in Los Angeles with twelve of our high school girls. We learned how to use a sewing machine and program Arduinos. One of my mentors is the founder of DIY Girls, Luz Rivas, who continues to connect me with resources and support my personal and professional ambitions. I couldn't have accomplished my goals without the support and mentorship of great women like Luz.

"As someone who has gone to some of the best educational institutions in the world, I see myself as a role model to other young Latinas in my home community, the northeast San Fernando Valley. My community has a rich Latino culture that is celebrated with great pride. Unfortunately, this same community is plagued by gangs, violence, and drugs. Many of our families live below the poverty line and typically attend underperforming schools. At my high school, the math and science achievement is below 50 percent, as measured by a passing score on standardized tests. Our schools lack basic resources, and the administrators are just starting to understand that we need to teach subjects like engineering and computer programming to prepare kids for twenty-first-century careers.

"I want to show young Latinas that anything is possible. I was not supposed to succeed. I was expected to fall into the same traps as many of my peers—dropping out of high school or getting pregnant at a young age. By telling my story, I share a story of hope with other young women who feel trapped by the stereotypes or their circumstance. As a female builder and maker, I help other young women consider traditionally male-dominated fields of study. I hope to show them that, together, we can change the face and future of technology and engineering.

"My advice for young girls: *Jump right in*! You don't have to wait for permission or wait until you're in college to be an engineer. Engineering is all about solving humanity's problems. Making and building gives everyone, especially young people, the ability, the confidence, and the tools they need to address problems that affect their own communities. Start now by coming up with simple solutions to everyday problems. Everyone has something to contribute!"

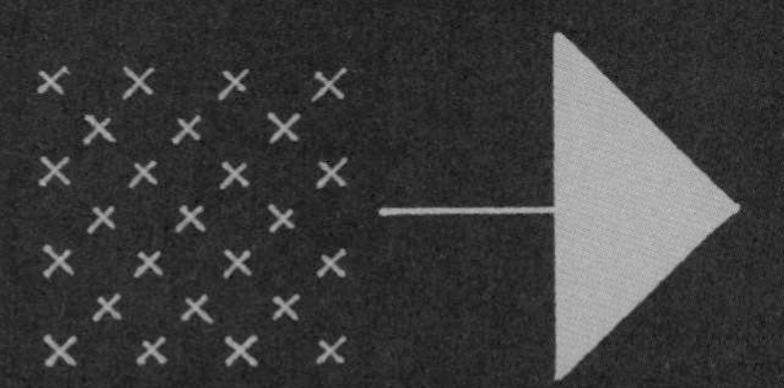

THE TOOLBOX

Now that you're well versed in safety gear, it's time to meet the materials and tools you'll use to actually build something. I've organized the major sections—Building Materials; Hardware; Measure, Lay Out, and Secure; Hand Tools; Saws; Power Tools; Sanding and Finishing; Metal Tools; and Clean Up—to mirror the logical progression, or workflow, of what you have to think about and then do in any building project.

Once you know what you're going to build, you'll first buy (or scavenge) your materials and hardware. Then you'll measure, lay out pieces and parts, think about connections, and select and use your tools, increasing in their complexity. Saws are their own category (outside of other hand tools and power tools) because cutting your materials is a feat in itself—be proud of the work you do with them. Activities like sanding, finishing, and cleanup are last because they come later in any project.

Within each section, you'll get to know the tools, starting with the most commonly used. For example, under Hammers, the claw hammer is the most common, so that's where we'll start.

The tools included here are mostly analog, low-tech, and tend to lean toward carpentry, metal, and other basic materials. Even as 3-D printing and all kinds of awesome technologies emerge, I see these low-tech tools as more directly gratifying when learning to build. Holding a circular saw, feeling its vibration as it cuts through plywood, and getting sawdust all over your jeans is a more instinctive and present experience than watching the robotic arm of a CNC (computer numerical control) router do the work for you.

For this reason, the analog tools included are a direct path to your growth as a builder! Working with these tools and materials will make you feel strong and will help you learn to negotiate problems in the real physical world.

I also acknowledge that this book in no way represents the entire universe of tools! I've made some decisions about what to include and what not, so please venture beyond the boundaries of this book and explore many more tools in the vast world of building.

Let's build our knowledge!

BUILDING MATERIALS

Over the course of your life as a builder, you'll likely encounter tons of different materials with which to build. Wood and metal are most common, but masonry products, manufactured plastics, and other materials are all in the ecosystem of amazing components that help us make our world. I am constantly curious about materials and how they work—what makes them break, how to take advantage of their strengths and properties, and how to use them in new ways. Here is a short list of some of the most common building materials you might encounter. Keep in mind that this is just the tip of the iceberg!

Lumber

Lumber is a piece of a tree! You might use dimensional lumber (a wood product cut directly from a tree at a specific dimension, like your friendly 2×4) to build a structural frame (think tree house or doghouse), and carpenters exclusively use dimensional lumber to frame the studs of houses or other structures. Two-by-fours are commonly used for house framing, and 4×4s are commonly used as posts for fences, mailboxes, or birdhouses. (Turn to page 274 to make your own!)

Even though you can physically see the rings of a tree in lumber, it can be hard to imagine the whole process from forest to your worktable. Most trees for common lumber are grown on tree farms, so the growth is fast, predictable, and very straight. When the tree is cut down, it goes to the mill to be cut up into lumber. Just like everyone has their own technique to slice an apple, there are different techniques for cutting a tree trunk into lumber.

Plain-sawn, also known as flat-sawn, is the most straightforward (and inexpensive) way to cut a log—straight across into parallel slices. A plain-sawn piece of lumber has grain that makes tall arching shapes across the face of it. Plain-sawn lumber is very common, especially for wider lumber, like 2×10s, but because of how it is cut across the tree's rings, it can warp or cup upward as it dries or ages.

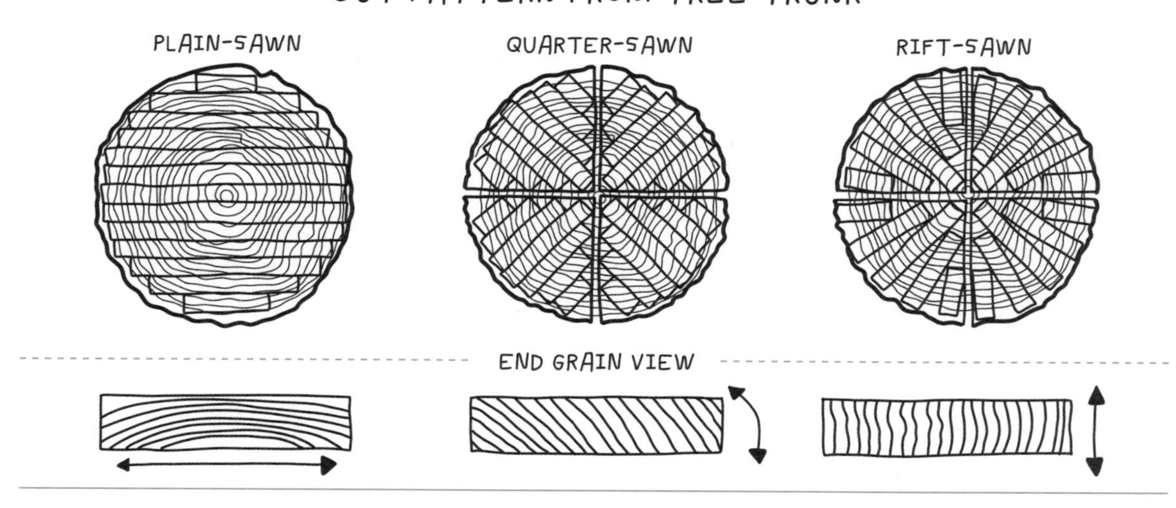

DIMENSIONAL LUMBER

NAME	ACTUAL
1" X 2"	3/4" X 1 1/2"
1" X 4"	3/4" X 3 1/2"
1" X 6"	3/4" X 5 1/2"
1" X 8"	3/4" X 7 1/4"
1" X 10"	3/4" X 9 1/4"
1" X 12"	3/4" X 11 1/4"
2" X 4"	1 1/2" X 3 1/2"
2" X 6"	1 1/2" X 5 1/2"
2" X 8"	1 1/2" X 7 1/4"
2" X 10"	1 1/2" X 9 1/4"
2" X 12"	1 1/2" X 11 1/4"
4" X 4"	3 1/2" X 3 1/2"

FUN FACT!

"Timber" is what you call the tree after it has been cut down (the log), but before it goes to the sawmill and is cut into pieces. Once it is milled it is called lumber.

Quarter-sawn lumber starts with the log being cut into quarters along its long axes. Then each quarter is cut into parallel slices. Quarter-sawn lumber usually has long lines of grain on its face, with some marks or flecks.

Rift-sawn lumber is the cream of the crop, the most expensive, and also the rarest kind of cut. Logs are cut into quarters, like quarter-sawn lumber, but rift-sawn pieces all point perfectly toward the center of the log, like a sunburst. This cutting pattern gives you pieces that have parallel grain along the edge of your lumber and consistent lines of grain on the face. The downside to rift-sawn lumber is that it has a lot of waste, because pieces are not cut parallel to each other.

Dimensional lumber comes in standardized sizes (see table) like 1×4, 2×4, 2×6, 4×4, and others. These dimensions refer to the width and height of the lumber in inches, and the length is determined by you and your project. So, you might have an 8-foot-long 2×4 and a 4-foot-long 4×4.

One extremely sneaky fact about dimensional lumber is that a 2×4 does not actually measure exactly 2 inches by 4 inches! That would be too easy. For example, a 2×4 in real life measures only 1½ inches by 3½ inches. If you're wondering why on earth lumberyards would do this, here's why: When the lumber is first cut from the log, it *is* cut at 2 inches by 4 inches (or whichever regular dimension it is supposed to be). But because logs are often still wet when they are milled, the cut lumber dries and shrinks. Once dry, it also goes through a planer tool that trims it to the 1½-inches-by-3½-inches dimension. So no, a 2×4 is not actually 2 inches by 4 inches, but you *can* depend on it always being 1½ inches by 3½ inches.

Plywood

Plywood is the most common "sheet good" in the wood family. It is a great choice for projects where you have to cover a large area with a wood surface. For example, you might use dimensional lumber to build the frame of a doghouse (see page 300 to build one for your dog!), but you would likely use plywood as your walls to sheathe the frame (and then paint it with paw prints and your pup's name).

Plywood comes in large sheets (a "full sheet" measures 4×8 feet) and is made of multiple thin layers of wood. "Ply" means layer, just like when you buy 2-ply paper towels. The purpose of the plies is to make the entire sheet stronger. The thin plies of wood are cut from tree trunks using one of a few different methods—my favorite being a rotary cut, where the tree trunk is spun like a paper towel roll and a thin layer of wood is shaved off as it unrolls. These thin sheets (plies), usually 1⁄32 inch to 1⁄8 inch thick, are then laid on top of each other, glued, and pressed.

If you look at the cross section of a piece of plywood, it almost always has an odd number of layers. Common plywood has 3, 5, 7, or 9 plies, but really nice plywood for fine furniture projects can have 13 or more plies. The plies are laid down in a specific order, with the wood grain running in alternating directions. Having the grain of one layer running perpendicular to the layer above and below it makes a stronger bond when glued, and also makes plywood very stable; it won't warp or distort in any direction. Some types of plywood have finished face layers that are smoother than the interior plies.

Common thicknesses of plywood are ¼ inch, ½ inch, ¾ inch, and 1 inch, and they come in various grades, depending on how rough or finished you want your edges and face. The highest grade is A, with both the face and the back layers nearly perfect and without knots or other imperfections. Grade A/B has a front face (A) without flaws, but its back layer (B) may have knots or discolorations. Grades go down from there, with C/D being a lower-end plywood, which likely has some knots and discoloration, but is cost-effective if you aren't using it for a highly finished purpose (like flooring or roofing that you'll cover with something else).

Plywood can be made from a variety of woods, but the most common kind you find at a hardware store or lumberyard is likely birch, fir, or pine.

PLYWOOD

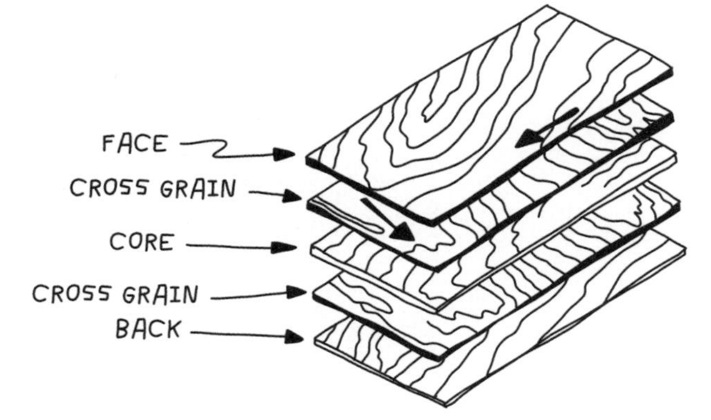

MANUFACTURED WOOD BOARDS

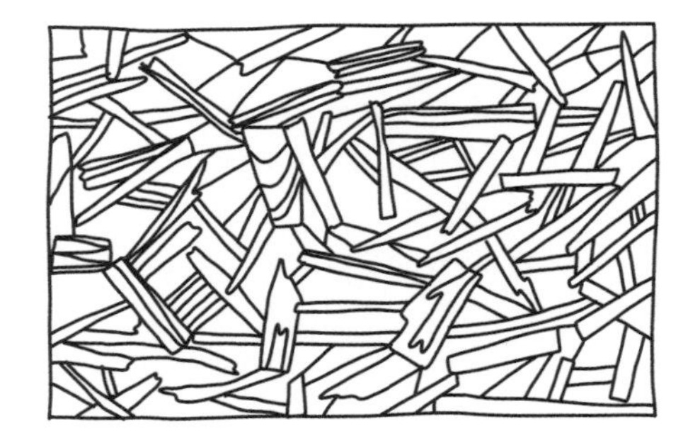

OSB (ORIENTED STRAND BOARD)

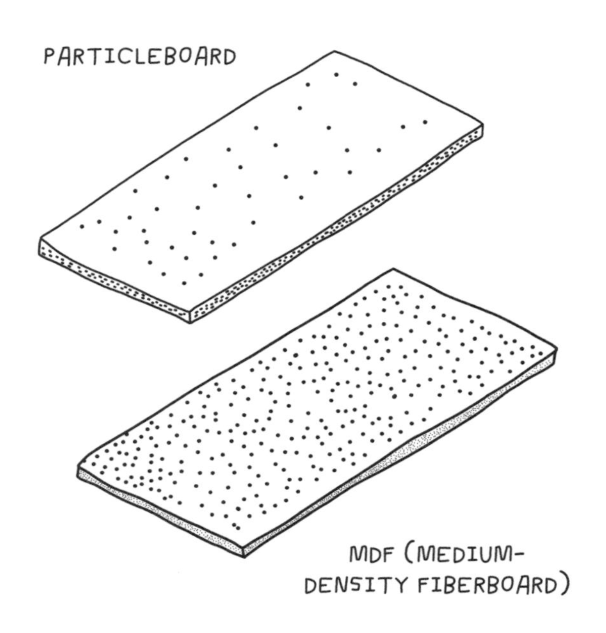

Manufactured wood boards

Oriented strand board (OSB) is made from flakes or strands of wood glued and pressed together with a resin glue. The wood is often spruce or pine plus other hardwoods, which are shaved off tree trunks in small strands. The strands are then laid on top of each other at various angles, which makes the board stronger. Using a sprayed adhesive, the strands are pressed together to form the board. Because of the varying angles of the strands, OSB doesn't warp or distort under different weather conditions.

Surprisingly strong, OSB is the go-to wood panel product used for construction, particularly as a wall layer in wood-framed houses. OSB is also used as a layer of flooring before wood or carpet goes on top. It comes in four grades for different applications and weather conditions, ranging from 1 (light duty for dry conditions) to 4 (load-bearing, heavy duty, or for humid conditions). It's also just cool to look at; at Girls Garage, we use OSB as the backboard for our tool wall, where all our tools have a particular place.

Particleboard is an inexpensive wood product made from leftover shavings, particles, and wood chips that are glued and pressed together. A lot of the affordable furniture we probably all have in our homes, schools, or offices is made from particleboard. Particleboard can be covered in a layer of wood veneer or plastic laminate to give it a nice surface. Most particleboard is made from wood byproducts and pressed using a resin glue. If you look at a cross section of particleboard, you can actually see the individual particles. Particleboard is not as dense as medium-density fiberboard, but has smaller particles than OSB.

Medium-density fiberboard (MDF) is similar to particleboard, but much denser and made from finer wood particles. It is made from small wood fibers—almost dust—mixed with an adhesive and pressed into boards. Because MDF uses smaller particles, it is much denser; when you look at a cross section, it looks much more solid and consistent. However, because of the amount of glue used to make it (and because that glue can contain bad chemicals, such as formaldehyde), **wear a mask when cutting and handling MDF.** Some manufacturers do make a formaldehyde-free MDF, so if you can find it, use it!

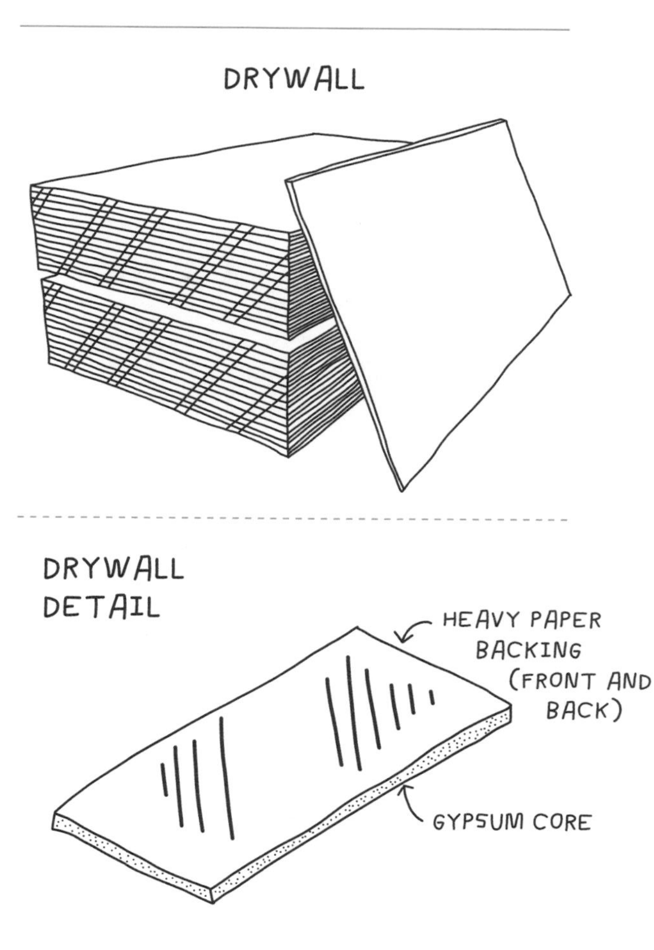

Drywall

Centuries ago, most interior walls of homes and other structures were made of plaster, which was expensive, and took a long time to mix and dry during construction. Drywall emerged as an easy-to-install sheet that worked as a substitute for plaster. By the mid-1900s it was the go-to wall covering for its easy installation and affordability.

In most wood-framed structures, drywall is hung in full sheets that measure 8 feet tall by 4 feet wide. This dimension allows them to be attached consistently with drywall screws to the wood studs behind them, hiding the insulation, plumbing, and electrical wiring behind the sheets.

Drywall consists of a gypsum core with a layer of heavy paper backing on either side, making a sort of gypsum sandwich. Gypsum is a mineral that can be formed into durable sheets (and can also be found in blackboard chalk). Unlike plaster, drywall sheets can usually be hung quickly, then seamed and painted soon after.

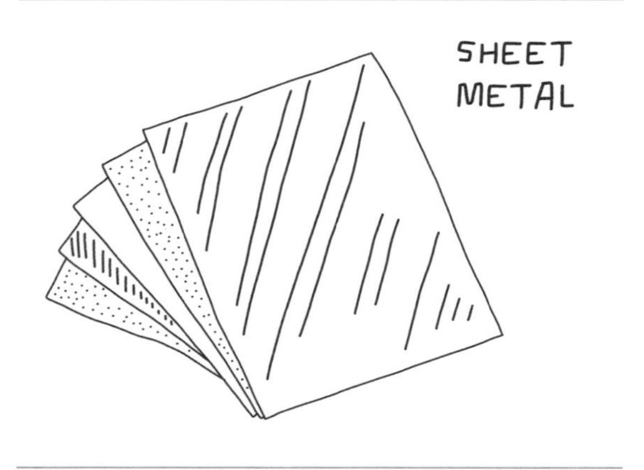

STANDARD METAL STOCK

FLAT BAR

ANGLE

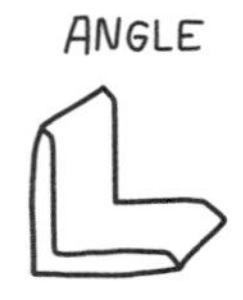

ROUND

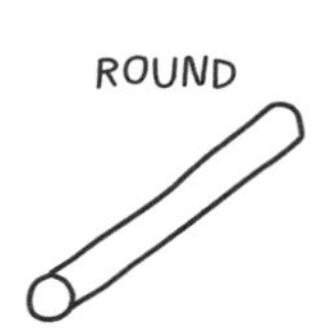

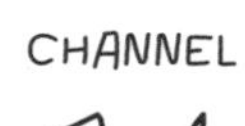

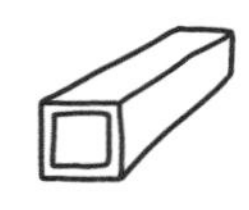

PIPE/ROUND
TUBE

Sheet metal

Sheet metal is so versatile—look around right now; I bet you can find more than a few items made from it. You can buy sheet metal of all types (aluminum, steel, stainless steel, copper, brass, and more), and in different sheet sizes and thicknesses. The thickness of sheet metal is measured in "gauge," with the most common gauges ranging from about 6 to 20. The larger the gauge, the thinner the sheet, which sounds a little counterintuitive. Sheet metal is made by heating and rolling metal between two compressing rollers at a certain thickness. You can cut, bend, stamp, form, and shape sheet metal into so many different things. Your car doors, a metal toaster, file cabinets, jewelry, and about a bazillion other products are made from sheet metal!

Standard metal stock

Metal you might buy from a hardware store or steelyard comes in long skinny lengths in various shapes and profiles, sometimes referred to as "sticks." If sheet metal is like plywood, then standard metal stock is like a 2×4, with a specific width and height dimension and a long length. The most common kind of steel stock you'll likely use is called mild steel, which is formed through a process of hot-rolling, wherein it is heated up and shaped into its respective stocks. You might also encounter cold-rolled steel, which is basically hot-rolled steel that has undergone further processing at room temperature. Some of the most common metal stocks are flat bar, angle (sometimes called "angle iron"), round, standard I-beam, channel, square tube, pipe, and rebar (for reinforcing concrete structures). You can find these types of stock in common dimensions at hardware stores or get custom-cut dimensions from steelyards. Steel stock is common for welding and can easily be cut using your abrasive saw or an angle grinder and then welded together to make awesome frames for furniture and more! Other metals such as aluminum and stainless steel are also commonly sold in standard metal stock dimensions and shapes.

Threaded rod

Sometimes called "all-thread," threaded rod is like a thick screw with no head that threads all the way down its length. You can buy long lengths of threaded rod and cut them down for specific projects, or buy them in specific lengths. Threaded rod is useful as a substitute for a bolt, where you might need to fasten both ends of the rod with a nut. Like bolts, threaded rods come in different diameters and thread counts.

Cement and concrete

One of my pet peeves is hearing the terms "concrete" and "cement" used interchangeably. They're not the same. Here's the difference: Cement is an *ingredient* in concrete, which is a mixture.

Cement is made from limestone, which is mined from the ground. It is a raw material, and the most common type is referred to as Portland cement.

Concrete, on the other hand, is the actual mixture you see most commonly in the world. Concrete is made from cement, water, and some type of aggregate (like gravel, sand, or crushed stone). You can get premixed dry concrete in bags for various uses: all-purpose, countertop (fine and smooth), and more. Concrete is often poured into a mold, called formwork, which holds the wet concrete in place to set, and is then removed. Often, formwork is made from wood screwed together so it can easily be taken apart. Concrete is heavy (a common cinder block weighs about 35 pounds!) but also very strong. Especially when it is mixed properly and poured with some kind of reinforcing structure, like steel rebar, it is one of the strongest and most durable building materials.

THREADED ROD

FUN FACT!

When I was a kid, I used to press flowers between two pieces of wood. The two pieces of wood were like a sandwich, with threaded rod running through them at each corner so I could tighten wing nuts on either side and squeeze my flower in place.

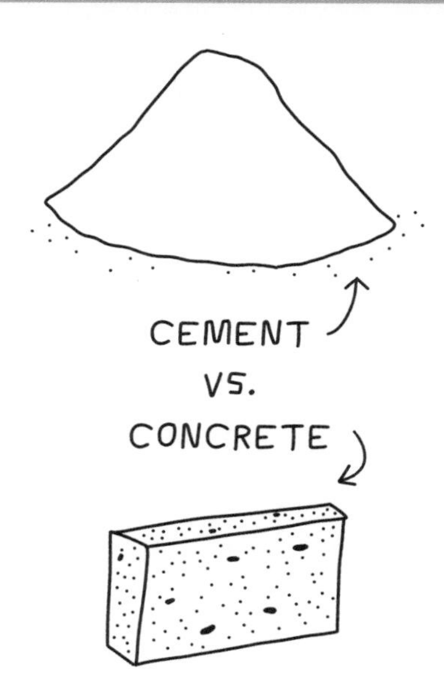

CONCRETE = CEMENT + AGGREGATE (SAND, GRAVEL, ETC.) + WATER

SIMONE PARISI

Girls Garage student

East Bay Area, California

I met Simone when she was eleven years old, as a Girls Garage summer camper! That summer, she helped build a playhouse for the children living at a local women's shelter. Since then, she has been part of Girls Garage every single year. As one of our most advanced builder girls, Simone taught Melinda Gates and Mandy Moore how to use a drill and driver when they visited Girls Garage in 2019! One of my favorite memories of Simone is when she accompanied me to an event hosted by a professional organization that was raising funds for the Girls Garage program. Even though she identifies as an introvert, she got up to speak in front of hundreds of people about her love of building, and every person in the room was inspired by her story. Every day, Simone practices bravery and pushes herself to grow in new ways as a builder and leader.

"As a young child, I loved art class! I loved painting, sewing, drawing, and working with clay. Then in second grade, I started taking a carpentry class after school, and the only thing I remember is that I was the only girl. It was the first time I had worked with wood or used hand tools, and I remember being proud of the work I created but also self-conscious and embarrassed that I was the only girl. After that, I did not continue in the class.

"I didn't start building again until I discovered Girls Garage. I was drawn to Girls Garage because it felt like an inclusive space where I could express my creativity and work with my hands. I had never used power tools before and was interested to try something new and exciting while also giving back to the community. As a Girls Garage builder over many years, I have learned how to use a chop saw, jigsaw, drill, driver, speed square, table saw, and how to weld and draft by hand. These are tools and skills that I never would have picked up and gone out and learned on my own. My absolute favorite tool is the impact driver. I love the feeling of holding the driver, squeezing the trigger, and using my own strength and force to drive a screw into the wood.

"I'm most proud of the 21-foot-long sidewalk bench and seating area that I built with twelve other teen girls at Girls Garage for a local restaurant client. The project combined all of our individual ideas into a complex, huge, and intricate design, and it pushed me as a builder and leader.

"Building provides a creative outlet for me that I am unable to experience at home or school. As a girl who has not always enjoyed or felt confident with math, building has increased my mathematical confidence because you get to use it in totally different (and real) contexts. It also allows me to express my ideas in a different way and bring them into the world. Through building, I have the power to design, problem solve, and create tangible objects that have use in the world.

"I would tell other young girls who want to learn how to build to not doubt yourself or your strength. Do not seek perfection; embrace your mistakes. It is a really important time for girls to learn how to build because we are at a time when girls can shed old stereotypes of what society thinks we should be and do.

"My own dream is to teach younger girls how to build. I want to be able to pass on my knowledge to help other girls use power tools and become strong, confident builders. It is time that girls and women are recognized for their full strength, power, intelligence, and talent. I hope that women will continue to push the limits and show the world that they are capable of doing anything. I hope that people come together to pave the way for new generations of young girls who will be fearless, unapologetic, and who know how to use a chop saw."

3 helpful kinds of tape

Duct tape is something of a fix-all and an icon of any DIY work. Since the early 1900s, duct tape has been used for everything from wrapping power cables to fixing shoes and assembling steel bridge cables—and for good reason. Duct tape can be used for pretty much anything. It is made from a cloth mesh coated in adhesive on one side and often a reflective pigment on the other.

While duct tape is every builder's favorite, the one use it's not great for is high-heat situations because the adhesive can get gummy and become hard to peel off. Duct tape has a slight stretch and can be torn easily.

Painter's tape, commonly colored green or blue, is basically masking tape with an adhesive that peels off easily without creating any surface damage or peeling off paint. It is a great light-duty tape for woodworking and painting.

I've used painter's tape to create awesome paint patterns by covering up the part I want to protect and painting everything else. You can also use painter's tape to tag or mark your work pieces, as it comes off easily without leaving residue.

Painter's tape does not stretch, but can be torn easily because it is made of paper.

Electrical tape is made of plastic or vinyl. I love electrical tape mostly for its color selection. It's great for taping or marking metal pieces, wrapping tool handles, or for its original purpose—insulating electrical connections.

Electrical tape tears easily and can be written on with permanent marker for labeling.

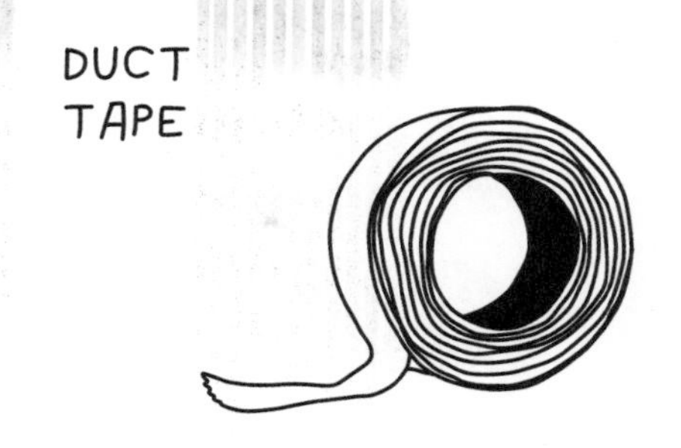

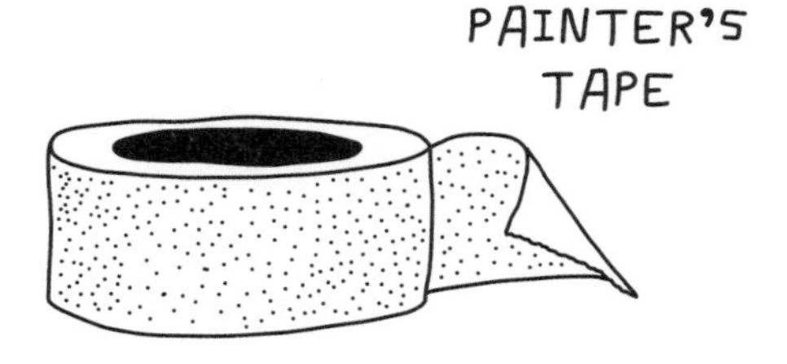

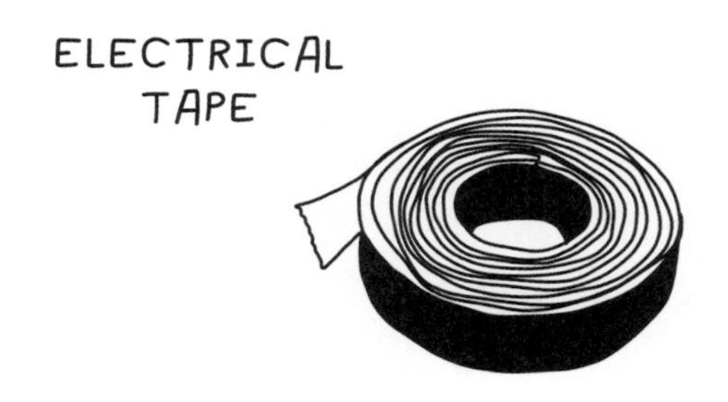

FUN FACT!

Duct tape is also sometimes called duck tape because of its original material—duck canvas.

GLUES

WOOD GLUE

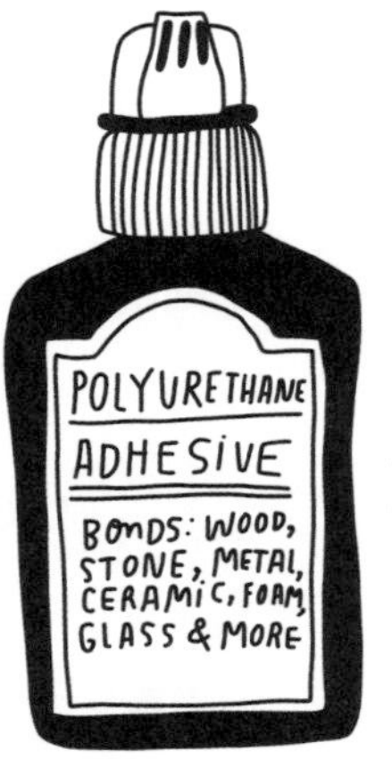

POLYURETHANE GLUE

SPRAY ADHESIVE

3 great glues

Wood glue is like white craft glue for the woodshop. It's usually a pale yellow color and has different use options—one for interior, and one that is waterproof, for outdoor use.

Use wood glue to join pieces of wood, but you'll need to clamp them together or hold them in place to dry. Wood glue is also what you use for a dowel-joined or biscuit-joined wood joint (see page 72). You can use wood glue to press thin layers of wood (veneer) together, like if you were making your own skateboard (which, while not a project in this book, is an excellent DIY project, on which many detailed books have been written!). Wood glue only works on porous materials (like wood or paper products) and does not work on metal, glass, or plastic.

Polyurethane glue is activated by moisture. It expands as it dries, dries very hard, and is waterproof. Polyurethane adhesive can bond to almost anything, including wood, metal, glass, and plastic, so it's great for fix-it jobs where you aren't too concerned with seeing the dried glue residue.

Spray adhesive is the way to go for lightweight materials or large surface areas. Spray adhesive is a multipurpose spray glue that works great on paper, fabric, cardboard, or other lightweight materials, especially when you have to spray a large area. You can get spray adhesives that are permanent, or a version that allows you to remove and reposition your materials (like a sticky note).

Polyurethane

For wood and metal projects you want to protect from staining, rusting, discoloration, or general damage, a polyurethane sealant is your go-to clear-coat product. Polyurethanes can be water-based or oil-based, with high-gloss or matte finishes, and they even come in different stain colors. One of my favorite polyurethane products is Wipe-On Poly, which you put on a rag and hand-wipe onto your project. Other polyurethanes brush on and come in larger paint cans, or convenient spray cans. For most projects, I go with a satin-finish water-based polyurethane, but for heavier-duty outdoor uses, an oil-based product is a better choice.

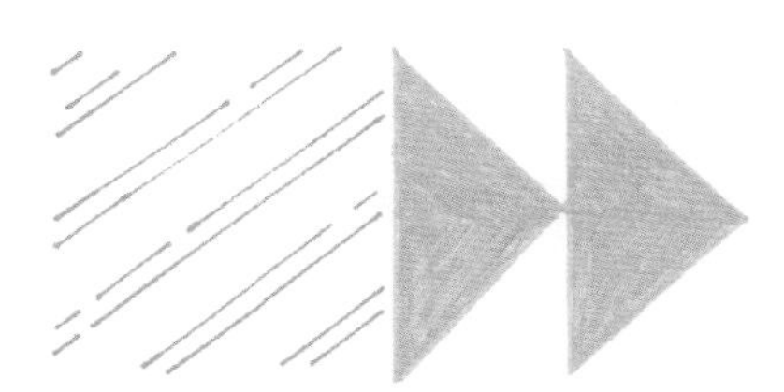

POLYURETHANE

HARDWARE

I clearly remember the first time I walked down the hardware aisle without needing help. I knew the exact size and type of screw I needed and where to find it. I was building a wave-shaped bench with a steel frame that needed to be anchored to a concrete floor without leaving giant holes once it was removed. So, when the store employee asked me, "Sweetie, do you need help?" I responded confidently, "Nope, just grabbing some 1¾-inch Tapcon masonry screws, thanks!" Knowing exactly what you need and how to name it, in life and in building, is powerful. Particularly in the hardware aisle—a sea of small metal objects—this can be particularly empowering. Just say, "1¾-inch Tapcon masonry screws" out loud a few times—and say it like you mean it!

When talking about tools and building, the word "hardware" generally refers to the metal fasteners or other devices that help us hold items together. Screws, nails, and bolts are among the most common kinds of hardware. There are so many kinds of hardware, you'll likely wonder, "How can there possibly be this many items that look mostly the same?" But, as you'll find when working on a project, there are times when you definitely need screws, not nails, or flathead wood screws you can countersink instead of round-head screws.

So having just surveyed our building materials (in the previous section), let's meet the wide world of hardware and, most important, learn when and why to use each type to fasten those materials into your projects.

Screws

Screws are a reliable fastener for almost all projects. They are incredibly strong and versatile—you can find a particular screw for virtually every job you can dream of. Screws are almost always made of steel, have great tensile strength, and are easy to remove at a later date with a screwdriver or an impact driver. Screws don't require any extra adhesive (like you might use with brads or nails), and they generally "bite" into wood with more grip because of their threads. In general, screws have greater strength than nails in squeezing materials together (strength along the axis of the screw), but weaker "shear strength" than nails, meaning they are more likely to break or snap when put under sideways pressure.

Screws and bolts are similar, but screws have threads that "mate" with corresponding internal threads, either by biting into the wood and creating them or threading into a preformed threaded hole. Bolts, on the other hand, have threads designed to connect to a corresponding nut.

Tips for use

To install screws, use either a handheld screwdriver or an impact driver (sometimes called a hammer drill or an electric driver; see page 152). This can also be a reason to use screws over nails: Because nails require the swing of a hammer, screws are ideal for smaller spaces where you don't have room to swing. Some screws work best with a "pilot hole," a small hole you predrill before installing the screw, especially when you are driving the screw into a small piece of wood and don't want to split the wood.

Parts of a screw

A screw has six basic parts that vary, depending on the type of screw and what it is used for: the head, drive, shank, shaft, threads, and point. The shaft is the whole length of the cylindrical part of the screw, while the shank refers only to the smooth, unthreaded part of the shaft.

For your building projects, you will need screws of a particular length and width: long and thin or short and thick or any number of other combinations. A good tip for buying screws is to choose a length approximately ½ to ¾ the thickness of the material it's going into. Or, if you're attaching two separate pieces of material, choose a length that is long enough to pass through the first piece and half to three-fourths of the way into the second piece. So if you are attaching two pieces of wood that are both 1 inch thick, you would need a screw about 1½ inches to 1¾ inches long.

Screw sizing

Screw sizes are shown on the box in a particular format: a number indicating the screw's thickness or width, and the length of the screw shown in inches, for example: #8 × 2½ inches is a very standard-size construction screw. The number of the screw's thickness (#8 in our example) is the width of the shank. The higher the number, the thicker the screw, and the lower the number of threads per inch.

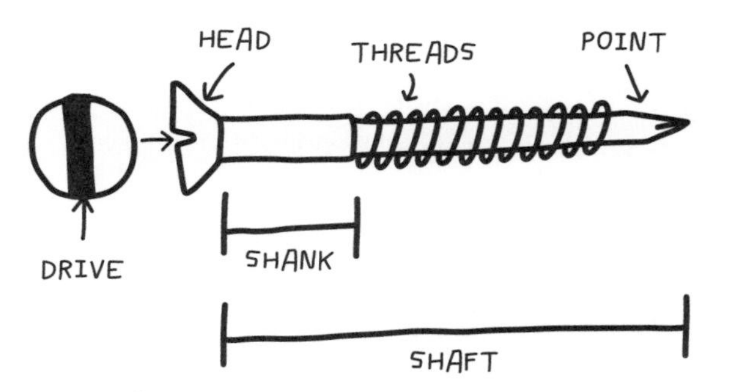

SCREW SIZE CHART

SCREW SIZE	SHANK DIAMETER	PILOT HOLE SIZE*
#2	3/32"	1/16"
#4	7/64"	1/16"
#6	9/64"	5/64"
#8	5/32"	3/32"
#10	3/16"	7/64"
#12	7/32"	1/8"
#14	1/4"	9/64"

* Pilot hole sizes listed are for softwoods, like pine and other woods you will likely use for the projects in this book. For hardwoods, pilot hole size should be 1/64" larger.

Screw head types

The screw's head is very important to its strength, performance, and installation. The head houses the drive (where you stick your screwdriver or driver bit) and is also the part of the screw that's visible and sits above or at the surface of your material, in case you ever need to remove it. Screw heads vary in size and profile, and whether they are meant to be countersunk—allowing the screw head to dive deeper into the wood and sit flat (or flush) with the surface of your piece.

Flathead screws are the most common head shape for wood screws. Most flathead screws have a head with an angled V-shaped underside designed for countersinking, so that when installed, they sit flat on the wood's surface. A countersunk wood screw is desirable for furniture projects and house framing, when you don't want screw heads protruding from the wood's surface.

Round heads are almost semicircular in profile. They are taller than most heads, and have a flat (not angled) underside, so they stick up above the surface when installed. Because they protrude from the surface, they are not very common in wood construction projects, and are better suited for metal pieces, sheet metal, or on machinery. They might also be used decoratively to accentuate the screws themselves.

Oval head screws, just like flathead screws, are designed to be countersunk, so they sink deeper into your material. But they have a slightly rounded top, making them a sort of hybrid between the round heads and flatheads. Oval-head screws are common when you need the deep, tight grip of a screw that is countersunk, but might want a more decorative, rounded screw on the surface.

Pan-head screws are also rounded, but they are shorter and have short vertical sides. They are similar to round-head screws, with a flat underside. They also protrude from the surface when installed, but have a bit lower profile.

Truss heads are wide and flat and have a flat washer-shaped base that helps distribute the weight or force over a wider surface. They are commonly used to secure license plates to their frame.

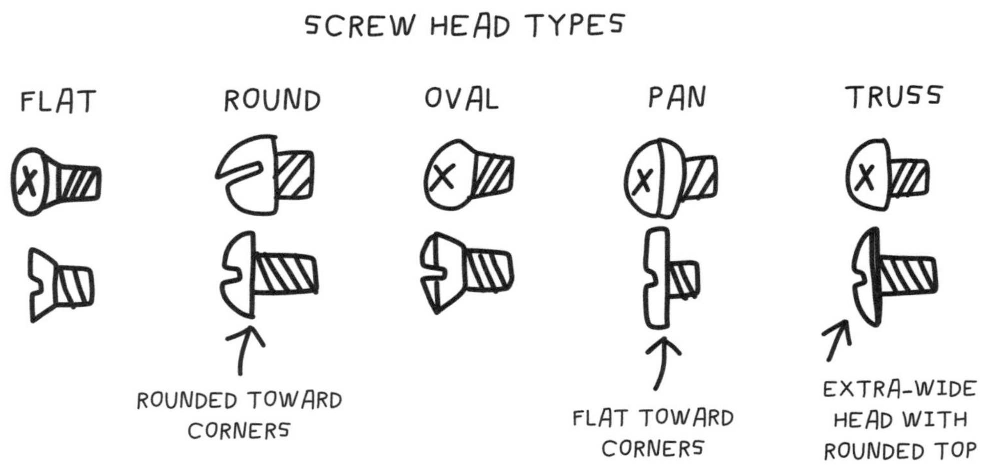

SCREW HEAD TYPES

FLAT
ROUND
OVAL
PAN
TRUSS
ROUNDED TOWARD CORNERS
FLAT TOWARD CORNERS
EXTRA-WIDE HEAD WITH ROUNDED TOP

Screw drive types

In addition to the shape of the screw's head, different types of screws have different types of "drives," which is the slot in the head where you insert your screwdriver or drill/driver bit to install it. The most common drive types are:

Flat/slot is a single slot across the entire width of the screw head. The sizing of a flat drive is measured in fractions of an inch, such as ⅜ inch or ¼ inch corresponding, to the width of the head.

Phillips is a cross-shaped drive. Phillips drives come in various sizes, with #2 (or PH2) being the most common. Phillips screwdrivers or driver bits are marked with the corresponding size—PH1, PH2, PH3, etc. The cross-shaped slot of a Phillips drive is actually not a perfect cross, though, as the intersection has rounded or angled corners, making it easier to grip. The Phillips drive was invented by John P. Thompson, who, after failing to manufacture it, sold his design for the screw drive to Henry F. Phillips, hence the name.

Star/Torx is a six-pointed star-shaped drive. Torx drives are great for wood screws because they have more contact area between the screw and the driver bit, giving you better grip when installing it. The most common size for Torx bits is #25 (or T25), and the others are T15, T35, and so on, with the smaller numbers corresponding to smaller-size stars.

Square (aka **Robertson**, after Peter Lymburner Robertson, a Canadian inventor and tool salesman who patented the design in 1909) drives are square holes of varying sizes, the most common being #2 (or S2). Square drives are often found in finish screws with small heads and have a similar grip to star drives. Robertson invented the square drive after cutting his hand when his grip slipped while

SCREW DRIVE TYPES

MOST COMMON

LESS COMMON

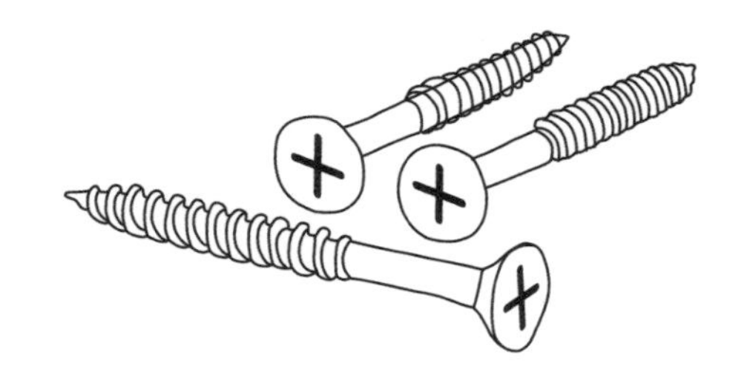

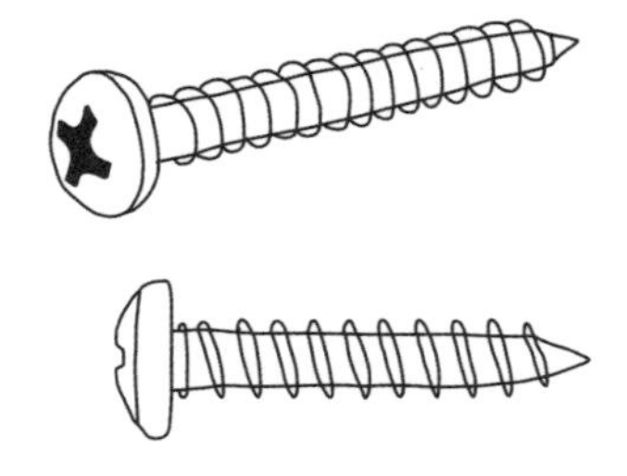

demonstrating a flathead screwdriver to a potential customer. Robertson's screws were also used on some of the early Ford car models produced at the Windsor plant in Canada.

Hex socket drives are sometimes called **Allen** drives, referring to the corresponding Allen wrench used to install them. Their drive hole is the shape of a hexagon. Hex socket drives can withstand high amounts of torque (rotating force) when installing, and are often used for bike hardware.

Types of screws

There are many types of screws, but these four are some of the most common you'll likely use over the course of your building career.

Wood screw: Wood screws are—you guessed it—essential to any wood project. They are characterized by a head that is angled underneath so that they can be countersunk, sitting flat against the surface of the wood. Many also have a shank that is unthreaded, which also helps the screw sink into the wood.

Most wood screws have fewer threads per inch than other types of screws. This creates less friction between the wood and the screw and makes it easier to install. Wood screws are made from steel that is tempered, a heat treatment that makes them less brittle. This means the screw will bend before it breaks, which is especially helpful because wood shrinks and expands over the course of its life.

Some helpful types of wood screws include deck screws (which are coated or galvanized, to resist rust in outdoor settings), drywall screws (designed to attach sheets of drywall to wood or metal studs in construction), and trim-head or finish screws (which have smaller heads, so they aren't as visible for decorative or furniture projects). Phillips and star drives are the most common drive types for wood screws.

Sheet-metal screw: These screws are used for many tasks involving metal, including attaching metal to another material, such as wood or plastic, and installing hardware, like hinges or brackets. They have pointed ends and are "self-tapping," meaning they cut their own little channels for their threads as they are inserted, giving them a precise fit. Some sheet-metal screws are also "self-drilling." They have a drill bit–shaped tip, so you don't have to drill a pilot hole before installing them.

Sheet-metal screws have fully threaded shafts and more threads per inch than wood screws, making them extra secure when installed in different types of materials. They generally have a round or an oval head that sticks up above the surface, with a flat bottom that distributes weight across a metal surface. Sheet-metal screws are usually made from stainless steel or zinc-plated steel to keep them from rusting.

Machine screw: Some might consider a machine screw to be a bolt because of its flat end, but because it has a drive in its head and needs to be installed using a driver or screwdriver, I consider it part of the screw family. Machine screws are used to join or fasten together two or more parts using predrilled (and sometimes pre-threaded) holes.

Because their ends are flat, not pointed, they are installed into threaded holes and can go entirely through the material and out the other side, where they are fastened using a nut and washer. Machine screws are small in size, have fully threaded shafts, and are used commonly in the assembly of equipment, machinery, and appliances. They often have a round or pan head, so they sit nicely atop their surface material.

Masonry screw: You'll use these to attach things to concrete, brick, or cinder block. Masonry screws require a predrilled pilot hole, but are self-tapping and cut their own thread channels into the masonry as they are installed. The Tapcon screws I mentioned in the introduction to this section are one of the most common brands of masonry screws and are recognizable because of their royal blue color.

Almost all masonry screws have one of two head shapes: a hex-head screw without a drive (using a bit that grabs the outside of the screw head and turns) or a head that is angled underneath for countersinking and screwed in with a drive.

SELF-DRILLING SHEET-METAL SCREW

MACHINE SCREW

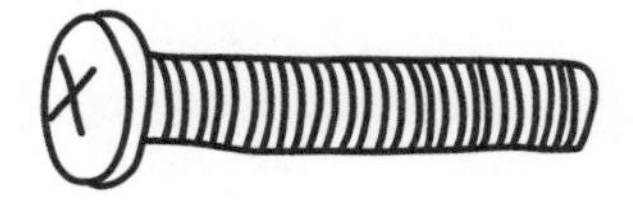

FUN FACT!

Machine screws are also sometimes called stove bolts, because, historically, they were particularly useful in assembling wood-burning stoves made of sheet metal.

MASONRY SCREW

TAMI GAMBLE

President and CEO of the Girly Shop Teacher

Dallas, Texas

One of the things I love most about Tami Gamble is her unapologetic use of the word "Girly." Known as the "Girly Shop Teacher," Tami is a skilled construction and trades educator who knows her way around power tools, and she also wholeheartedly embraces her femininity (red lipstick and a chop saw, yes please!). Tami has an infectious smile that her fans, students, and coworkers cannot deny, and I've been a fan of hers for years.

"I am an African American woman with a psychology degree from Howard University, a mom of three sons, and a teacher of the building trades! I teach Construction and Building Trades at Duncanville High School on the south side of Dallas, Texas, and I am known for my energetic comedic spirit and never-quit attitude.

"My first moment of construction inspiration came at age six, when I helped my dad set and pour concrete steps. I remember spending the bulk of my childhood pressed over his shoulder watching his hands create, mold, and shape raw materials. It was his guidance that gave me the knowledge of the building trades, and I have gone on to utilize the skills he taught me my entire life! Oddly, it was the recognition of these skills by my hairdresser that ultimately propelled me into the classroom—teaching the trades to high school students. Family and friends soon encouraged me to share my work on social media, and I became known as the 'Girly Shop Teacher.'

"My favorite project I've done with my students has to be the design, construction, and installation of much-needed habitat structures for the bird exhibit area at the Dallas Zoo. As I began planning and budgeting, a friend who works at the zoo suggested (and requested!) the possibility of an all-girl construction crew of students.

"The scope of work was huge! The zoo asked for thirty large steel-framed structures for the birds to inhabit. With my team of young women, we started working and made great progress, and garnered some attention in the local media. And then, the young men in my class who heard of our massive assignment stepped up, wanting to contribute and help the birds as well. This opened up a valuable teaching moment: Working behind the scenes, the young men appreciated the young women more and saw true value in their work. And the young women came to comfortably rely on and collaborate with the male students on other projects. From that day forward, the idea of gender imbalance in my shop class disappeared.

"While I know how capable I am, I still ask for help from others in my areas of weakness and offer others help in areas of my strength. To young women who are interested in building, don't allow gender to be a crutch or a weapon and instead use your most valuable tool, your brain, to make wise decisions for yourself. If someone is demeaning or objectifying you, know that you were placed in that moment as a teaching tool for your own growth as well as those around you."

PARTS OF A NAIL

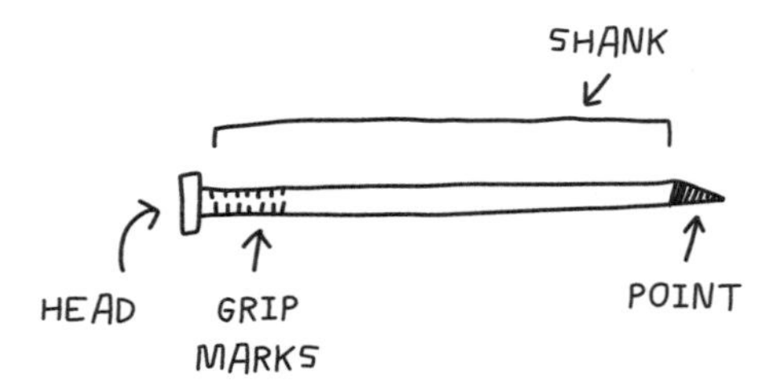

FUN FACT!

Nails have been around since ancient Egypt! Nails today are almost always made from steel, but historically, they were made from bronze or iron. The early nails were forged by blacksmiths who heated wrought iron to a high temperature and physically shaped and cut *each individual nail* using hammers and anvils. Imagine how much time and care this took, and how hard it must have been to make two identical nails. During the industrial revolution in the United States near the end of the eighteenth century, nails could be made more efficiently by cutting small slivers off a long metal sheet. Modern-day nails are made from spools of wire cut to length and shaped mechanically.

Nails

Nails get their holding power by displacing wood fibers as you hammer them in, which creates a tight squeeze against the shank of the nail. This pressure is what holds the nail firmly in place.

Tips for use

Like screws, nails come in many shapes and sizes—all are made for a specific purpose. You might be wondering, *When should I use nails instead of screws?* Different craftspeople or builders might give you different answers. At a basic level, nails have weaker pull-out strength but have much better sideways (shear) strength and flexibility.

In plain English, nails will pull out of materials easily but can withstand a lot more force horizontally, and will bend before they break (unlike most screws). Any house with a wood frame made of 2×4s and other dimensional lumber is nailed—not screwed—together because nails hold the wood studs together more firmly under shear forces. If you screwed your house studs together, chances are that over time, as your house moves and vibrates and withstands earthquakes, the screws would snap in half! Most building codes have specific requirements for the type of nails and even the specific pattern in which they are installed for houses and other structures.

Parts of a nail

One of the most fundamental fasteners in any shop, nails are characterized by a pointed tip at one end, a long, unthreaded shank, and a flat head on the other.

Nail sizing

Just like screws, nails have a unique sizing system for their lengths. The length of a nail is measured using a unit called a penny, which is indicated by a lowercase *d*. If you see a box of nails labeled "8d," those are 8-penny nails. "Penny" dates back to the pricing of nails in medieval England, with the *d* standing for the Roman coin denarius.

In practice, the larger the penny size of a nail, the longer it is (and thicker, too). Smaller nails, usually 1 inch or smaller, are called brads and do not usually have a penny size, but instead just their dimension in inches. Extra-large nails, usually longer than 4 inches, are called spikes, like the railroad spikes that hold down railroad tracks. There's no need to memorize the penny-to-inches conversion, because most boxes show you both the penny size and the size in inches. A 16d nail is the most common one used for framing houses, measuring 3½ inches.

Nails are also available with various coatings or finishes to help your specific purpose. "Bright" nails are uncoated (so they are prone to rusting in outdoor settings) and are bright and shiny, as the name implies. Cement-coated (CC) nails are coated with resin that heats up from the friction created when they are installed, serving as an adhesive for a more secure fit. Galvanized nails are dipped in a zinc coating to keep them from rusting or corroding.

NAIL SIZE CHART

PENNY SIZE	LENGTH (INCHES)
2d	1"
3d	1¼"
4d	1½"
5d	1¾"
6d	2"
7d	2¼"
8d	2½"
9d	2¾"
10d	3"
12d	3¼"
16d	3½"
20d	4"
30d	4½"
40d	5"
50d	5½"
60d	6"

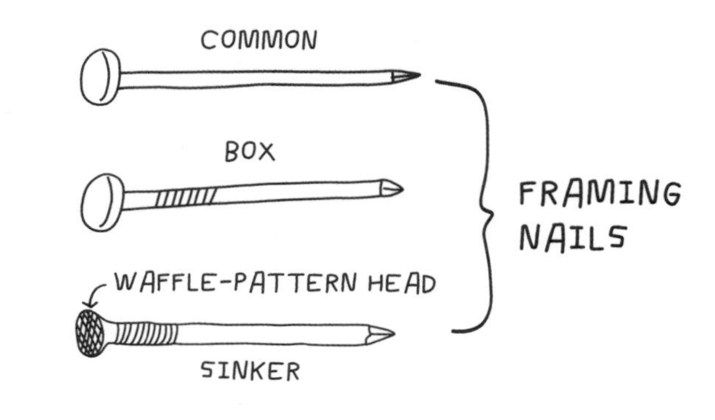

Types of nails

There are many more types of nails than the ones listed here, but these are the most common for projects you'll likely be building.

Framing nails—common, box, and sinker: If you've ever driven by a construction site with a half-built house, the wood studs you see are being held together by framing nails. Framing nails are used for framing houses and other wooden structures and are pretty much the go-to nail for most carpentry projects. Common, box, and sinker nails are fairly similar, and are all used for framing, but each has slight differences that can be helpful in certain situations.

- All three framing nails have smooth shanks (though some have grip marks for greater grabbing power) and flat heads about three to four times the width of the shanks.
- Common nails and box nails are nearly identical, but box nails are slightly skinnier than common nails. Historically, box nails were used to assemble small wooden boxes like apple crates, so the thinner shank made them easier to install on skinnier pieces of wood. Common nails, being slightly thicker, are stronger than box nails.

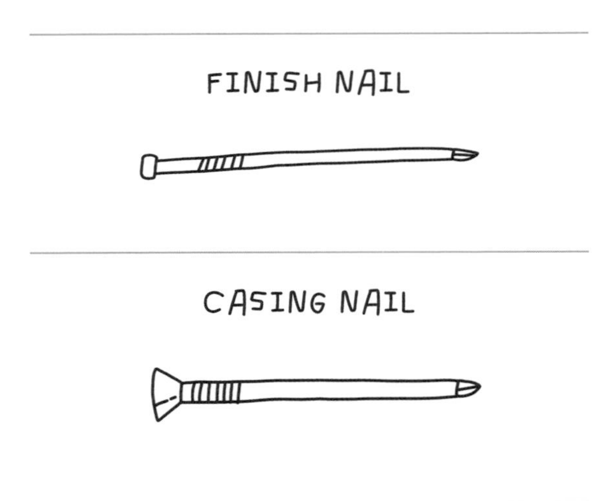

- Sinker nails serve the same purpose as common and box nails, though they have three features that give them a slight advantage in some construction situations:
    - Sinker nails are often vinyl coated, which gives them an extra level of adhesion when installed.
    - They often have a textured head that provides a little extra grip, making them easier to install with a hammer at odd angles.
    - The underside of the head is sometimes angled underneath, helping them sink, or dive, a little deeper into the wood's surface.

Finish nail: If you're working on a project like a handmade piece of wooden furniture in which you don't want anyone to see your nails, finish nails are your best friend. Unlike framing nails, finish nails have a small barrel-shaped head that is not much bigger than the nail shank, so they can be easily driven in below the surface of the wood and basically disappear. Furniture designers use finish nails and then cover up the nail holes with wood putty, making their finish nails even more unnoticeable. Finish nails are also used in applications like molding and trim in residential settings, where visible nails would be an eyesore.

Casing nail: A "big sister" to the finish nail, casing nails are similarly stealth with slim heads that are less visible when installed. Casing nails are slightly larger than finish nails and therefore stronger, making them great for applications like heavier molding, window frames, or door casings.

Duplex head nail: You may find yourself building something that you know you're eventually going to take apart. For example, if you're pouring concrete into a wooden mold, which is called formwork, you will need to remove that wooden formwork to release your concrete piece. In this case, a duplex nail is the perfect choice. Duplex nails have a second "head" that sits on the shank below the actual nail head. This second head stops the nail from going into the wood all the way, so part of the nail remains sticking out. When you need to take apart your brace or formwork, you can easily pull the nail out without prying and digging out the head.

Annular ring-shank nail: Ring-shank nails have a series of rings on the nail's shank. When installed, the rings interlock with the wood to create an extra-snug hold that is resistant to the nail pulling out of the wood. Ring shank nails are commonly used for shingles or to install siding or paneling. Most are also galvanized and resistant to rust.

Brad: Nails that are 1 inch or shorter are designated as brads. Similar to finish nails, brads are designed to be mostly hidden and are helpful in making small frames or attaching paneling. Brads are easily installed by using a brad gun, which uses pressurized air to shoot it into the wood. Because they are so tiny, they're not usually hammered.

DUPLEX HEAD NAIL

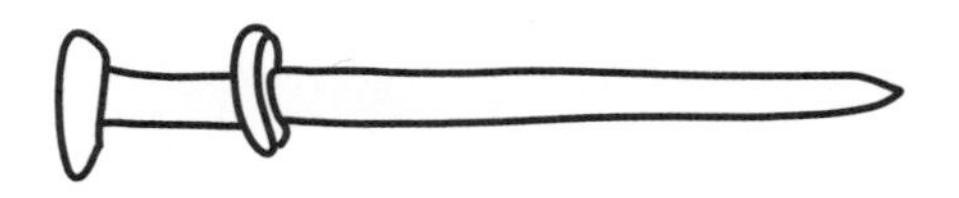

ANNULAR RING-SHANK NAIL

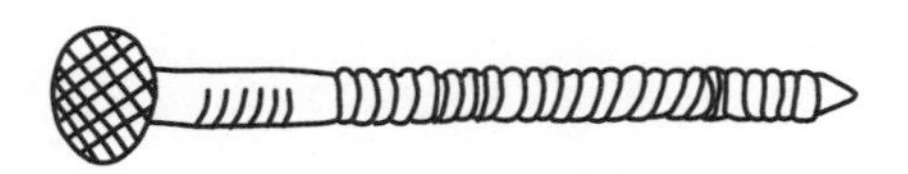

BRAD

KARI BYRON

TV Personality: *MythBusters* and *White Rabbit Project*

San Francisco Bay Area, California

When I was a young girl, I used to watch *MacGyver* and was enthralled by the way he would solve problems in creative ways with the most constrained materials. As an adult, I felt the same way about the *MythBusters* crew as they hacked and built and blew objects up with the sole purpose of figuring out how they worked (or didn't). Some of my favorite episodes were those in which Kari Byron, the show's lone female expert, was pregnant with her daughter AND using power tools and setting items on fire. Kari is a modern-day MacGyver and, as a person in the public eye, has carried the torch for many young women who aspire to build (and blow stuff up).

"I have always been a maker/builder. I used to sit in the garage watching my dad's woodworking projects in awe. The smell of freshly cut wood still fills me with warm memories. I was too little to use the big saws, but he let me hammer nails and sand away splinters. When I wasn't helping him, I was deconstructing items I would find in the trash to make models. Around the fourth grade, I used to make finger skateboards from wood scraps. Paper clips would be axles, pencils cut into sections with the graphite scraped out made perfect wheels. Once painted, I sold them to my classmates for candy money.

"The tough part about my TV role on *MythBusters* is that some of my most beautiful creations were made to be blown up or crushed. One time I built an arsenal of rockets shaped like dragons that deployed arrows. I had so much pride wrapped up in making them work, and in making them look like the historical artistic renderings. I stayed up late nights creating small-scale replicas to make sure the final experiment would look right. The rockets worked, but because they were meant to be destroyed, there wasn't much left of them at the end of the day.

"There is a special feeling of satisfaction that comes from making something with your own hands. It is a tangible result you can admire and feel proud of. I have gained so much confidence from completing a tough build. I carry that confidence into all parts of my life. Being a maker/builder has made me a stronger woman. The best way to learn is just to go for it. Mistakes help you learn. I have ruined so many projects and started over to make them even better!"

Bolts

A bolt is like a screw's cousin: they can look similar but have distinctly different characteristics and uses. Bolts have some key features that distinguish them from screws and make them great for specific jobs.

Bolts are almost always used with a nut and a washer (or multiple washers). The nut and washer help secure the bolt in place.

Bolts usually have an external drive—meaning instead of inserting a screwdriver into a slot in its head to tighten it (as with a screw), the entire head of the bolt is grabbed from the outside and turned by using a wrench. Bolts often have a hexagonal (or other geometric shape) head that makes this external turning easier.

Bolts can be threaded through a predrilled hole, all the way through the material and out the other side, where they are connected to a corresponding nut. Because the nut-bolt combination can be tightened to squeeze the materials in place, bolts have incredible holding power.

The threads on bolts are very different from those on screws. While a screw's threads are great for gripping and biting into wood, a bolt's threads are meant to thread into a corresponding nut, not the material itself. You can see this in the threads: bolts often have threads that are closer together and not as sharp as screws. Of course, there are exceptions to all this!

BOLT ASSEMBLY

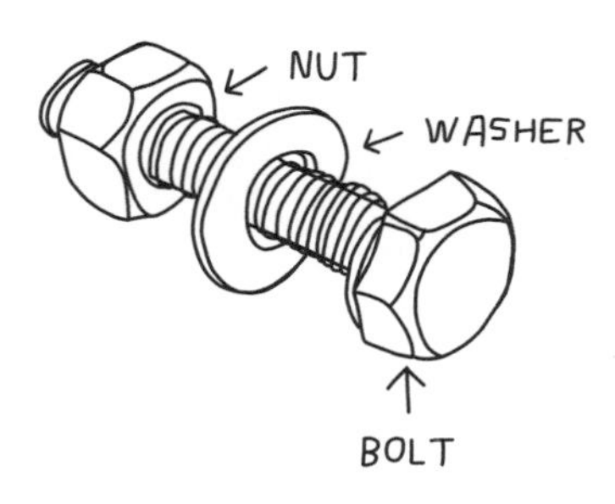

PARTS OF A BOLT

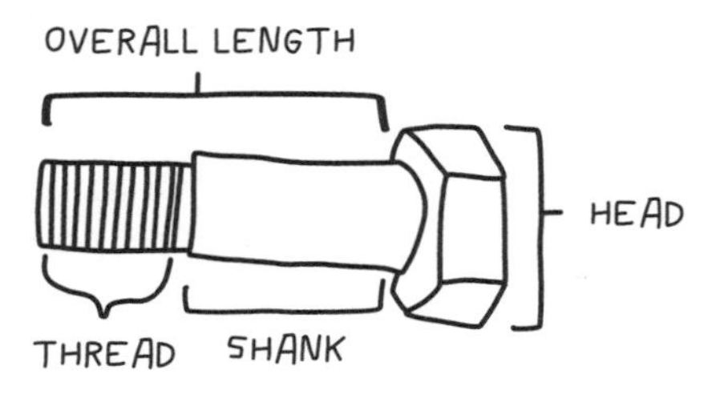

Tips for use

Bolts are fabulous for projects you might need to disassemble at some point or for sandwiching multiple materials together with great strength. Bolts are also generally (but not always!) longer or thicker than screws and come in sizes that are enormous enough for heavy-duty construction projects.

Parts of a bolt

The main parts of a bolt include the head (the geometrically shaped top that is grabbed and turned), shank (the smooth, unthreaded length of the bolt), and thread. The threads on a bolt are measured in threads per inch (TPI). So, for example, with a 20 TPI bolt, you could count 20 individual threads in 1 inch.

Bolt sizing

When you shop for bolts, just like with nails and screws, the sizing system can be confusing. Generally, bolt sizes are shown in the following format:

shank diameter - thread count (TPI) × length, or ½"-20×2"

These numbers are super important, especially if you're using a bolt with a corresponding nut and washer. Nuts and washers are sized according to the bolt they go with, so a ¾-inch bolt needs *both* a ¾-inch washer and a ¾-inch nut. The trick with nuts, however, is to also make sure *the thread count matches the bolt*. If your bolt has gigantic threads spaced far apart, it won't thread into a nut with tight threads. Using the preceding example, that ½"-20×2" bolt would need a ½"-20 nut.

One last helpful tip for installing bolts: When predrilling holes for bolts to fit into, drill at a diameter that is *slightly wider* than the diameter of the bolt (maybe an extra 1⁄32 inch or 1⁄16 inch). This makes it easier to get the bolt all the way through the material, and because the bolt's threads do not grip the material itself, it's fine if it's not snug-as-a-bug-in-a-rug tight.

Note: This doesn't apply to lag bolts, which DO bite into the wood, so they actually need a hole that's *slightly smaller* than their diameter!

Types of bolts

Just like with screws and nails, there's a bolt for every job, and a job for every bolt! Hex, carriage, and lag bolts are some of the most common. Depending on the grip strength, material, and whether you're sticking your bolt all the way through to the other side of your material, here are some of your best options.

Hex bolt: Probably the most common bolt, hex bolts have a hexagonal head that can be tightened using a wrench or a nut-driver bit. Hex bolts might be fully threaded or only partially threaded with a shank. When installing hex bolts using a corresponding nut on the other side of your material, it is important to tighten the nut while holding the head of the bolt stationary with a wrench; otherwise, the bolt just spins itself around and around in the hole. I like to hold the head of a hex bolt using a wrench with my nondominant hand and tighten the nut using a wrench or nut-driver bit with my dominant hand. Hex bolts are a great go-to bolt and are available in a huge range of sizes. They are great for sandwiching materials, like plywood, to a wood frame.

Carriage bolt: These are a funny breed of bolt, characterized by an "ungrabbable" round domed head with no drive (you can't use a wrench or screwdriver to grab the head) and a square "collar" just below the head. This square collar is designed to fit into square holes in metal pieces so that once installed, it cannot rotate.

Because the head cannot be grabbed and tightened, carriage bolts can only be installed by tightening a nut at the end of the bolt, so they are commonly found in security settings like safes, where the exposed head can't be unscrewed. The end of the bolt is flat, not pointed, and threads into a corresponding nut to keep the bolt securely installed.

HEX BOLT

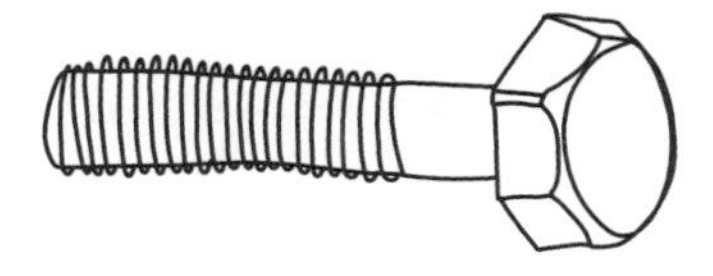

CARRIAGE BOLT

FUN FACT!

Carriage bolts were originally designed to connect the metal frames of carriages to wooden axles—yes, that's where their name comes from!

LAG BOLT

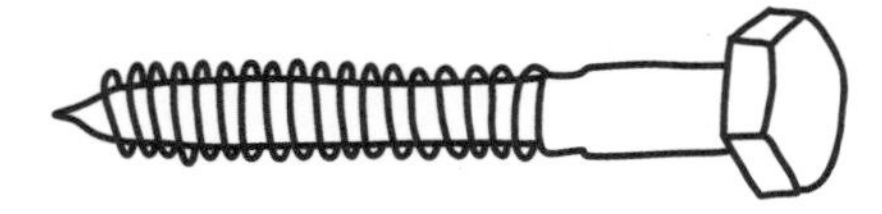

SOCKET BOLT

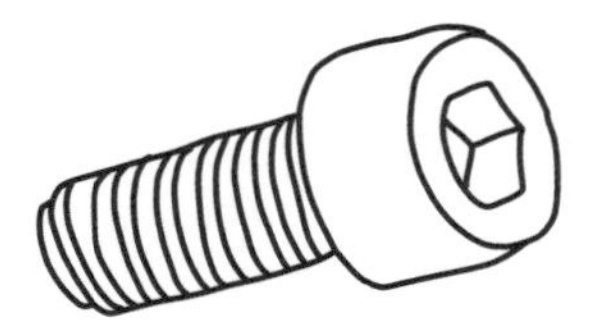

FUN FACT!

You might find socket bolts holding your bike together!

Lag bolt: I had a long internal debate about whether lag bolts belong in the bolts or screws category, because these hybrid fasteners have characteristics of both! Because of their size and hexagonal head, I'm inclined to call them bolts, though I'm sure some may disagree, and you'll likely hear them referred to as "lag screws" at some point, too.

Lag bolts, a burly version of a wood screw, are often used for jobs like deck building or for roof beams that hold a very heavy load. Lag bolts have a pointed tip like a screw and require a pilot hole that is slightly smaller than their diameter so the threads can bite into the wood. Unlike most other bolts (but like most other screws), their threads are meant to cut into wood, rather than thread into a nut. Lag bolts can be fully or partially threaded, but they are almost always used with a washer between the head and the surface of the material. They are easily installed using a socket wrench or a driver with a nut-driver bit.

Socket bolt: Socket bolts, also known as Allen bolts, are ideal for projects where you don't have much room to maneuver. Whereas other bolts require wrenches or drivers to install, socket bolts need only a hex wrench (aka Allen wrench) to tighten. This makes them ideal for assembling machinery and equipment, as they also have fine threading and incredible holding strength. Because the bottom side of the head is flat, these bolts almost never require the use of an additional washer.

Eyebolt: If the name of these bolts isn't enough of a hint, eyebolts have a circular ring (or eye, like the eye of a needle) on the head end. Eyebolts are useful for creating a secure hanging point for a rope or chain. You might use an eyebolt to hang something from the ceiling or to create a pulley to lift a heavy load. There are different types of eyebolts to hold different weights: some eyebolts have an open loop, while others are solid loops with a built-in flat washer surface to withstand pulling forces from all angles. Eyebolts generally thread through a material and into a corresponding nut on the other side, though there are also eye lag bolts with a pointed end that are installed like lag bolts into the material itself. The bike hooks used in the Wall-Mounted Bike Rack project (page 280) are similar to eyebolts, but have an open hook instead of a complete loop.

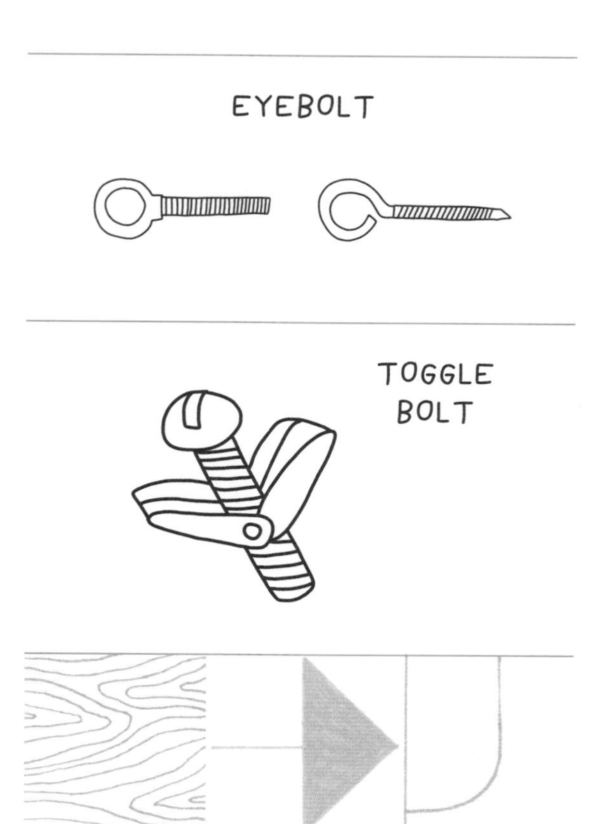

Toggle bolt: For anyone who has tried to hang something heavy on a wall only to watch it come tumbling down, the toggle bolt is your new best friend. Toggle bolts are thin, usually round-headed bolts that come with a corresponding toggle, which looks like a set of wings. They are installed through drywall by folding up the wings and inserting the bolt and wings into the drywall. Once the winged toggle is through to the other side, the wings open and brace against the wall to hold the bolt (and object!) in place. Toggle bolts are great for installing items like shelving or for bracing a bookshelf to a wall.

The toggle bolt's sister is the Molly bolt (which happens to be my actual sister's name!). Molly bolts work in a similar way, but instead of using wings, they have a metal sleeve around the shank that expands against the walls of the hole when installed. Molly bolts can be a little trickier to install than toggle bolts, but some have a pointed end that you can tap into the wall without having to predrill a hole.

Nuts

Nuts and washers go with bolts like peanut butter and jelly go with bread: they *need* each other. Unless you're using a lag bolt or other type of bolt that doesn't go all the way through to the other side of a material, you'll need nuts and washers to secure your bolts in place. The job of the nut is to secure the end of a bolt. Its threads match the threads on the bolt so that they fit together perfectly.

Tips for use

After you fit the nut (and probably its corresponding washer) onto the end of your bolt, turn the nut to tighten it down the length of the bolt. When tightened, the nut sits on and squeezes against the surface of the material. Depending on the type of nut, you'll likely use some type of wrench to tighten it: a socket wrench, a nut-tightening bit on a driver, or a combo wrench that matches the size of the nut.

Also, you might be tempted to tighten your nut as much as your muscles will allow, but remember that nuts will likely have to be unscrewed at some point, so don't go too crazy and overtighten.

Nut sizing

Make sure the thread count of your nut matches the thread count of your bolt! The worst thing in the world is trying to force a nut onto a differently threaded (and therefore unhappy) bolt.

Remember that bolts have both a diameter and a thread count, so you need a nut with a hole with the same measurements. Nut sizes are shown as **hole diameter - threads per inch**—for example, ¼"-20 (which would fit on a ¼"-20 bolt of any length).

Types of nuts

Just as you might have a favorite nut in the kitchen (hello, hazelnut!), you might also explore the wide world of nuts in the hardware store and find your own favorite. As with so many other types of hardware, there's a nut that is perfectly designed for specific tasks. Sometimes you might need a nut that tightens and stays put forever, or one that can easily be removed, or one that will distribute weight evenly across a thin sheet of metal. Not to worry—there's a nut for that!

Hex nut: When my first class of high school students "graduated" from my design/build program, I awarded them each a commemorative hex nut with their name engraved on the side. I also wear a stainless steel hex nut on my right middle finger. All this is to say that the hex nut holds a special place in my heart.

When in doubt, go with a standard hex nut! Probably the most common type of nut, a hex nut has a hexagonal profile that makes it super easy to turn using a wrench.

Hex nuts come in many sizes, types of metal, and finishes. You'll find giant galvanized hex nuts for outdoor projects, such as decking, and tiny hex nuts for smaller assembly. Depending on the type of bolt you're attaching the hex nut to, and also the type of material you're securing, you may want a washer to help distribute the load of the tightened nut. Use a flat washer under a hex nut for softer woods or sheet metal in particular, so you don't damage or dent the surface.

FUN FACT!

The world's largest hex nut measures 10 feet tall and sits in front of the Packer Fastener plant in Green Bay, Wisconsin.

Jam nut: These are just hex nuts with a shorter height, usually about half the height of a traditional hex nut. Despite their short stature, they have a very clever purpose: You can use two of them and jam them up against each other to lock them into a specific place on a rod or bolt. This is helpful should you ever need a stationary nut on a bolt or rod as a stopper of some sort. A jam nut secured against a standard hex nut can also be an easy way to create a tighter hold in instances that call for only a hex nut.

Square nut: Square nuts are not that common anymore, but some people prefer them to the hex nut. They serve almost the same purpose, but can provide a little more holding power than the hex nut. Why, you ask? Geometry. Think about a square and a hexagon of the same diameter. If you were to calculate the areas of those shapes, the square would have a larger area. For a nut, this is helpful, because more surface area in contact with the material equals more ability to hold it in place. The downside of the square nut is that because they're less common, they can require more obscure tools to install. Square nuts are usually paired with square-head bolts.

Nylon-insert lock nut: You might hear this type of nut called "nyloc" for short. A nylon-insert lock nut is like a hex nut but with a nylon-lined collar inside that helps keep the nut from loosening. The lock nut is tightened onto the end of the bolt, with the very end of the bolt tightened into the nylon collar. The nylon in the collar compresses around the bolt, giving it that extra squeeze! Because of their holding power, nylon-insert lock nuts are not usually used with washers.

HEX NUT

JAM NUT

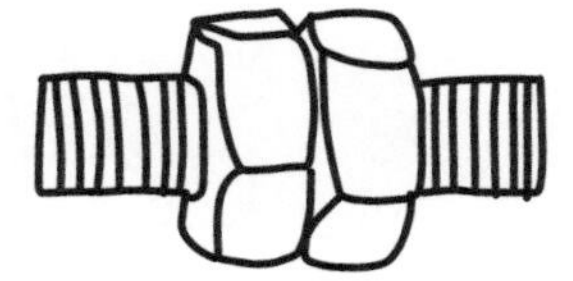

SQUARE NUT

NYLON-INSERT LOCK NUT

CAP NUT

FLANGE NUT

WING NUT

FUN FACT!

Wing nuts were commonly used to secure wheels to old bike axles for quick removal and are used on drum sets to attach cymbals and other drums to the stand for easy adjustment and disassembling.

Cap nut: A cap nut is a hex nut with a domed "hat" over the top. The rounded cap covers the end of the bolt, so you can't thread your bolt all the way through it. They are often used for decorative purposes like details on furniture or residential decor. Keep in mind that when you use a cap nut, the bolt must be the precise length to pass through the material and thread into the cap nut snugly, because unlike a hex nut, the bolt cannot pass through a cap nut. Cap nuts have a sibling, the acorn nut, which has a taller and sometimes pointed cap. Cap nuts, because they are mostly decorative, are also not typically used with a washer.

Flange nut: This is a hex nut with a built-in washer. One side of the washer has a flattened flange that functions like a washer, creating a wider area that sits against the material. This can be very useful for applications like auto maintenance or assembly when using a washer in hard-to-reach places would be more difficult and time-consuming. Most flange nuts have a textured serration on the flat side, giving the nut even more grip and resistance to loosening. Another secret advantage to the flange nut is that because of its wide flange, you can insert it into a socket wrench and install it with one hand—without having to hold the nut in place. The wide flange keeps the nut from falling into the socket!

Wing nut: The wing nut has an adorable quality to it, like bunny ears, and is a highly functional fastener. Wing nuts are characterized by their two large "wings," which function as mini handles, so you can easily tighten and loosen them over and over.

Washers

Washers help nuts stay put and hold strong. They are flat and usually round disks of metal with a hole in the middle—almost like a coin—that help distribute the pressure of the tightened nut across a wider area.

Tips for use

Washers protect the surface of a material from pressure damage or wear. Washers fit around the bolt and sit between the nut and the material (or in some cases between the head of the bolt and the material), creating a flat surface against the material. Without a washer, when nuts and bolts are tightened they can often chew away at or bite into the material, or come loose. In general, it's a good idea to always use a washer with a nut and underneath the head of bolts that have small heads. Especially if you're working with a material that you don't want to damage, like a soft wood, sheet metal, or something that has been painted, a washer will help protect your surface.

Washer sizing

Just like other hardware, the sizing of washers can leave you wondering whether you're buying the correct item. Washers have an outside diameter (the width of the entire washer) and an interior diameter (the diameter of just the hole). The interior diameter is the dimension that matters most, because you need your bolt to fit through it! Buy washers that list the same dimension as your bolt diameter, so a 1/4-inch-thick bolt needs a 1/4-inch washer.

Types of washers

Used in combination, nuts and washers make bolts even stronger and less likely to come loose over time. Just like bolts and other hardware, there's a washer made for every job.

Flat washer: These are your go-to standard washer and are great paired with hex nuts. They help distribute the load of a nut evenly across a wider surface. You can use flat washers underneath a nut and/or underneath a bolt head to distribute the load on both sides. There are even some situations where you might see two washers stacked together. This additional washer provides even more protection for the material surface and prevents a single washer from bending or buckling under extreme force. In most cases, though, you'll only need one.

Fender washer: Fender washers are just like flat washers, but they have a wider outside diameter, giving them a wider area to distribute the load. They get their name from the automotive industry, which uses them to secure sheet-metal fenders. Because sheet metal is relatively prone to bending or denting, a fender washer helps distribute the pressure of the bolt head so it doesn't dent the surface.

Finish washer: For projects with visible screw heads or bolt heads, a finish washer can help protect the surface of your material while providing a decorative finish. Finish washers are only used on the screw or bolt head side of an installation and, because of their shape,

COMMON WASHERS

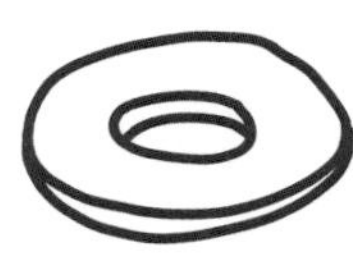

FLAT

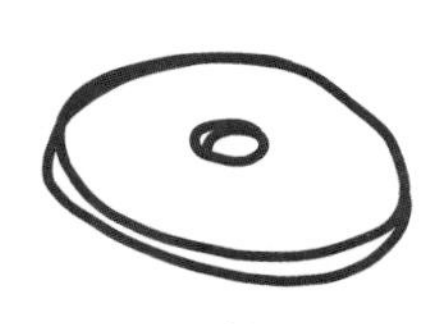

FENDER

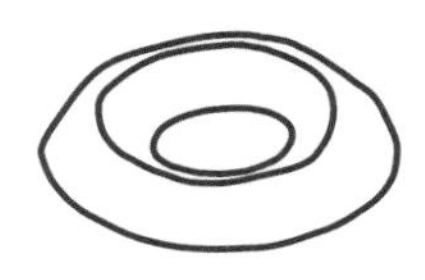

FINISH

LOCK WASHERS

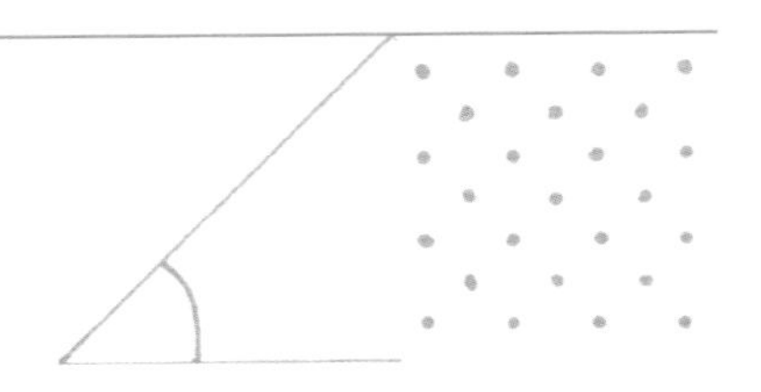

accommodate an angled bolt head or screw designed to be countersunk. The result is a finished look with the washer sitting on top of the material, and the screw or bolt head sunken into it. Finish washers are great for furniture- or cabinet-making projects where screw or bolt heads will be visible, but you still need a strong hold.

Lock washer: Lock washers are a whole category of washers, sometimes also called "spring washers." When these washers are sandwiched between a material and a nut, their shape allows them to exert a spring force that keeps everything from loosening. The most common type of lock washer is a split washer, which is simply a ring washer that is cut at one point and bent slightly out of shape. When installed, it compresses flat, but it wants to revert back to its original shape and presses against the nut and the material surface to create an extra-snug fit that's resistant to the bolt unscrewing. Other lock washers include a tooth-lock washer, with external or internal teeth (an external tooth-lock washer looks a bit like a throwing star with a hole in the middle), and a wave washer, with a wavy profile.

Miscellaneous hardware

There are a few other common types of hardware that are great for attaching diverse types of material (like paper to wood) or for hanging heavy objects on walls, but they don't fall neatly into a category like screws or nails. You'll definitely find them at your hardware store, though! Staples, rivets, anchors, dowels, and biscuits are all fairly specific in their use, but you'll probably find yourself needing one or the other at some point in your adventures.

Staple: Just like office staples, but burlier, industrial-use staples, for construction and building, can be an effective and quick option for securing thin materials to wood. Staples are made from steel wires lined up and glued together and then bent into the shape of a staple and cut into the strips you load into your stapler. Industrial staples are usually installed with a staple gun or hammer tacker, and they come in various widths and lengths. You can use staples to attach chicken wire to a wood frame, and more.

Rivet: While there are lots of different types of rivets in the world (like the tiny metal ones on your jeans), rivets used in building are most commonly a particular type, called blind or pop rivets. These rivets are installed using a rivet gun and are most useful in trying to connect two thin pieces of material when only one side is accessible.

I renovated an old Airstream trailer once. When installing the new sheet metal as the interior walls, I needed to attach it to the structural ribs of the trailer, but couldn't access the ribs once the sheet metal was set over it. In this instance, thousands of pop rivets were just the trick!

A rivet has a thin post (called a mandrel) that feeds into a wider sleeve at the bottom (called the rivet pin), which has a flat collar (the rivet head). The rivet pin is inserted into a predrilled hole up to the rivet head, which stops the rivet from going in farther. Then, you use a rivet gun to grab and pull the mandrel, which compresses the rivet pin against the back of the surface. This compression keeps the rivet in place, and the remaining length of the mandrel can be cut off. Rivets are mostly used for sheet-metal work but can sometimes be used to attach thin wood plywood or veneer. Buy rivets with a rivet pin length that will clear through to the other side of your materials (even though you won't be able to see it). Drill a hole to match the diameter of the rivet pin.

STAPLES

SINGLE STAPLE

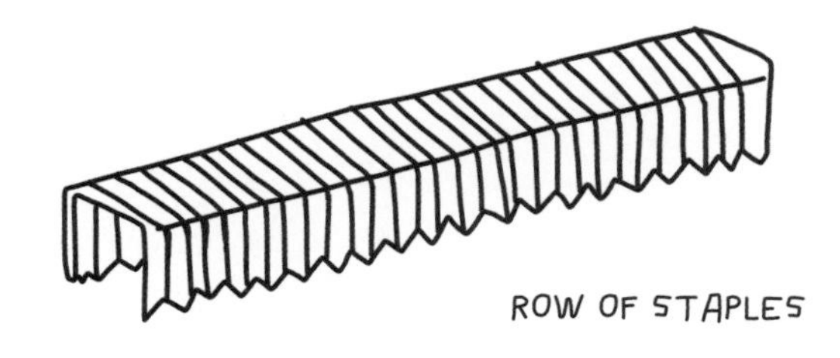

ROW OF STAPLES

RIVET

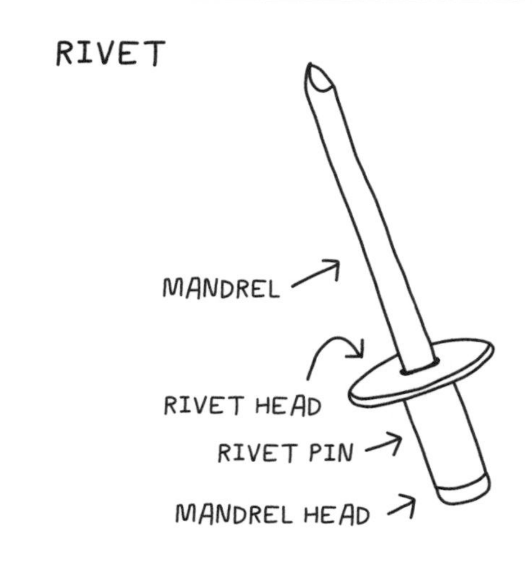

ANCHOR

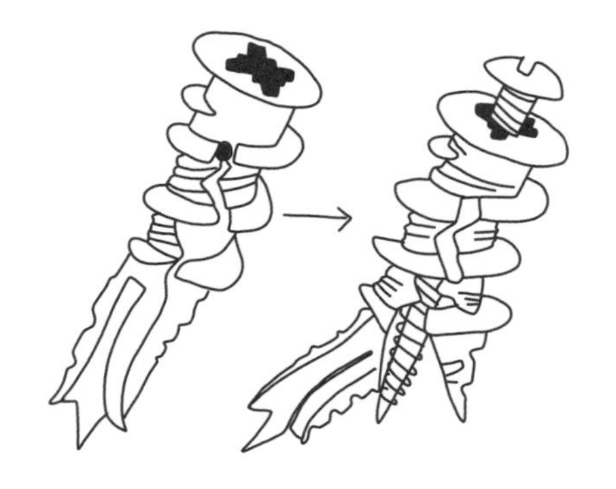

DOWEL

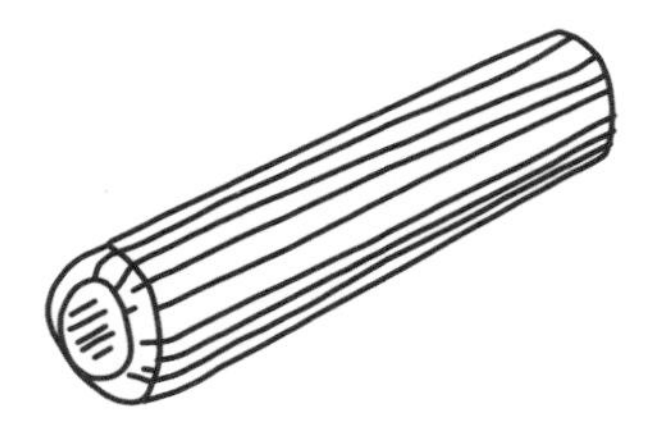

BISCUIT

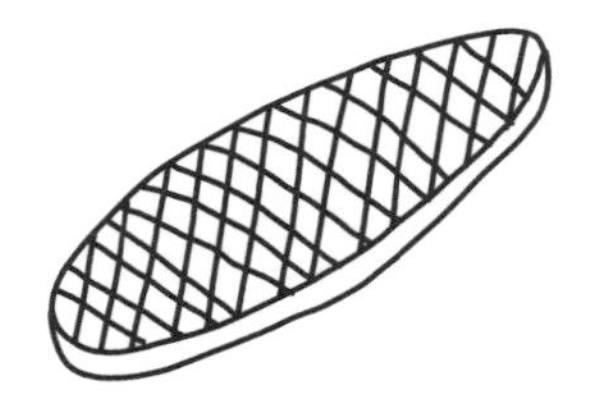

Anchor: The word "anchor" can have different meanings in the realm of building—from concrete anchors (like the Tapcon masonry screws we learned about on page 53) or toggle bolts that help you hang items. For our purpose here, I'm giving a shout-out to the simple but mighty plastic wall anchor, which looks like a plastic screw but functions as a hyper-helpful plastic sleeve for screwing into drywall.

These come in handy anytime you're trying to hang something heavy on a wall or securing a vertical element, like a bookshelf. After inserting the plastic anchor, you then drive an actual screw into the anchor, creating an extra-strong hold in your drywall that can typically support up to 75 pounds or more.

Dowel and biscuit: If you've ever assembled simple furniture, you've likely used dowels. Both dowels and biscuits are small pieces of wood that help internally fasten pieces of wood.

Dowels are cylindrical. They work by inserting one end into a shallow hole of one wood component and the other into a hole in another wood component.

Biscuits are flat, football-shaped wooden wafers that work the same way as dowels but, because they are thinner, can help attach slimmer pieces of wood. Biscuit holes, however, must be made using a biscuit joiner, which carves out a perfect slot in both pieces of wood to be joined. You insert the biscuit into both sides to connect them.

Dowels and biscuits are used commonly by cabinetmakers or furniture makers with wood glue or adhesive and are ideal for attachments in which you want to hide all signs of fasteners.

MEASURE, LAY OUT, AND SECURE

Now that you've returned from the hardware store victorious, with building materials and hardware in hand, you're ready to get started! Before you cut or attach any pieces, it's important to spend some time measuring, laying out, and learning how to secure your material in place so that you can safely work on it. Whether you are making a cut with a saw, attaching two pieces, drilling a hole, or anything else that manipulates or connects materials, you will want to make sure your material is (a) the exact size, (b) in the correct position, and (c) secured and held in place in the right spot. These next three sections highlight the best tools to measure materials, lay them out straight and square, and secure them snugly so you can work your magic.

Measure

There's an old saying: "Measure twice, cut once." As a math lover and craftsperson, I say, "Measure thrice, cut once!" Back in 2010, when I was building a farmers' market pavilion with my first class of high school students, I was working with my student Jamesha to cut dozens of deck joists (beams), which ran across the width of the structure's floor. These gigantic 2×12s were specified on the drawings at 96 inches, but in real life, they had to be cut perfectly to fit the width of the deck as it was built. Some were actually 95¾ inches, and some were 96⅛ inches—just a tiny bit off from the drawing. Jamesha measured the width of the first and told me, "Ninety-six . . . ish." I told her, "There is no 'ish' in building, Jamesha!" We remeasured together and cut the piece to fit perfectly into the joist hangers.

It may seem like overkill, but I always try to take and mark measurements as accurately as possible, at least accurate to the nearest 1⁄16 inch, and sometimes to an even smaller tolerance. This means the smallest tick marks on your tape measure matter, and instead of measuring ⅝ inch or ¾ inch, it might actually be 11⁄16 inch. These tiny differences and fractions of an inch matter, especially if you're attaching multiple pieces—if they're all off, your error will be compounded.

The moral of the story: Precise measurements are always the best way to ensure a well-built end product. Before any piece of wood, metal, or other material gets cut, measure it precisely. The following tips and tools help you do just that, no matter the material, size, or project.

The caret

The caret isn't a tool (or a vegetable!), but a useful way to mark precise measurements. When using any measuring device, you need to make a mark on your material where that measurement falls. Most people put a dot or a small line to mark their measurement. But a single dot is so tiny that you may not be able to find it later, and a line can be confusing when you move your material to the saw and wonder, "Wait, was my 16-inch mark at *this* side of the line or *this* side of the line?" The small V-shaped caret will serve you far better because it works like an arrow that points exactly where your measurement falls.

I like to use an extra-sharp pencil for making carets so that lines can be erased later, and because the pointy tip gives me that much more accuracy. With your tape measure (or other measuring device) pulled, locate your spot, and place the pencil right next to the tape measure at your precise measurement, on the material. This is the point of your V-shaped caret. From that point, draw one diagonal line to the right, then one to the left. Now you've got a precise mark to reference when you make your cut!

Tape measure

A 25-foot measuring tape is an absolute essential. I have a miniature 3-foot tape measure on my keychain and use it literally every day.

A standard measuring tape consists of a metal tape marked with linear measurements. The tape is stiff enough to remain straight when extended, but also flexible enough to rewind itself back into its case. At the end (zero mark) of the tape, a small metal hook helps you latch the end onto whatever material you are measuring. Most tape measures also have a clip on the side of the case (so you can wear it on your belt—with authority!) and a locking mechanism you can engage to keep the tape from retracting when it is extended. The trick with tape measures is to always hook the end of the tape securely to the end of your material. Then pull the tape to extend it, lock it in place with the mechanism, and either take your measurement or make your cut mark.

Here's a helpful hint: If you are taking a long measurement where one end is far away or out of reach (like if you were measuring the height of a ceiling), use the "bent tape" technique. A tape measure will bend back on itself. If you were measuring the height of a tall

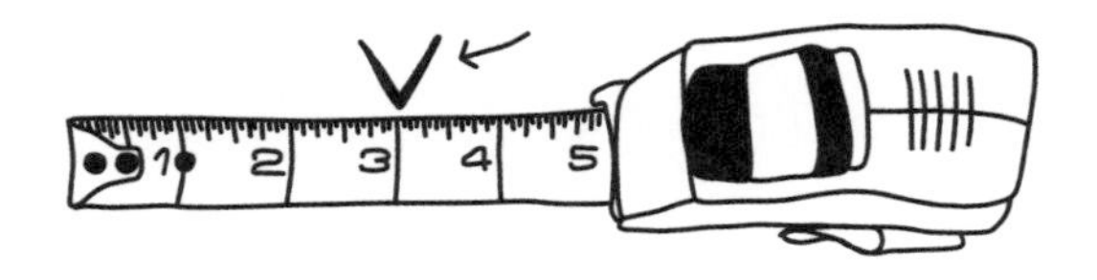

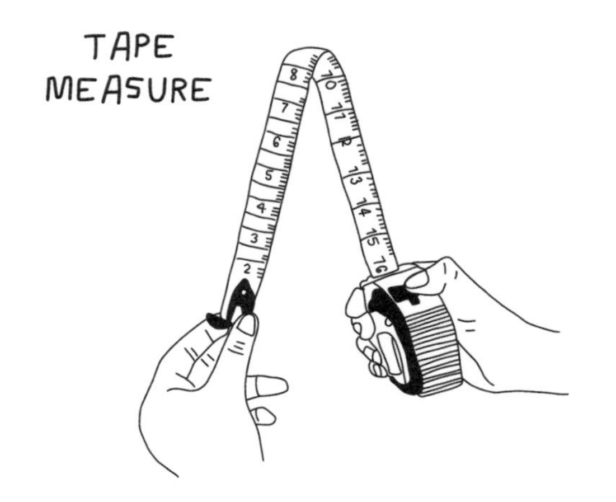

ceiling, it would be very difficult to extend the end of the tape vertically 10 or 12 feet without it teetering and falling. But you can hold the end of the tape at the floor (use your foot to hold it in place) and then run the tape up the wall, and then bend the tape back onto itself toward the zero point. Because the tape is stiff, you can now stand at the zero point and "push" the tape away from you, and it will extend away from you easily. Once the bent point has reached the top of the ceiling, you can read the measurement.

Long tape

Long tapes are commonly used by surveyors marking out an area on the ground or measuring the footprint of a structure to be built. Unlike retracting metal measuring tapes, long tapes are usually 100 feet or longer, made from a flexible coated fabric, and unwind from a spool that can be manually wound up by hand. Because long tapes are floppy, they are very helpful for measuring across uneven terrain. For example, you might use a long tape if you were measuring out the base of a playhouse you wanted to build in your backyard.

Ruler

We've all used rulers, ever since they first appeared on our school supplies list in elementary school. While there are all kinds of rulers and straightedges—made from wood, metal, plastic, and more—having an 18-inch or 24-inch metal ruler with a cork backing is a great asset for any project. This type of ruler is particularly helpful as a straightedge, to make straight pencil lines connecting two marks. For building projects, the cork backing keeps the ruler from slipping across your surface and the 18- or 24-inch length is long enough to measure most pieces without being too hard to store.

You might use a ruler instead of a tape measure when making nonperpendicular measurements or connecting lines. For example, in Your Own Go-To Toolbox project (page 249), a step requires you to cut a diagonal corner off your end pieces. You are asked to measure 3 inches in from one side and 6 inches in from the other, and then draw a line to connect the points. This would be a tough task with a tape measure, but it's a perfect job for a flat metal ruler!

FUN FACT!

Many tape measures have different-colored numbers (usually red) every 16 inches, which is the standard spacing of wall studs in wood-framed houses and buildings. This is particularly helpful for carpenters, as those differentiated numbers help them easily find and measure the distance between studs.

LONG TAPE

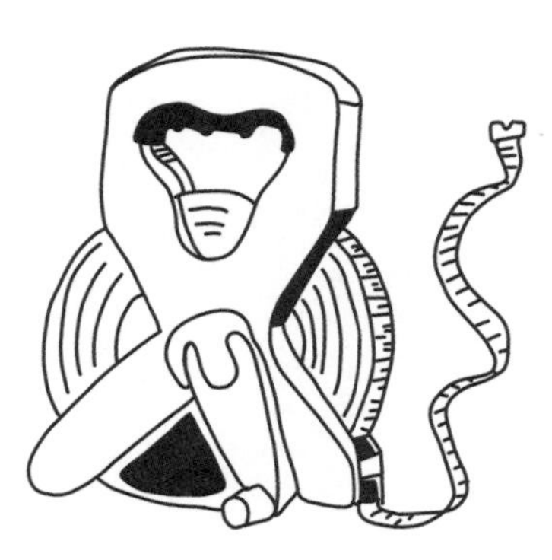

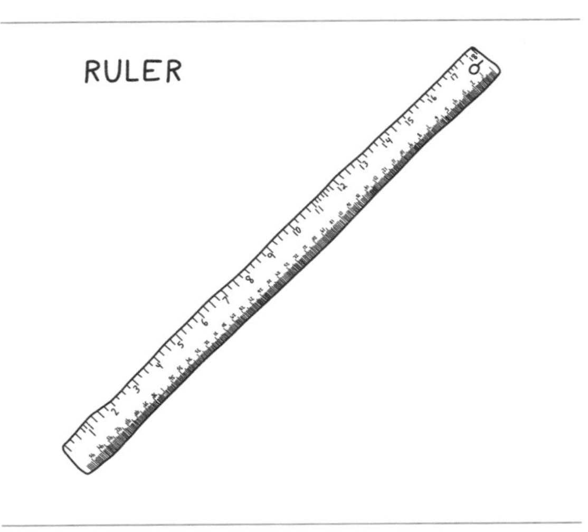

Most rulers used in building projects have a zero point that starts about ⅛ inch from the edge, leaving a small gap between the zero mark and the end of the ruler. Be careful when taking measurements with your ruler: *confirm you are starting at the zero point* and not the end of the ruler itself! This gap exists to allow for damage or chipping on the end of the ruler (which happens more frequently with wooden rulers), without losing any measuring length.

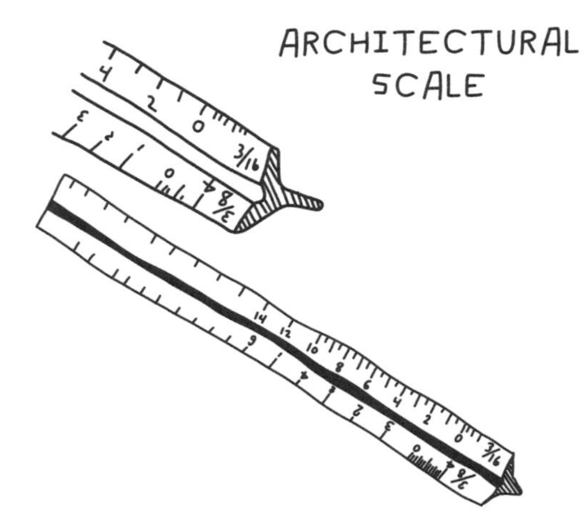

Architectural scale

When making drawings of a building or a project, we can't draw that structure at full scale because it won't fit on the page! So we shrink it at a specific mathematical proportion, or "scale." Most architectural or building projects show that scale as a number of inches on the page equal to 1 foot in real life. An architectural scale is a measurement tool that helps you measure a line on the paper and immediately figure out its dimension in real life. Most architectural scales are three-sided with a triangular profile, and each edge shows a different scale—for example, 3⁄16 inch = 1 foot. Instead of inch markings, markings are 3⁄16 inch apart, so when you use it to measure a line, you can accurately measure the feet in real life.

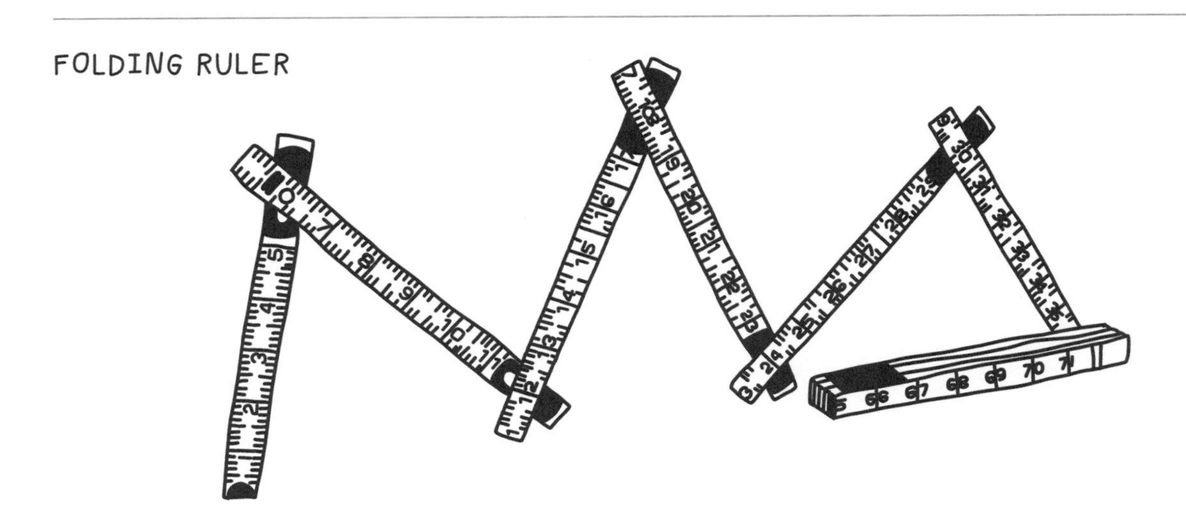

Architectural scales have a standard set of scales that are most commonly found on construction drawings so that you can measure the real-life dimensions off a drawing. You might also use an architectural scale to create your own set of drawings. If you were drawing a project you designed, in which you needed 1 foot in real life to be drawn as ½ inch on the paper, you would use the ½ inch = 1 foot scale edge to draw accurate lines.

Folding ruler (aka carpenter's ruler)

Before the retractable tape measure came into existence, the folding carpenter's ruler was the go-to measuring device. Usually measuring a total of 6 to 8 feet when extended, a folding ruler is most commonly made up of 6-inch wooden lengths connected together. These 6-inch sections pivot, unfolding into a long zigzag. Most high-quality folding rulers have an additional brass extender to measure smaller dimensions between 0 and 6 inches beyond the last 6-inch section. While not as common anymore, they still have some advantages. Because they are more rigid than a tape measure, they can be helpful in measuring anything overhead, upside down, or in compact spaces. And because they pivot, they can also bend around corners or other oddly shaped spaces.

Calipers

A pair of calipers is a highly precise measuring device, used to measure inside or outside dimensions of a small object (like a piece of hardware or the thickness of wood). There are typically two arms that form a jaw, which, when clamped around or in between an object, gives you a precise digital or analog reading of the object's measurement. For example, you may have a hex nut you need to measure so you can buy the appropriate bolt or other hardware (and you lost the original box with the dimensions on it!). You could use calipers to measure the interior diameter of the hole, as well as the exterior diameter of the entire nut. For a builder's purposes, vernier calipers with a digital readout is an ideal choice, as it has two sets of arms—one set that can be placed inside a hole or object to measure its interior dimension, and one set that measures from the outside, to give you an outside dimension. While I'm usually a fan of analog tools (and there are vernier calipers that come with analog rulers), one with a digital readout can save a lot of time and give you even more accuracy.

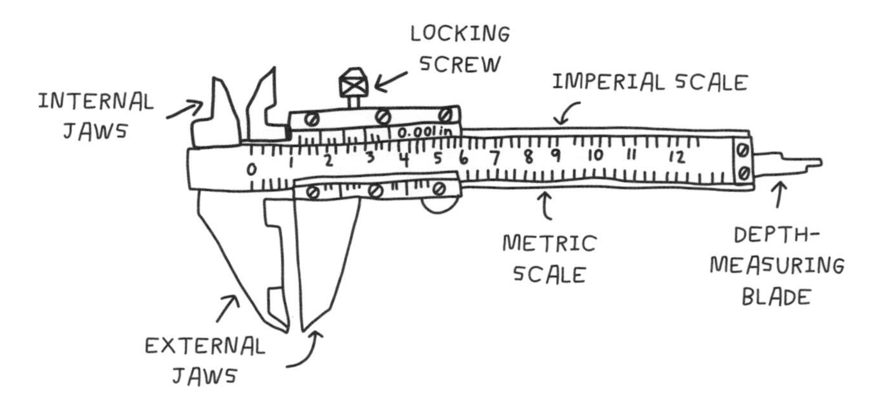

SPEED SQUARE

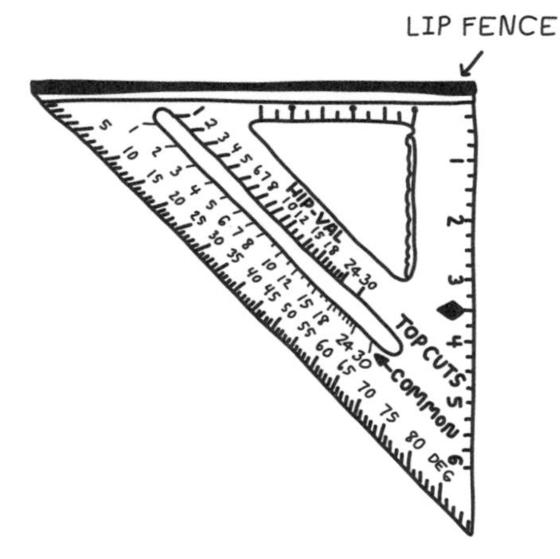

FUN FACT!

The speed square is a trademarked tool, invented by Albert J. Swanson in 1925, produced exclusively by the company he founded, Swanson Tool Company. Some people refer to the speed square as a "Swanson" or a "Swanson Speed Square."

COMBINATION SQUARE

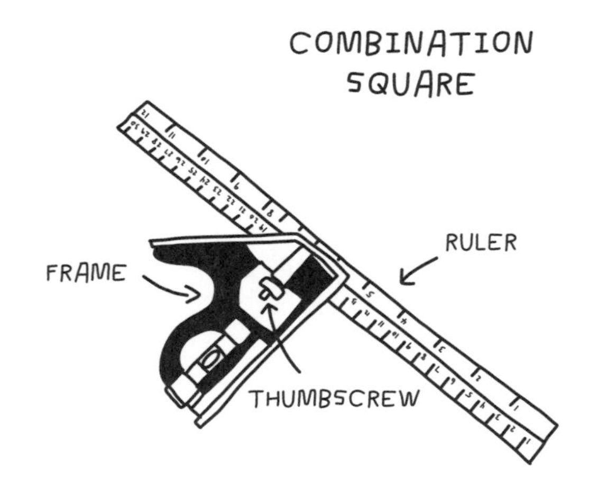

Lay out

Now that you have mastered the art of precise measurements, you need to lay out your materials so you can attach them properly. The tools in this section can be used to lay out components at precise angles, but they can also be used as measuring devices to gauge the squareness of existing spaces or objects. Most of these tools exist to help builders keep pieces "square," "plumb," or "right," meaning at 90-degree angles—perfectly vertical or horizontal—because as it's often said in building, "If it isn't a right angle, it's a wrong angle!" Layout tools are similar to measurement tools: they require a level of precision. But layout tools are generally most helpful in checking the geometry of how multiple pieces fit together, rather than the individual measurements of one.

Speed square

A speed square is a triangular-shaped tool you can use to lay out work pieces at perfect square (right) angles or to mark cut lines on your material perpendicular to an edge before cutting with a saw.

The speed square was designed for carpentry work, in framing houses, and to lay out accurate cuts of lumber, 2×4s, and other wood materials (and to do so more speedily!).

The speed square is a right isosceles triangle (45-45-90 degrees), with two equal sides and a hypotenuse (the side of the triangle opposite the right angle). One of the two equal sides has an extra lip, or "fence," designed to slide along the side of a piece of wood. When this fence is hooked on to the side of your wood, the other side lies across the face so you can draw perpendicular lines to guide your saw cut. Speed squares are also very helpful tools for "checking square," or making sure that multiple pieces are meeting at a right angle. Speed squares are usually metal or a hard plastic.

Let's say you have a 2×4 and need to cut a 36-inch length off it. Measure 36 inches from one end using a tape measure and make your caret mark. Now you have one accurate point, but it would be more helpful to have a full line across the entire face of the wood at the place you'll cut. This will help you later when you go to line up your saw blade. Hook the fence of your speed square along the long edge of the 2×4, and slide the perpendicular edge up to meet your caret mark. Now draw your straight line across the width of the 2×4. This is exactly where you'll make your cut.

To check square using a speed square (for example, if you are building a box but are not sure your corner pieces are meeting at perfect right angles), set your speed square on the inside corners. If both of your speed square's sides sit flat against the frame sides, congratulations, your box is square! If there is a gap between your speed square and the frame, or if your speed square doesn't quite fit, your frame is not square and needs adjustment.

If you're wondering what all the other marks on a speed square mean, most of them are designed to help carpenters find precise roof and staircase angles when framing houses. They aren't used except for this super-specialized purpose, but it's nice to know what they're for. You'll notice these markings on the speed square:

- **Degree:** Helps you mark and lay out degrees from 0 to 90 in relationship to the edge of the speed square.
- **Common:** Helps you mark the vertical "rise" in inches, in relationship to a 12-inch horizontal run of a roof rafter or staircase.
- **Hip/val:** Helps you mark the rise in inches over a 17-inch run of a roof rafter or staircase.

Combination square

Like a speed square, a combination square (or "combo" square) is a helpful tool for checking right angles or marking perpendicular cut marks. The combo square, though, also has a sliding ruler, which gives you more flexibility to lay out different orientations of pieces and take measurements at the same time.

The combo square has a metal frame and a sliding ruler that is adjusted using a thumbscrew in the frame. You can orient the frame to check 90-degree and 45-degree angles easily, and some also include a small level to check for the flatness of a surface. Both the angled 45-degree edge and the 90-degree edge of the combo square work as a fence, so you can latch it on to the edge of a piece of wood easily. The ruler can then be used to mark dimensions, check angles, and more.

To check "square" on a piece that has already been built, slide your ruler all the way to one end of the combo square frame so that the end of the ruler is in line with the flat edge of the frame. This way you can use the L shape to check right angles.

To draw perpendicular lines like you would with a speed square, slide the ruler halfway down the frame, so you can use the flat edge of the frame as a fence and the ruler sticking out as a perpendicular edge.

Framing square (aka carpenter's square)

Framing squares are mostly used by carpenters to check the layout of wood pieces that make up the roof, walls, and stairs of a building. It acts as an oversize speed square, letting you check the squareness of multiple pieces you will join together.

The framing square has helpful markings that can help carpenters calculate the angle or "rise and run" (a certain measurement up and a certain measurement over, for staircases and

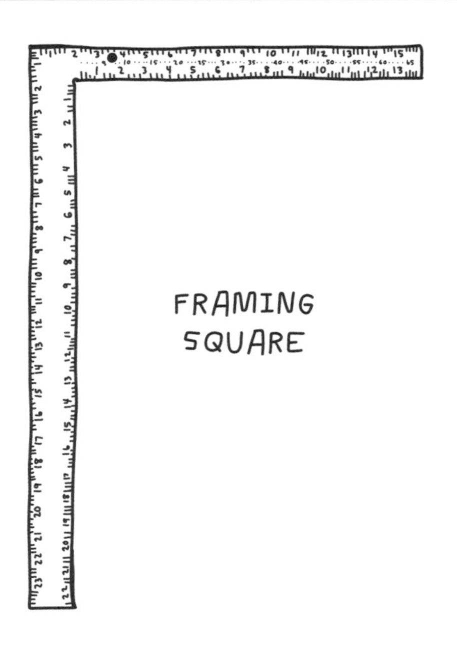

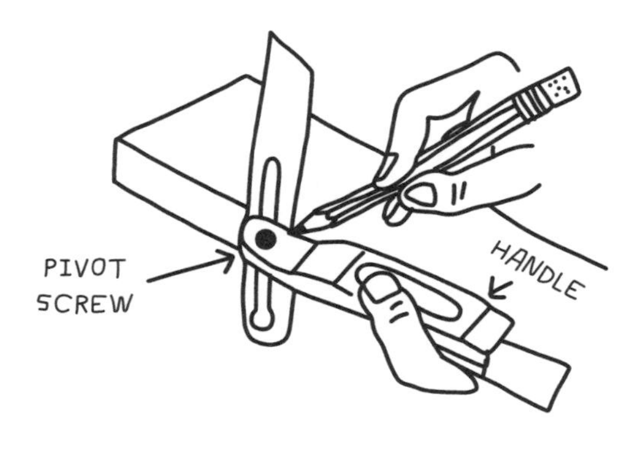

slanted elements). A framing square can be a good substitute for a speed square when you don't have an edge to latch on to (like you can with the fence of a speed square).

Most framing squares are made of a durable metal and come in sizes ranging from 1 foot to 3 feet on the long side (generally one side is shorter than the other).

To check square, lay the framing square over a corner, or set it into the corner of a frame or box to make sure the two perpendicular pieces meet at a 90-degree angle. You might also use the framing square to mark a diagonal line to cut off a corner, because you are able to measure from the corner in two directions.

Bevel square

Also known as a sliding T-bevel angle finder or bevel gauge, this is a magical tool for transferring specific angles from one location to another. You might use a bevel square to match an angle on a project that's already been built, and then transfer that exact angle to a new piece of wood to cut. The bevel square is great because it can match angles without the use of a protractor or knowing what the actual angle measurement is.

A bevel square is made of two parts: a handle (usually wood) and a metal blade or straightedge. This metal part is connected to the handle with an adjustable screw that lets you change the angle relative to the handle. Some bevel squares come with a built-in level, protractor, or digital reading of your angle.

Use the adjustable thumbscrew (sometimes a wing nut) to loosen or tighten the grip on the metal blade. Pivot the blade to match the angle of your work piece, then retighten the thumbscrew to lock the angle in place. Now you can transfer that angle over to other pieces of material.

Miter box

This tool helps guide your handsaw blade to make precise angled cuts. Use a miter box to hold your wood in place and the built-in slots in the box to guide your saw along precise angles. A miter box is a great tool for cutting two pieces with 45-degree edges that attach to form a 90-degree corner, like for a picture frame.

You can find plastic miter boxes at most hardware stores. They have precut slots for 90-degree, 45-degree, and other common angles. A miter box may come with its own handsaw, typically a backsaw with a rigid spine.

To use, clamp your miter box securely to a table or work surface. Place your wood inside the box. Use the built-in hold-down pegs to keep your piece stationary. Locate the appropriate slots for your saw based on the angle you want to cut. Slide the saw into the slots and begin cutting, with short scoring cuts at first, and then longer cuts to complete the cut.

Level

No one likes a tilted table! No matter what you build, you'll want your flat surfaces to be flat—or "level"—meaning truly horizontal in the world. When a surface is level, you can place a marble on it and the marble stays put. If the marble rolls off, your surface isn't level.

A level is a tool that helps you gauge whether a surface is actually level and, if it isn't, can show you which side is high or low.

While many kinds of leveling tools exist, the most common and easiest to read is a spirit level, which uses a small vial of liquid that's not quite filled, so there's a small air bubble telling us how level a surface is. The vial has a very slight curve upward in the middle, so if the surface is truly level, the air bubble sits perfectly within two small lines in the vial. Some spirit levels have more than one vial: one to help you find the horizontal level, and another to help you gauge the plumbness (exact verticalness) of an object. Others also have a 45-degree-angle vial. Fancy!

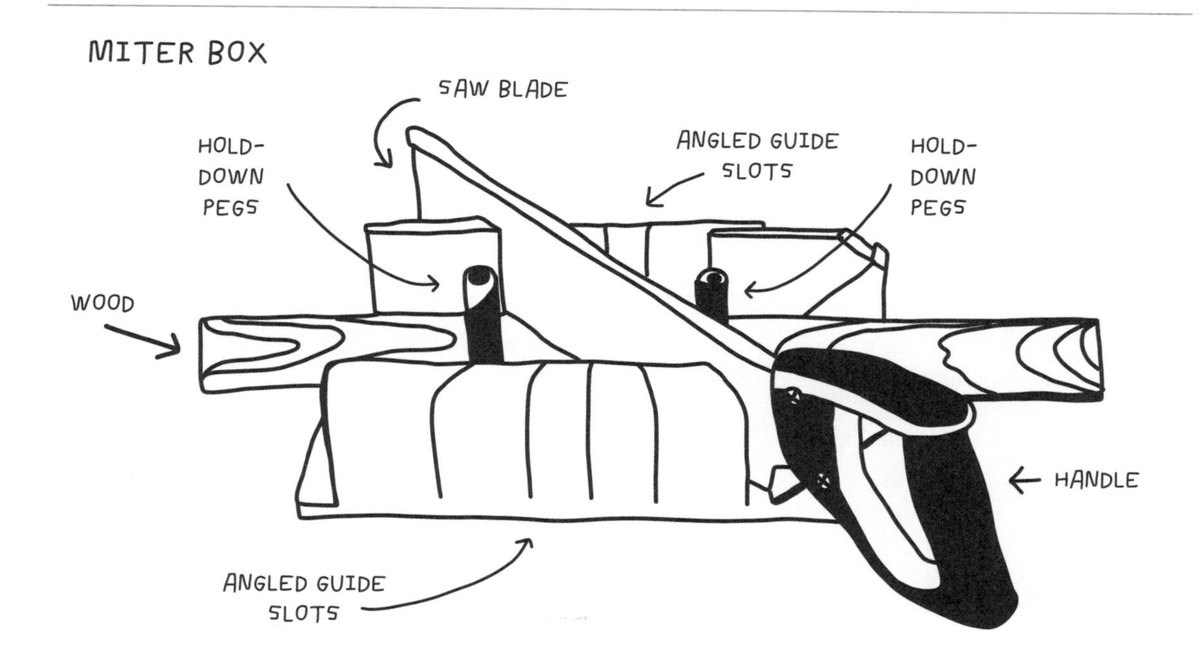

LEVEL

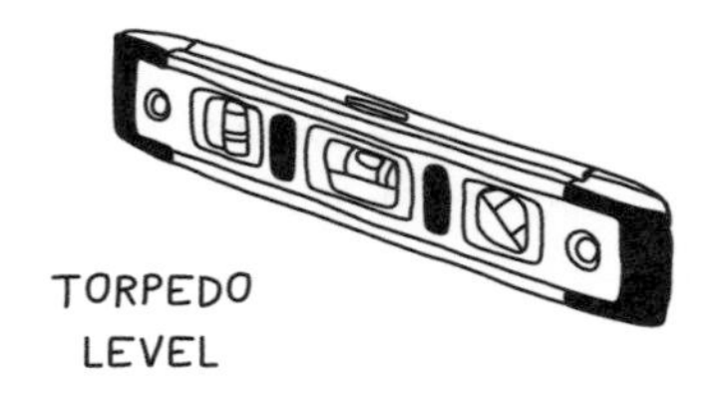

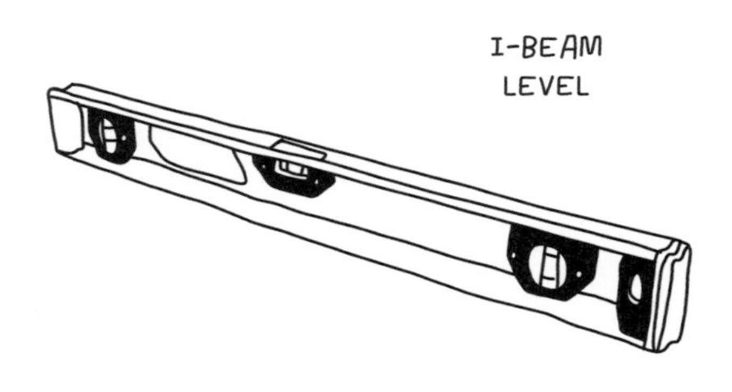

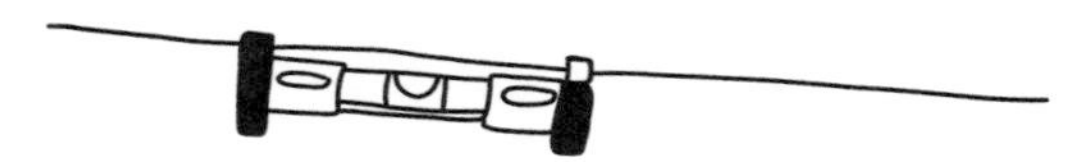

FUN FACT!

The spirit level gets its name from the olden days, when the vials were filled with colored alcohol (or spirits) instead of water.

There are three types of spirit levels that are highly functional and can help you check the level on nearly any project:

Torpedo level: This compact spirit level (usually measuring approximately 8 inches long) has three leveling angles. A torpedo level can quickly tell you how horizontal or vertical a surface is, as well as the accuracy of a 45-degree-angle slope.

I-beam level: This longer spirit level (usually 2 to 6 feet long), is usually made of aluminum or another metal and is used for more accurate level readings across long distances.

Line level: This small spirit level can be strung on a line or string to indicate the levelness of that line across a distance. You might use a line level between two fence posts, for example, to make sure the fence boards in between remain in a straight line. Brick masons also use line levels to check the levelness of one line (or "course") of bricks along a long wall as they are laid.

Place your level on the surface and then read the spirit level by looking at the position of the bubble with your eyes at the same level as the vial. Hopefully, the bubble is exactly between the two lines. If yes, voilà!—your surface is level. If the bubble sits to the right of the lines, the right side of your surface is higher than the left. If it sits to the left of the lines, the left side of your surface is high. Adjust accordingly!

Plumb bob

If there's one absolute you can always trust, it's gravity. A plumb bob is a bullet- or radish-shaped, heavy metal weight (usually made of lead or iron) that can be hung from a long line or string. With gravity, a perfectly vertical line is created along that string. You can use this string line like a vertical level. You might use a plumb bob for tasks like hanging large paintings, when you need to line up an edge along a vertical line. Plumb bobs are used by carpenters to check how perfectly upright their wood studs or beams are. The tip of the plumb bob usually has a point, which can also be helpful for marking a specific spot on the ground, directly below a corresponding point above.

To attach the plumb bob to the hanging line, remove the screw at the top. The top has a hole in it where you insert a string, and you knot it below the surface. Reattach the top and you have a hanging line that's perfectly centered with the center of your plumb bob.

To hang a plumb bob, use a hammer and nail to create a hanging point in the wall or ceiling. Then tie a looped knot in your string and hang it from the nail. Adjust the length of the hanging string so the plumb bob extends to the floor. Wait until the bob stops swinging, then use your line as your vertical guide, or mark the spot under the point of the bob.

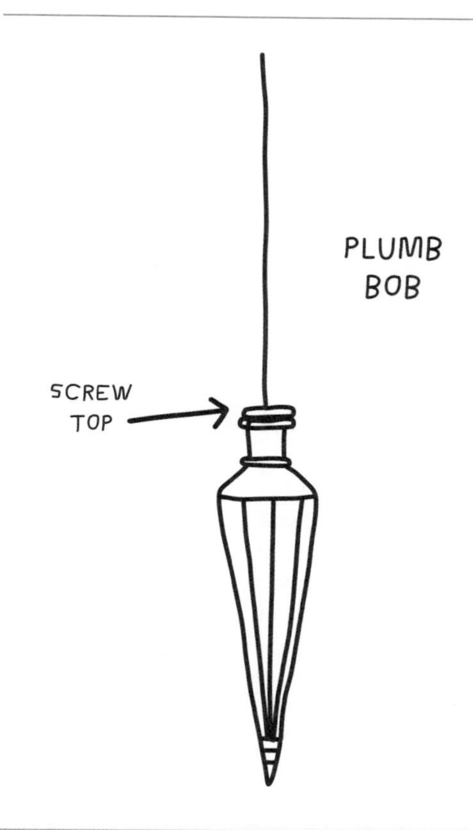

FUN FACT!

The word "plumb" means vertical, in the same way that "level" means horizontal. "Plumb," as well as "plumber," come from the Latin word for lead: *plumbum*. And the symbol for lead on the periodic table of elements is Pb!

MIRIAM E. GEE

Architect and Cofounder, CoEverything

Boston, Massachusetts

Miriam is the real deal. She walks a construction site like a boss—because she is the boss—and runs her own design/build organization called Build Lightly. I admire her so much because she takes initiative to make things happen, rather than waiting for them to happen to her. She is also a gifted architect with an eye for structures that are beautiful and incredibly functional.

"I am an architect, a design/builder, and a design/build instructor. I build super-efficient homes and projects, such as pavilions, bike carts, and parks for nonprofit organizations. I always wear a fanny pack if I'm not wearing a tool belt, and I love my bevel square and worm drive circular saw!

"I was not brought up in a family of carpenters, so my first exposure to building was in an Architecture 101 shop class at Cal Poly State University at age nineteen. The 'status quo' was what initially held me back. But I got over that quickly!

"I'm most proud of the pedestrian bridge I built in Asheville, North Carolina, as it was the first part of an expansive regional greenway in the area. It was also my first permit set, and first project with my collaborator, Luke Perry, and we had all hands on deck from community artists, builders, students, volunteers, kids, and metalworkers to finish the job. Teamwork makes the dream work.

"I think it's important that women in the trades continue to challenge that only men and boys would or 'should' be interested in physical work. Know that hands-on work can be extremely satisfying to your body and soul!

"If you want to build, get out and try it! Find classes and resources in the community that will teach you the basics so you can have confidence and awareness. Then it's all about putting in your time and effort."

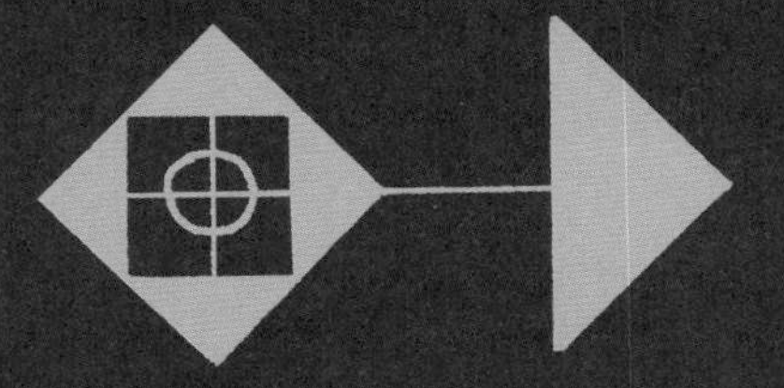

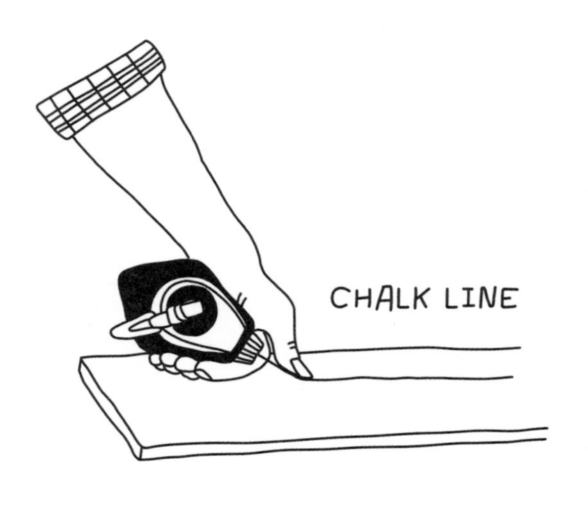

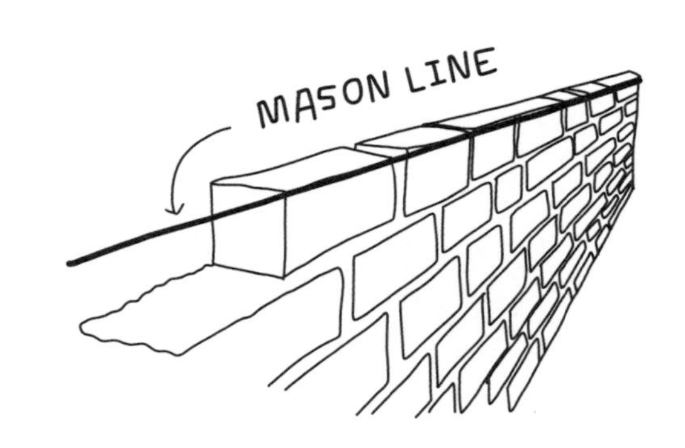

Chalk line

A chalk line is a spool of string coated in brightly colored powdered chalk that you use to mark a line across a long surface. You might need a chalk line if you're cutting a large piece of plywood and need a straight line, or if you need a straight layout line to follow across a long distance.

Chalk lines consist of a canister, or container, with a retractable spool inside. You can pour powdered chalk, which is bought separately, directly into the container to coat your string. Chalk comes in bright colors, usually red and blue, so you can see it clearly when you make your line.

A chalk line is simple to use: just pull the string across your surface, wherever you need your line marked. Have a friend help—one of you holds the end of the string and one of you holds the canister. Pull the string taut, hovering just over the surface of your material. Then pull the center of the string slightly upward and let go to snap it against the surface. This leaves a very satisfying chalk line on your material (and makes an equally satisfying *thwack!* sound).

Mason line

This very helpful type of string is named after the bricklayers and masons who use it to build walls with perfectly straight rows of bricks. You can tie mason line to connect two points, like fence posts, to help you complete the fence with an even line all the way across. You might also use mason line to lay out the footprint of a project, like if you have to dig a trench in your backyard and want a line to follow.

Mason line is usually a fluorescent pink, orange, or yellow color, making it easy to see. You can also use mason line in combination with a line level to ensure a horizontal line, leaving the mason line in place to help you complete your project.

When tying a mason line, use a fixed loop on one side (like a bowline knot; see Essential Skills, page 210) and a taut-line hitch knot (page 211) on the other, so you can adjust the tension until it is taut, then tie it off. Use your mason line like a laser to guide your work.

Scratch awl

Use this tool to scratch or scribe a line into wood. The scratch awl works as a helpful marking device in situations when you don't want to leave any ink or pencil marks.

A scratch awl is a simple metal spike with a sharp point and a wooden handle. You can also use a scratch awl to mark a specific point, making it easier to drive a nail or screw into the same spot.

Scratch awls work best when you are marking a line that follows the grain of wood. You can use a scratch awl across the grain, though your line may not be as consistent.

Sawhorse

Use two sawhorses as supports when cutting a long piece of material. If you're cutting a 10-foot-long 2×4 in half with a saw, you'll need to hold up the two ends of your piece so you can work on the middle. You can build your own sawhorses with instructions from the Building Projects section of this book (see page 254)!

Sawhorses are often sold in pairs but can be bought individually as well. They can be made from wood, plastic, or metal, and some models fold up, with the legs collapsing into the top.

When you need to use sawhorses, place them at the appropriate distance apart, slightly less than the length of your piece. Set the material on top, using each sawhorse as a set of legs at either end. Use clamps to secure your material to each sawhorse, and make your cut.

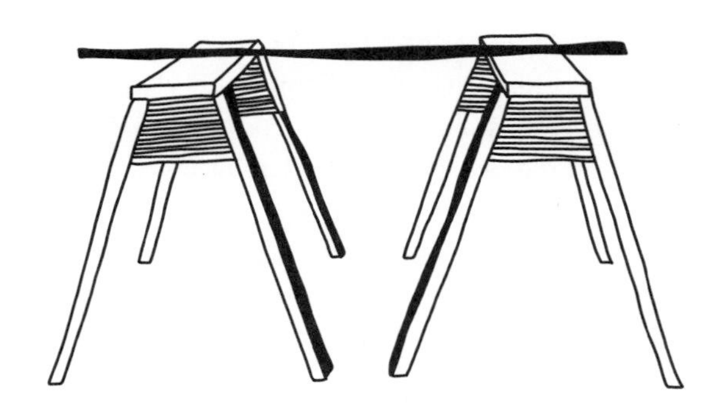

FUN FACT!

Many furniture designers use sawhorses as table bases. Because they are so sturdy, you can place a large sheet of plywood, a reclaimed door, or other table surface on top of two of them, and you have an instant work surface. Keep an eye out for very expensive modern furniture pieces that are really just two sawhorses with a top.

Secure

I have a deep affection for clamps—big ones, small ones, all shapes and sizes. Clamps literally bring order and stability to a chaotic world. Sometimes when I am working with a young girl at Girls Garage on a project that needs an extra set of hands, I'll tell her, "I'm your human clamp!"

When working on any project, after measurement and layout, you need to hold your material securely in place so you can attach it or manipulate it. This is not just to make your work easier, but also for safety. Imagine trying to cut a piece of plywood that isn't secured to a table, flailing about as you try to make a straight cut! Depending on the tool, clamps and vises can exert huge amounts of clamping pressure, measured in pounds per square inch (psi). Some heavy-duty vises can apply thousands of pounds of pressure, while bar clamps exert just a few hundred psi. The tools in this section ensure that you're ready for action, with everything secured to a work surface or its neighboring parts.

Bar clamp

Bar clamps are the all-star MVP of clamps—in my opinion. Use them to secure work pieces to a work surface (clamp a piece of wood you're about to cut to your workbench, for example), to clamp multiple pieces to each other (say, two parts of a project that need to be screwed together), or to glue up a project that needs sustained pressure while the glue dries. Bar clamps are easy to secure and adjust, making them great for projects where you need to move and readjust pieces frequently.

A bar clamp consists of a vertical metal bar with two parallel jaws that squeeze a material between them to create clamping force. The "head jaw" at the top end is fixed and does not move. Think of this as the top row of teeth. A bottom jaw slides up and down the bar, adjusting to the width of the material that needs to be clamped.

The two most common types of bar clamps are an F-clamp, with a swiveling screw in the bottom jaw that tightens the clamp, and a quick-release clamp, which uses a squeeze handle and

BAR CLAMP (F-CLAMP)

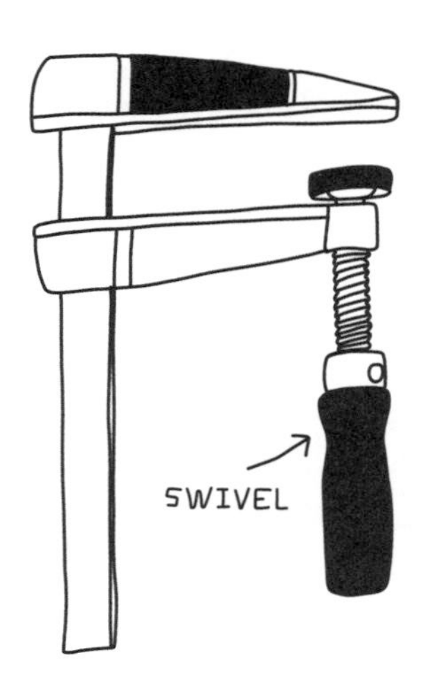

BAR CLAMP (QUICK-RELEASE)

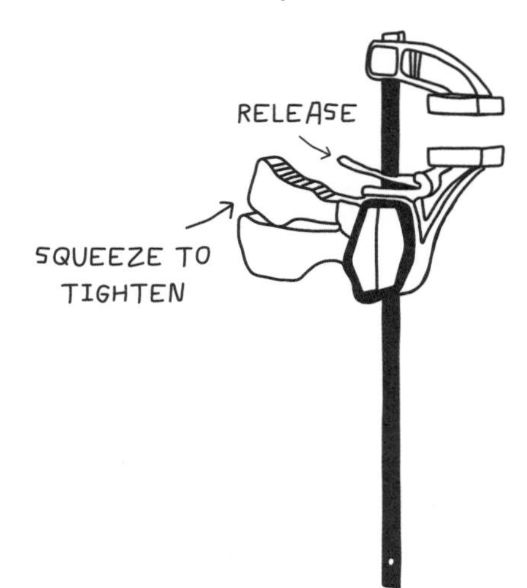

release lever, used with only one hand. Having four to six clamps around, ranging from 12-inch to 24-inch bar lengths, gives you tons of holding power for a variety of projects. Another dimension to pay attention to is "throat depth," which is the distance from the bar to the end of the jaw arms, and determines how far in from the edge of a material your clamp jaws can reach.

Whether you use an F-clamp with a swivel screw or a quick-release bar clamp, here's a good order of operation:

1/ Set your material or pieces in the position you want to clamp them. You might need a friend to hold the piece for you.

2/ Open the clamp's jaws wide enough to fit around your material (or your material and the tabletop, if clamping to a table).

3/ Set the top jaw on the material.

4/ Raise the bottom jaw until you make contact with the other side of the material.

5/ Now tighten! For an F-clamp, turn the swivel until it fits snugly. For a quick-release clamp, squeeze the large handle to tighten (the smaller lever is your release). Don't overtighten, as you might leave dimples, or marks, in your wood.

6/ To remove, turn the swivel on an F-clamp in the opposite direction to loosen. On a quick-release clamp, squeeze the smaller release handle.

If you're clamping a particularly soft material, use a piece of thin scrap wood between the clamp and your material to distribute some of the pressure and to prevent the clamp jaws from leaving an imprint on your surface.

FUN FACT!

In the United Kingdom and Australia, clamps used for woodworking projects are often referred to as "cramps."

PIPE CLAMP

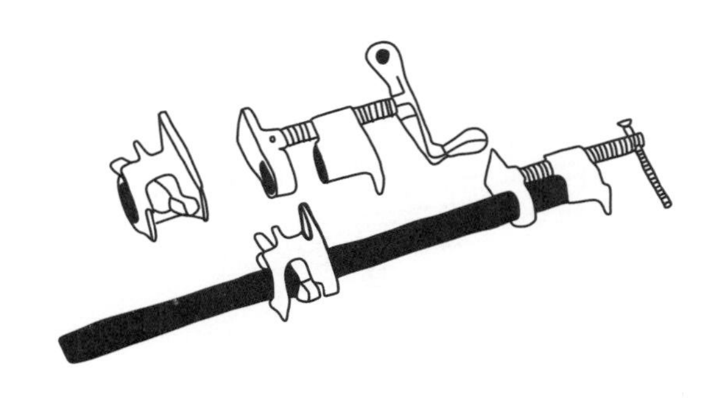

Pipe clamp

A pipe clamp works the same way as a bar clamp by tightening its jaws on two sides of your work material. Unlike bar clamps, however, pipe clamps have jaws that slide along a standard pipe. The benefit of the pipe clamp is that you can attach the clamp head and jaws to any pipe length. You could attach the clamp head to a 10-foot pipe if you wanted to, and glue up an entire tabletop!

The pipes used in pipe clamps are usually ½ inch to ¾ inch in diameter and have threads at both ends so they can screw into the clamp head. You can buy a pipe clamp that comes with a set size of pipe (24 inches, etc.), or just the clamp head to attach to your own galvanized pipe, which you can buy in the plumbing section of most hardware stores.

Just like a bar clamp, place the jaws of your clamp on each side of your work material or

WOODEN HAND-SCREW CLAMP

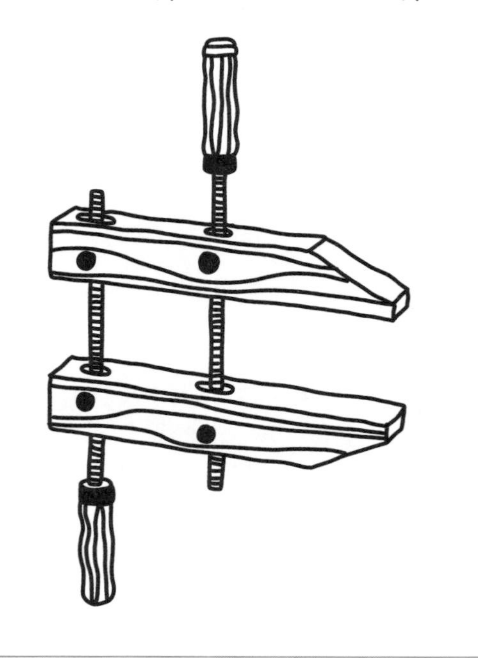

surface. To tighten it, squeeze the release lever on the bottom jaw and slide it up the pipe to meet your material. The top jaw, at the end of your pipe, has an integrated screw, which you can turn using the perpendicular bar until both jaws are tight against the surfaces.

Wooden hand-screw clamp

The wooden hand-screw clamp is something of a clamp ancestor—it is one of the oldest clamping devices used by woodworkers and craftspeople. It's a simple contraption, with two parallel pieces of wood connected by two long screws, which, when tightened at the same time, bring the two wood jaws together to squeeze the material between it. Because of the larger surface area of the wooden jaws, these hand-screw clamps distribute clamp pressure more widely, so they are great for projects with soft wood or materials you don't want to damage.

Wooden hand-screw clamps come in many sizes, with jaws most commonly opening from 4 to 12 inches.

To use it, open the wooden jaws wide enough to fit around your materials. Turn the two screws simultaneously (both clockwise, "righty tighty") so the jaws close while remaining parallel. Hand-tighten to your desired snugness!

C-clamp

It's easy to remember the name of this clamp because it's shaped like a C! C-clamps work similarly to bar clamps by applying pressure using two opposing jaws, or metal "shoes." C-clamps are similar to bar clamps in their strength and mechanics, but they still have special advantages. They're great for metal-work and, because they are made from solid steel or other metal, they are also great for welding, as they can withstand the heat of the weld. And you can use a C-clamp in more compact spaces where you can't fit the long bar of a bar clamp.

You can buy C-clamps in many sizes, with jaw sizes ranging from 1 inch up to 12 inches or more. Larger C-clamps also provide an advantage over the bar clamp if you need to clamp a point on your material farther in from the edge than a bar clamp can reach. In terms of clamping pressure, C-clamps are comparable to bar clamps.

The C-clamp is even easier to use than a bar clamp because it has only one moving part. Place the clamp's C shape around your material with the jaws open. The clamp's vertical screw has a perpendicular metal bar you can easily use to tighten the screw, moving it upward toward the surface of your material. Twist until tight!

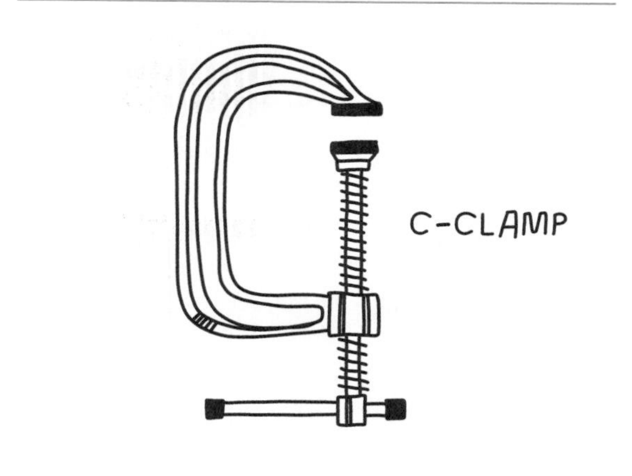

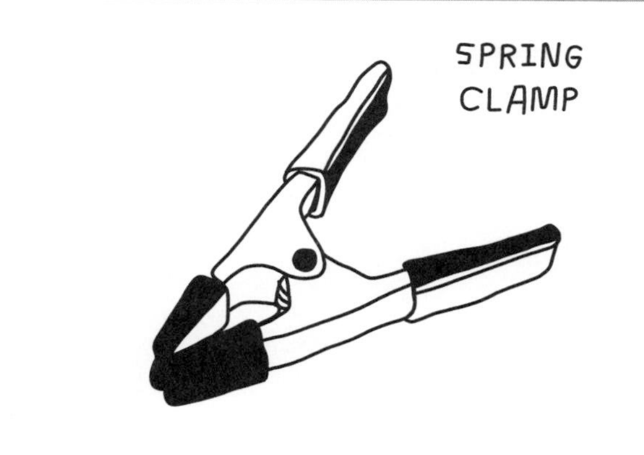

Spring clamp

Spring clamps have relatively low clamp pressure but are incredibly handy and simple to use. They exert pressure in the same way a binder clip holds together a stack of paper. What the spring clamp lacks in strength, it makes up for in user-friendliness. They are great for quick holds, where you just need an extra set of "hands" to hold something temporarily.

Spring clamps are made of plastic or metal, with rubber-coated jaws so they don't damage your material. And they come in a range of sizes ranging from itty-bitty (with a jaw opening of just 1 inch) to 4 inches or more.

Simply squeeze the handles to open the jaws, then release to tighten around your material.

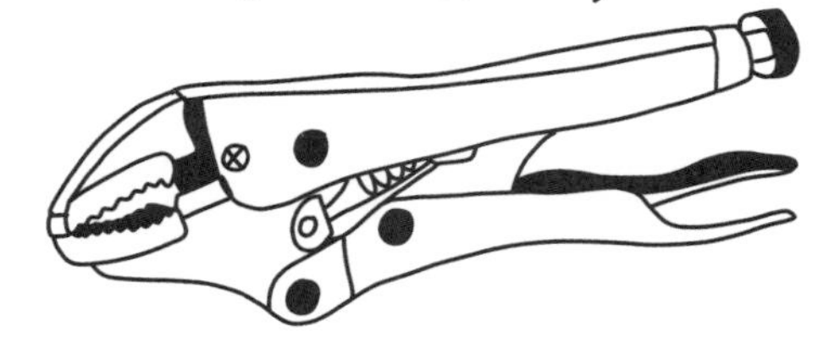

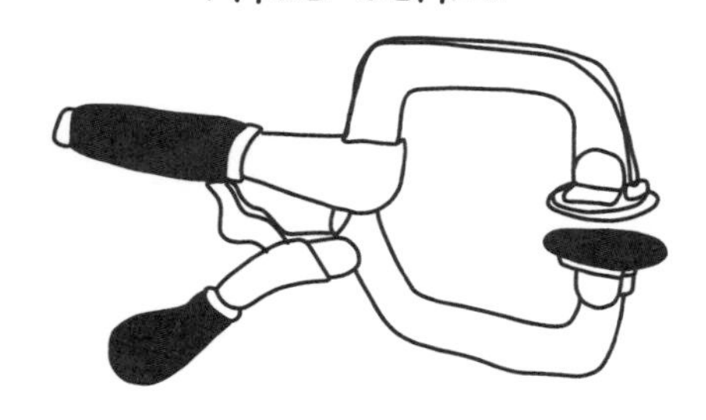

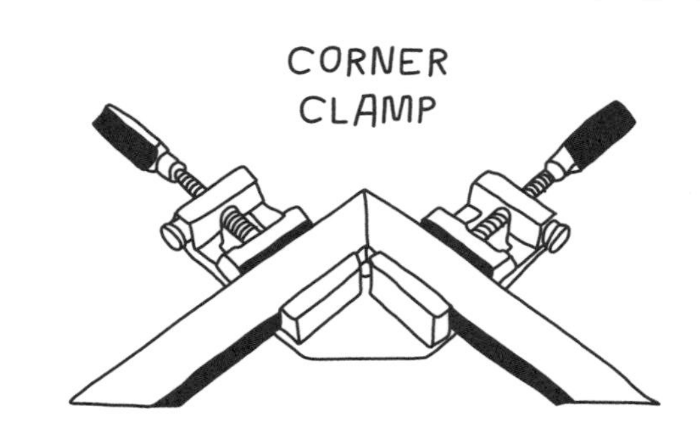

Locking pliers (Vise-Grips) and face clamp

Locking pliers (also called Vise-Grips) are like a cross between a clamp and pliers. They work great when you need extra holding strength or to hold something in place without your hand. Locking pliers have a dual-lever action that is like an extra muscle, locking down the already tightened jaws of the pliers. The greatest advantage of locking pliers is that you can grip something, lock the pliers, and walk away. The locking pliers stay in place until you release them. You can also use Vise-Grips to remove stripped screws or nuts that can't be removed using a screwdriver or wrench.

To use locking pliers, open the jaws wide by squeezing the release lever. Then fit the jaws around your material or hardware, and turn the adjustment knob at the end of the handle until the jaws fit snug around the object. Before they get too tight, release the jaws and turn the knob one more half-turn. Now place the jaws back around your object and squeeze the handle until you hear a satisfying click. To release, pull the release lever on the inside of the handle. Vise-Grips have serrated jaws and are *very strong,* so be careful not to use them on soft materials that could crack or dent.

A face clamp works similarly, but has flat plates as jaws, which are helpful for gripping materials you don't want to dent, like thin plywood or sheet metal.

Corner clamp

Use a corner, or miter, clamp to hold or connect two pieces that meet at a right (90-degree) angle. All the other clamps we've learned about so far are fabulous, but they don't help much with corners. A corner clamp lets you glue and join two perpendicular pieces of wood by holding them snugly at a 90-degree angle.

A corner clamp can connect two perpendicular pieces in an L shape, or pieces cut with 45-degree edges that meet at a 90-degree angle. These 45-degree cuts are called miter cuts (and maybe you cut them using your miter box!). Most corner clamps are adjustable to accommodate varying thicknesses of pieces. Some corner clamps also accommodate a T-shaped joint, with one piece running through the clamp and the other coming in to meet it at a 90-degree angle.

If you're using corner clamps, you'll probably want to apply glue to your pieces. Set both pieces into the clamp channels. Use the screw handles to tighten each side. Keep in mind that a corner clamp does not squeeze the two pieces together with tons of force, but rather holds them in place. This gives your glue time to dry, or you can insert a screw or nail to attach the two pieces. When you're done, just unscrew the handles and remove your piece.

Ratchet strap

Use ratchet straps (sometimes called tie-downs) to secure cargo or loads of wood, or to bundle material in the shop, or for about a thousand other tasks. Ratchet straps are so useful! While ratchet straps are unlike most clamps because they are fabric, they are incredibly strong and provide great holding strength. They also have a sister called a strap clamp, which tightens a circular strap around an object, but I find the ratchet strap to be much more versatile.

Ratchet straps consist of either one or two pieces of webbing (a strong woven strap made from a fabric such as nylon) and a ratchet mechanism that cranks and tightens, pulling the strap tightly around your load. You'll often see these tied around loads of lumber in the back of pickup trucks; just look for the neon orange or yellow straps.

You can use a ratchet strap in a one-strap or two-strap configuration. If you're wrapping a bundle of wood, for example, you'll need only one strap to wrap around the bundle, looping the end of the strap back into the ratchet. You can also use two straps, each with a hook at

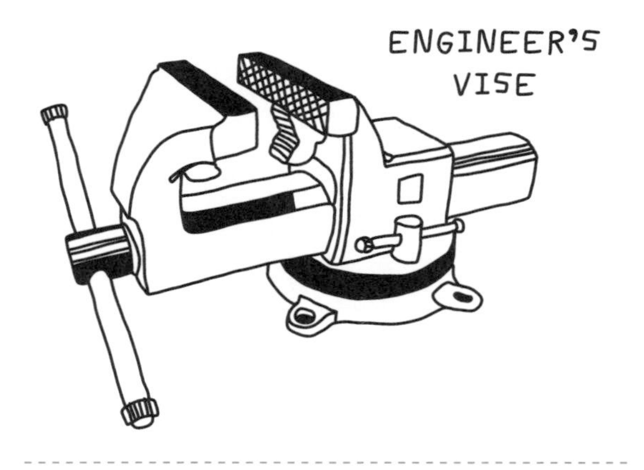

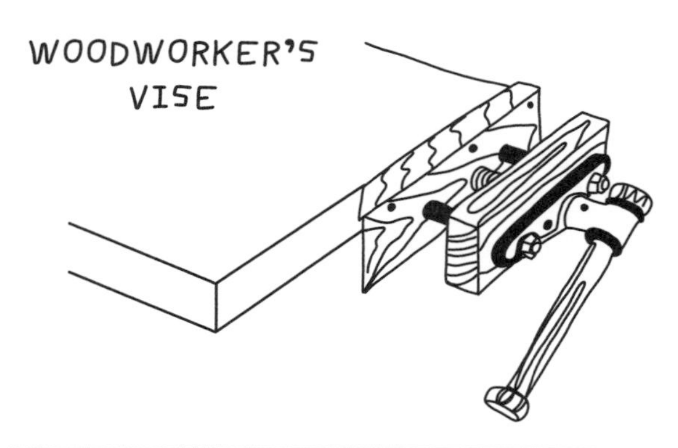

FUN FACT!

There are vises for so many different uses! In New England during the 1700s, the Shaker community was well known for producing high-quality round brooms (like the kind we picture witches riding). At the end of the century, a man named Theodore Bates designed a vise specifically for broom making—one that would press the broom's corn stalks into a flatter stack so they could be stitched flat across their width. This made Shaker-broom making far more efficient—and lucrative, as a result. Practically every broom we use today is flat, and we can thank the Shaker broom vise for its shape and functionality!

the end, to create a taut line attached to two separate points. The ratchet mechanism works because it rotates in one direction but not the other. Feed the end of your strap into the small slot in the spool of the ratchet, pulling the slack all the way through until the strap is tight around your material. Then crank the ratchet back and forth to your desired tightness; it will make a very satisfying ratcheting sound as you do this! When tight, lock the ratchet handle down so it lies flat. To release your strap, open the ratchet handle and squeeze the release lever in the handle (a small metal plate that slides upward toward the end of the handle).

Vise

A vise is a superstrong stationary clamp. It has a set of jaws, just like a clamp: one is stationary and the other moves along a threaded screw to tighten and loosen the jaws. The jaws' sides are parallel, with a larger surface area than a clamp. Most importantly, vises are attached to a work surface, typically a workbench. The two most helpful types of vises for building projects are an engineer's vise (a cast-iron device with wide metal jaws) and a woodworker's vise (attached to a wooden workbench).

A woodworker's vise is usually made of wood and has wide, smooth wooden jaws that can secure wood without nicking or denting it. Depending on the vise and workbench, the jaws of a woodworker's vise can open up to 12 inches or more. An engineer's vise, sometimes called a machinist's or metalworker's vise, is far better for clamping metal. If you try to clamp wood with an engineer's vise, the teeth on the jaws' surface will likely leave a mark. Engineer's vises can be incredibly strong, capable of exerting thousands of pounds of pressure!

Similar to a clamping mechanism, the jaws on a vise should fit snugly around your material. Insert your material between the jaws and rotate the bar lever clockwise until tight. To release, rotate the bar counterclockwise until you can remove your piece.

Bench dogs and hold-downs

Bench dogs are as helpful as their name is endearing. Unfortunately, they only work if you have a workbench with a built-in woodworking vise on the side of the table and bench holes in the surface (which I highly recommend).

Bench dogs are small plastic, wood, or metal pegs, about 1 inch in diameter and 4 to 6 inches long, which can be set into the holes in the table and vise, essentially creating two sides of a clamp's jaws. The bench dogs act as backstops, one in the movable vise, and one static in the table. They stick up just enough to hold your material in place; using the adjustable vise, you can securely hold your piece on the table's surface. This is particularly helpful if you're chiseling, carving, or doing surface work on a piece of wood.

Most bench dogs are cylindrical, with one end cut into a half cylinder. This creates a flat side that can stick up above the surface and rest against your wood. Some are square, though, and everyone has a favorite kind of dog—plastic, metal, wood, homemade, etc. The bench holes in a workbench can also be used for other types of dog devices, like hold-downs, which exert downward pressure on the surface, too. Holes for bench dogs are typically 6 to 12 inches apart in the table's surface, which, combined with the adjustability of your vise, should accommodate all lengths and widths of projects.

Set the piece you want to secure on top of the workbench, with one end hanging off the table, over the vise. Insert one bench dog in the vise, adjusting the vise as necessary so the bench dog sits against the end of the material. Then insert another bench dog into a hole in the table, close to the other end of the wood. Tighten the vise as needed until the wood is held firmly between the two bench dogs.

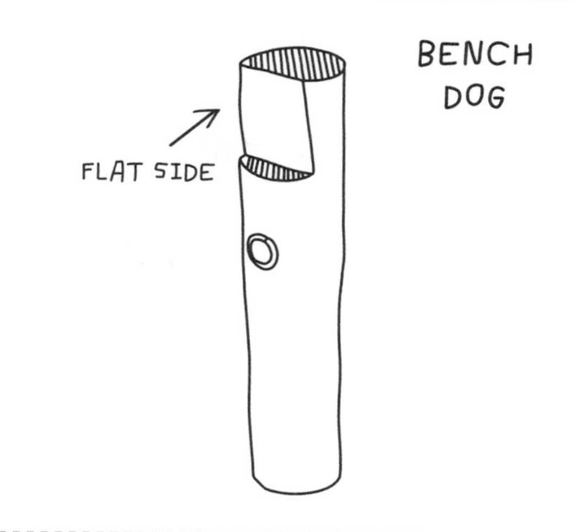

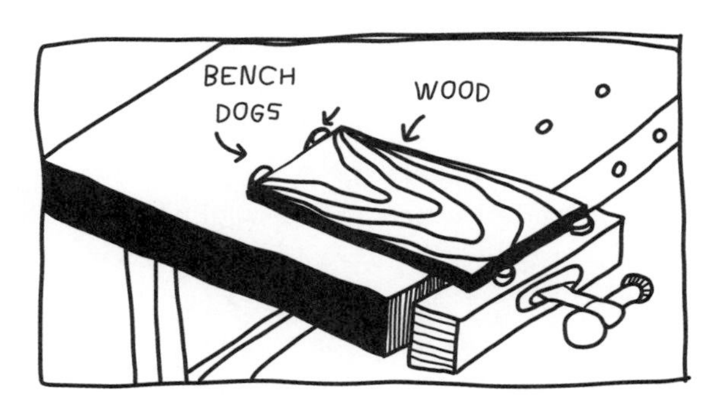

FUN FACT!

In engineering, in general, the term "dog" is any tool or part that physically obstructs movement. Apparently this comes from the image of a dog biting down with a locking jaw, but it might also be inspired by dogs like mine, who are opposed to movement when comfortably asleep on the sofa.

HAND TOOLS

By now you've got your materials purchased, measured, laid out, and secured. Now you're ready for action! The next step in any project is cutting or manipulating your materials using a variety of hand or power tools. We'll start with the hand tools for their simplicity and versatility, because who needs electricity? With a little muscle, and the appropriate hand tool, you'd be surprised at how much you can accomplish. This next section is a "best of" list of some hand tool choices for any job.

Hammers

About 2.6 million years ago, our ancestors started banging stones against other, larger stones to crack nuts, cut up food, break open skeletons, and later, to shape wood or other stones into more tools. Around 500,000 BCE (there is lots of debate about the exact date), humans started attaching a piece of wood or bone to these stones to create a handle. Game changer! The handle essentially extended the human arm, making the force of any blow much greater. And thus begins the lineage of the hammer we still use today!

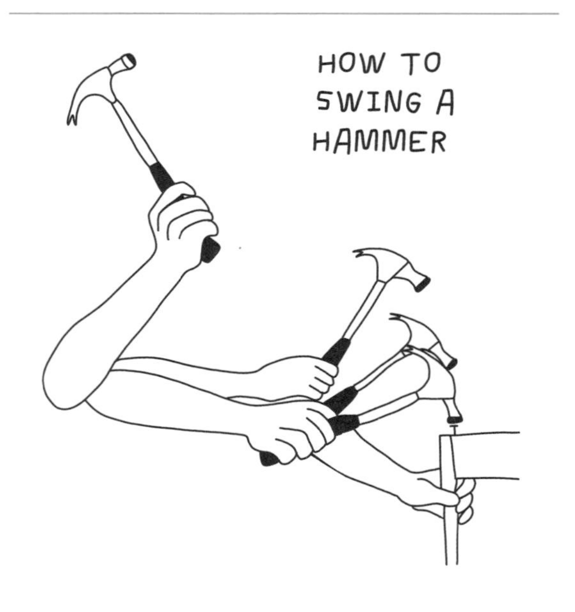

While there is a hammer for every job, in general, a hammer is simply a tool designed to strike a powerful blow. You might need one to drive a nail, or shape metal, or knock a dent out of your car.

Tips for use

For first-time users, it can be tempting to grab the top of the handle, near the hammer's head, and punch with it, like a boxer. This isn't very effective, as it basically cancels out the benefit of having the hammer's "extra arm length." Instead, follow these steps to correctly swing a hammer:

1/ **Always wear safety glasses when using a hammer!**

2/ Position your nail with your nondominant hand and tap the nail head with your hammer, just enough to hold it in place. Then move your nondominant hand out of the way.

3/ Place your hand at the end of the handle where there is often a rubber grip or grip marks.

4/ Keep your eye on the nail head, then swing from your elbow, then your wrist. Think of the motion you'd use when using a flyswatter. Your elbow and your wrist are both rotation points, so use them to your advantage to maximize force!

Types of hammers

The following are six very helpful hammers (though, of course, there are many more).

Claw hammer: If you have a hammer in your house, it's likely a claw hammer. As the jack-of-all-trades hammer, the claw hammer is great for driving nails, removing nails, and general jobs. While it is generally associated with wood and woodworking, it is a medium-duty tool that can be used on metal, plastics, and other materials as well.

The claw hammer is usually made from two parts: a wooden handle and a metal head. The head has a flat, untextured "face" on one side and a split "claw" on the other. Claw hammers come in different weights, usually ranging from 10 to 20 ounces.

Hold your nail in place with your nondominant hand and tap it lightly with the hammer, just enough to hold it in place. Then remove your nondominant hand. Swing the hammer from the end of the handle, making sure to strike the head of the nail squarely with the face of the hammer. To remove nails, use the top of the head as a pivot point, and slide the nail you want to remove into the split in the claw. Using a block of wood under the claw for leverage, pull the nail out.

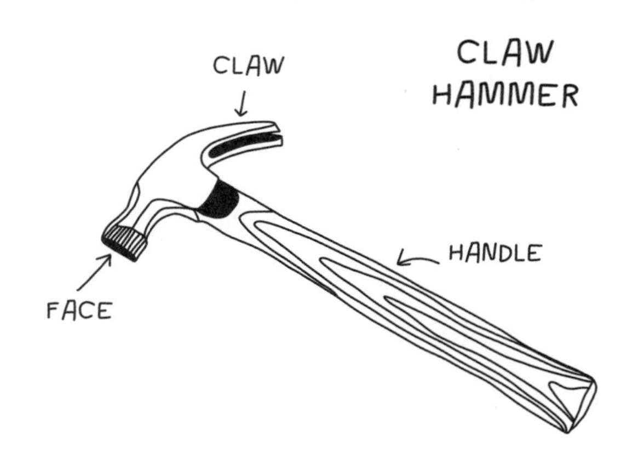

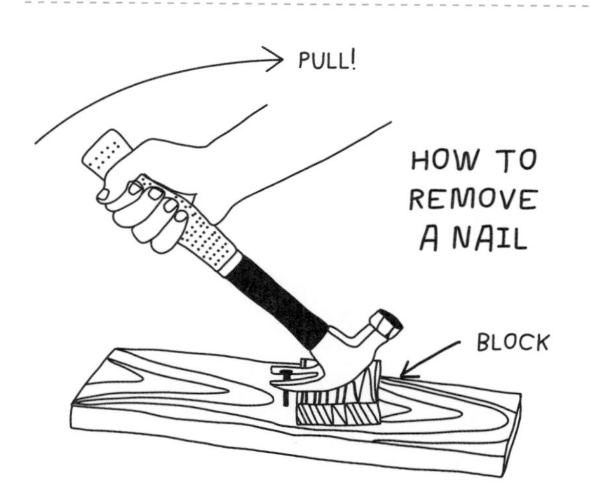

FUN FACT!

Pierson Building Center, a local hardware store in Eureka, California, claims to have the world's largest claw hammer. It measures 26 feet tall and is an exact-scale replica of the Vaughan company's No. D020 model claw hammer.

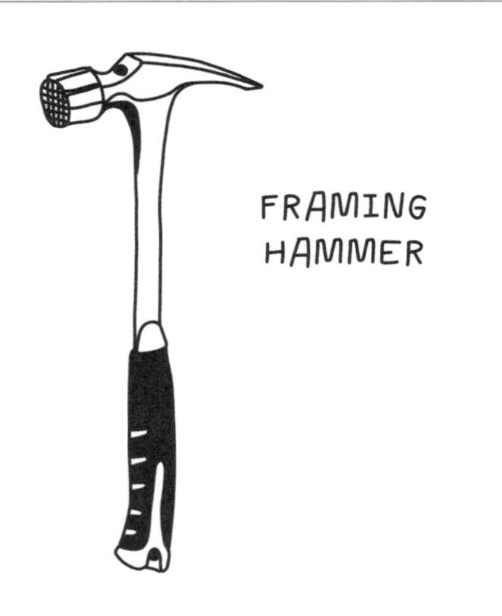

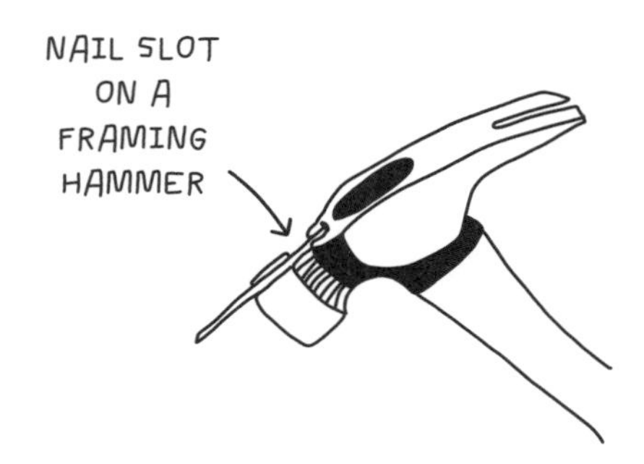

FUN FACT!

Some framing hammers have a magnetic head with a slot to help set a nail in place. The magnetic head and slot let you set and drive nails with only one hand!

Framing hammer: A framing hammer is similar to a claw hammer but is even more specialized, making it the ideal tool for wood framing. In particular, carpenters use framing hammers to drive nails into the 2×4-framed walls of houses. Use a framing hammer if you have a project that requires hammering a lot of nails!

When compared to a claw hammer, a framing hammer is generally longer and heavier, with heads weighing from 12 to 32 ounces. This provides more force with each blow, making it easier to drive many nails all in a row, with fewer swings per nail. Framing hammers also have straighter claws and a waffle-textured head for extra grip, which can help drive a nail even if the striking angle isn't perfect.

Swing a framing hammer just like a claw hammer (see page 97). However, because of its straighter claw, the framing hammer does not work all that well for removing nails.

Ball-peen hammer: The ball-peen hammer is also referred to as a machinist's hammer, and is mostly used for metalworking. Because of its dual head, it can be a good hammer to have around for rounding out metal edges or knocking small dents out of sheet metal. Small ball-peen hammers are also used in jewelry making to work wire or sheet metal into their final forms.

The ball-peen hammer has a double-sided head—one with a flat face, and one with a rounded ball. Historically, blacksmiths would "peen" their metal, hammering it over and over to harden it. While this process isn't as commonly done by hand anymore, the round end of a ball-peen hammer can be helpful in situations when you don't want a flat head, like if you were shaping metal and didn't want pronounced dents.

Tack hammer: Sometimes also called an upholstery hammer, the tack hammer is used to install small nails, tacks, or brads.

Because of its slim profile, the tack hammer is a lightweight and precise tool and particularly helpful in upholstery, when many nails or tacks must be installed to secure fabric or leather. Most tack hammers have a magnetic head that helps set the small nails and tacks.

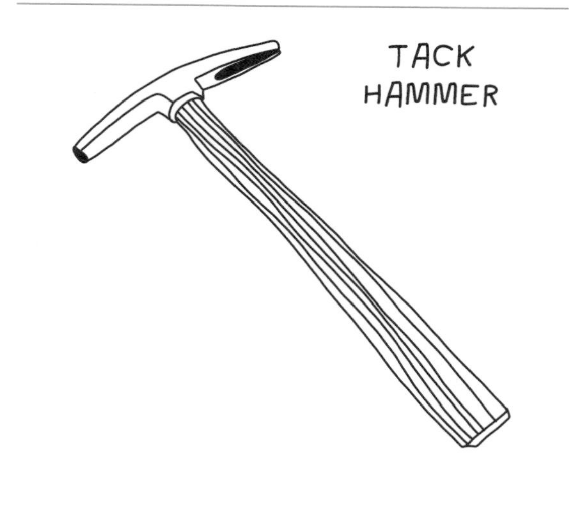

Sledgehammer: A sledgehammer, with its long handle and large and heavy metal head, is often used for demolition work: to tear down walls or break apart large blocks of masonry. We can also thank the sledgehammer for our railroads! During the construction of railroads, sledgehammers were used exclusively to drive the spikes into the ground, which held the rails and wood in place.

Sledgehammers' long handles range from about 20 inches to more than 3 feet long. And their metal heads can weigh up to 20 pounds! The head has a large face on either side, with a wider surface area to distribute the load more evenly.

Because of a sledgehammer's weight, you definitely need two hands to lift it. The good news is that gravity does most of the work, because of the sledgehammer's substantial weight. You can lift a sledgehammer over your head and let it drop, or start at waist height and swing downward.

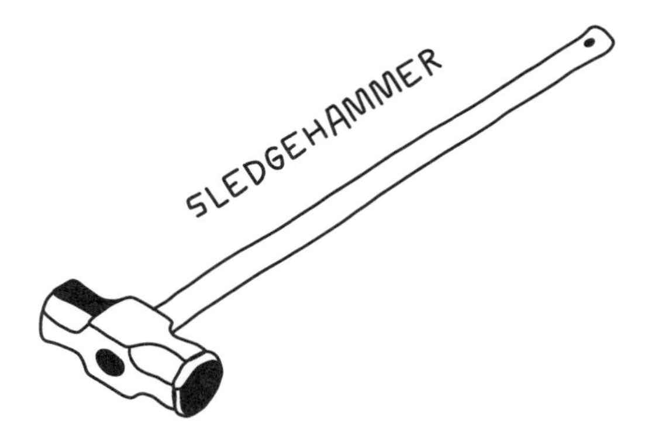

FUN FACT!

In the fall of 1989, the demolition of the Berlin Wall began with an iconic action symbolizing the fall of communism and end of the Cold War. Many German citizens (and tourists) were famously photographed swinging sledgehammers, stone-cutting hammers, and masonry hammers to chip away at the wall.

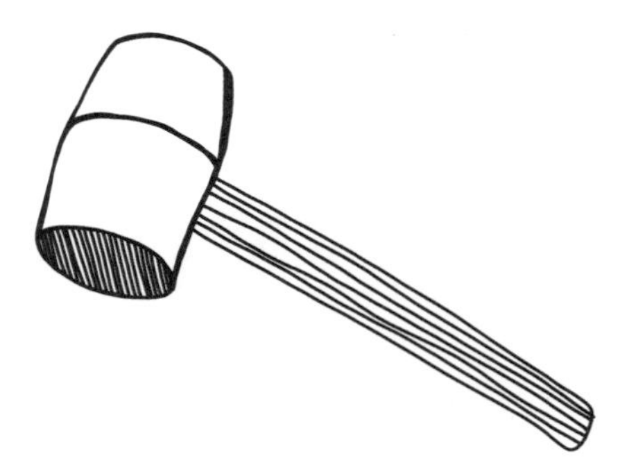

FUN FACT!

Mallets are used in many other industries beyond just building! They are used with percussion instruments (such as xylophones), to tenderize meat, in the game Whac-A-Mole, and as a source of constant comedy in cartoons like *Looney Toons*.

Mallet: A mallet has a large head that is often made from rubber or sometimes wood (something softer than steel). For building projects, a rubber mallet is incredibly helpful when you need to knock something in place with a softer blow than a hammer, or when you don't want to dent your material. You might use a mallet to tap a chisel when woodcarving, to tap wood components into place during assembly, or to strike finished metal without denting or marking it.

Most mallets come in weights of 8 to 32 ounces, with lengths that measure between 12 and 24 inches. The rubber versions are often white or black, and I prefer the white rubber because it is less likely to leave any markings on light-colored wood.

Mallets don't require a large arcing swing like hammers. Because mallets are typically used for lighter-duty force, you can tap them more lightly and with less precision because of their larger head.

Screwdrivers

Just as a hammer *knocks* items into place, a screwdriver *twists and tightens* them into place. Typically, as the name would imply, screwdrivers are for driving screws, rotating them along their threads into wood, metal, or a predrilled hole. Because screwdrivers are not electrically powered and require human muscle, they are generally more helpful for light-duty jobs, like installing small machine screws into predrilled holes or assembling furniture.

Tips for use

No matter what type of screwdriver or screw-drive type you're using, here are a few helpful tips:

- Righty tighty, lefty loosey! Turn your screw-driver clockwise (to the right) to tighten, and counterclockwise (to the left) to loosen.
- Push into your material as you turn. Your screwdriver tip should fit snugly in the drive slot in the head of the screw; apply pressure into your screw and material as you turn it.
- Keep your screwdriver in line with your screw. The shank of your screwdriver should be perfectly in line with the line of your screw, ensuring a good grip and that the screw is going into your material at the correct angle.
- Don't overtighten. Once you get started, you might be inclined to tighten like a superhero. In general, you're better off tightening until snug, but not overtightening to the point where you'd struggle to loosen it. For certain uses like bike repair, this is especially true because you'll likely need to remove and adjust these screws. A torque screwdriver can help you regulate this!

STANDARD SCREWDRIVER

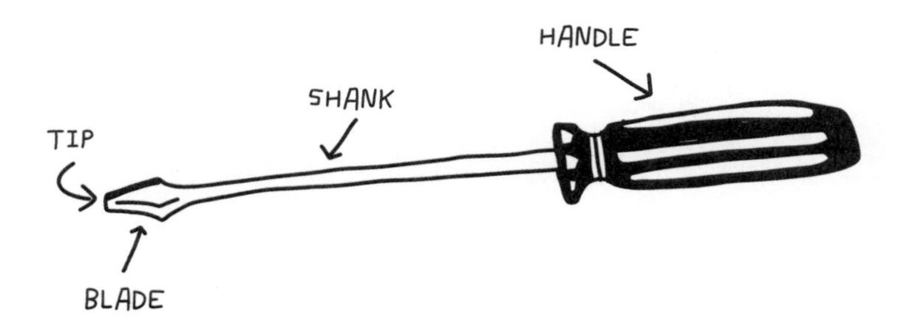

MOST COMMON DRIVE TYPES

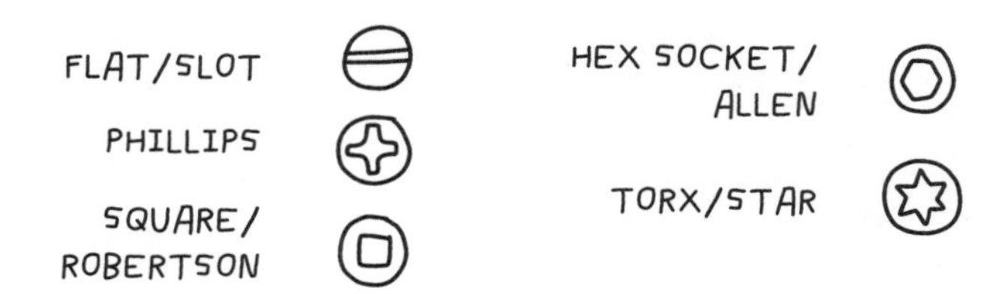

FEAR
LESS.
BUILD
MORE.

ERICA CHU

Girls Garage Fearless Builder Girl, Girls Garage student

Berkeley, California

At age ten she came to the very first camp we offered in 2013, and then we couldn't get rid of her (kidding!). Erica is an incredible young woman and, after attending nearly every Girls Garage class for four years, was the first girl in the program to earn all ten skill badges and her Fearless Builder Girl certificate. Erica is also an exceptional student, volleyball player, and now a junior instructor and leader to younger girls who want to learn how to build.

"My grandfather was an architect and was very handy, so I started building and fixing things with him when I was only four years old. We patched holes in the wall and installed a new handrail along the stairs. I was four years old, using power tools and plaster! My grandfather taught me how to read a blueprint, too. Building was something special I would do with my grandfather, and we bonded over building because we were the only ones in the house who knew how.

"Since then, I've had the privilege to continue my building experience in the Girls Garage program, which I started when I was ten years old. Learning from my grandpa and extending my skills at Girls Garage has inspired me to become a fearless builder woman, and this passion shows around my house—a whole area of my room is filled with furniture and projects I've built at Girls Garage!

"Of all the tools I know how to use now (and that's a lot!), I love the good old hammer. Something about hammering in a nail perfectly, with only a few swings, is very satisfying and makes me happy.

"I'm super proud of the nightstand I designed and built at Girls Garage summer camp. I drew it, engineered it, and built the whole piece from start to finish, with help from my fellow campers and female instructors. I built it exactly as I envisioned it, and painted it, too. My nightstand is still by my bed and is a physical representation of my unforgettable memories creating my first big project. Since then, I have earned all ten skill badges from Girls Garage (I was the first girl ever to do so) and am now a certified Fearless Builder Girl. I love helping other girls have these experiences, too, and as a junior counselor, I try to support them in every way.

"Being a Fearless Builder Girl has empowered me to push myself and be even stronger than the day before. When I am building, I am reminded that I have the power to go against female stereotypes. I want to be a structural engineer, and I know that won't be an easy path, but I know I can get there. I must work for what I want and persevere through any problems that may get in my way. I will never give up my 'building warrior' talent!

"I'd like to tell other girls and women: Do not be afraid to try something new, and find that spark that will help you persist through any complications. You don't know what you can and can't do if you don't try, and anything can be accomplished when you *believe* it can be done."

FUN FACT!

Some screwdrivers have a darkly painted tip, which is purely for visual contrast, to help you distinguish the tool from the screw.

STANDARD SCREWDRIVER

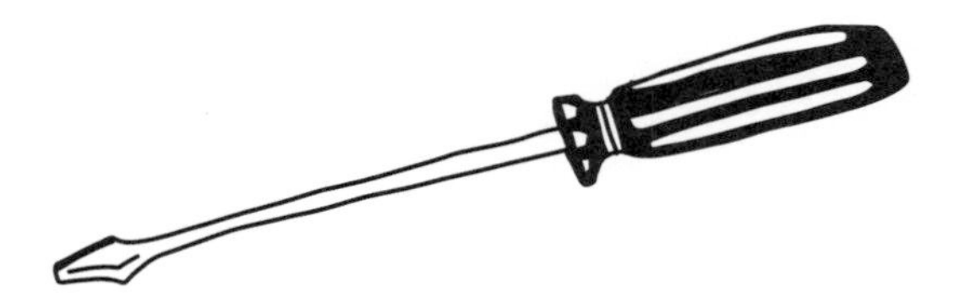

TORQUE SCREWDRIVER

FUN FACT!

Auto-mechanic shops use a pneumatic version of a torque wrench to tighten the lug nuts that hold your car's wheels in place. If your lug nuts were too loose, your wheels would come spinning off. And if too tight, the nuts can damage, fracture, or bend your wheels.

Types of screwdrivers

Screwdrivers have tips that correspond to the drive type of the screw (see page 51 for more on this), so a Phillips-head screw requires a Phillips screwdriver, and so on. These drive types also correspond to drill bits for power drills and drivers. The following are the most commonly used screwdrivers on many projects:

Standard screwdriver: Use a standard screwdriver to tighten and loosen screws. Because a screwdriver has a tip designed to match the drive type, slot, or recess in a screw's head, make sure that your screwdriver matches both the drive shape and the drive size of your screw (for example, Phillips #2 or Torx #25). Use a screwdriver instead of an electric impact driver for smaller or lighter-duty jobs, like tightening screws on furniture or machinery, where extra turning force (torque) is not needed.

Screwdrivers are available with as many types of tips as there are types of screws. Just like screws, you can buy screwdrivers with flat/slotted, Phillips, Torx/star, square/Robertson, or hex tips. Even among Phillips screwdrivers, there is a wide range of sizes. Screwdriver handles are usually rubber or plastic, and some have a hexagonal collar at the top of the handle, letting you use a wrench to turn it when you need extra force.

You may want to try turning the screwdriver by holding the handle and twisting with your dominant hand, and using your nondominant hand to keep the long metal shank of the screwdriver in line with your screw.

Torque screwdriver: For certain jobs, having screws that are too tight is as bad as having screws that are too loose! This is particularly true when working with metal parts, like in an auto-repair shop or on a bicycle. When you overtighten these screws, they can damage the parts they are attached to, making the whole object or structure weaker or more prone to breaking or bending. You might even break off the head of your screw. You can set a torque screwdriver to a specific tightness, which will prevent you from tightening your screw beyond that point.

What the heck is torque, anyway? Think about the arm strength it takes to open a stubborn jar of pickles. Torque is the force required to rotate or turn an object. Using a torque screwdriver, we can regulate how much force is applied to turning a screw. With a manual torque screwdriver, you use a small gauge to set the maximum torque force, usually between 2- and 40-inch pounds. Once you reach your desired tightness, the torque screwdriver either spins in place and does not allow you to tighten any further, or it makes a clicking noise to tell you to stop.

You can use a torque screwdriver in the same way you use a standard screwdriver. For any application where specific tightness is important, the manufacturer or manual will generally indicate a recommended torque so you don't have to guess.

Hex-head screwdriver: Some screws, like specific sheet-metal or machine screws, have hex-shaped heads. Most of these hex-shaped screws also have a flat or Phillips head slot so you can tighten them with a standard screw-driver. However, you can also use a hex-head screwdriver to rotate these screw heads exter-nally. In some cases, this can give you a lot more turning force.

Hex-head screwdrivers, unlike standard screwdrivers, turn screws externally, by grabbing the outside edge of the screw head instead of fitting into a hole or slot in the screw's head. Hex-head screwdrivers come in all the same sizes as screw heads, with dimensions that correspond to the diameter of the screw head.

Use a hex-head screwdriver in the same manner and direction as a standard screwdriver. You can also use them to tighten hex-head bolts!

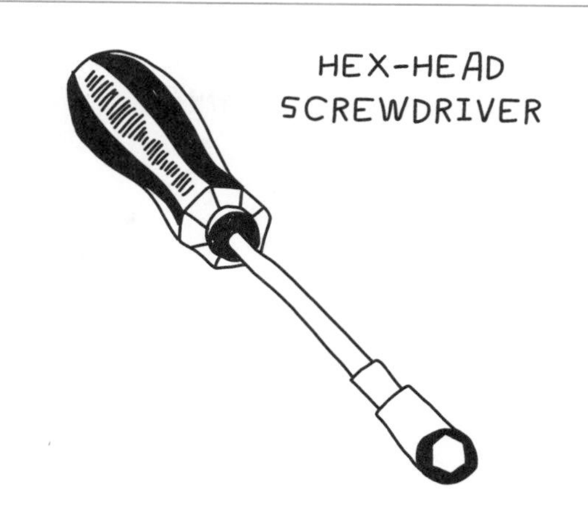

FUN FACT!

Don't confuse a *hex-head* screwdriver with a *hex-key* screwdriver. The hex-key screwdriver has a tip shaped like a hexagon, which is inserted into the hexagonal slot in the head of the screw.

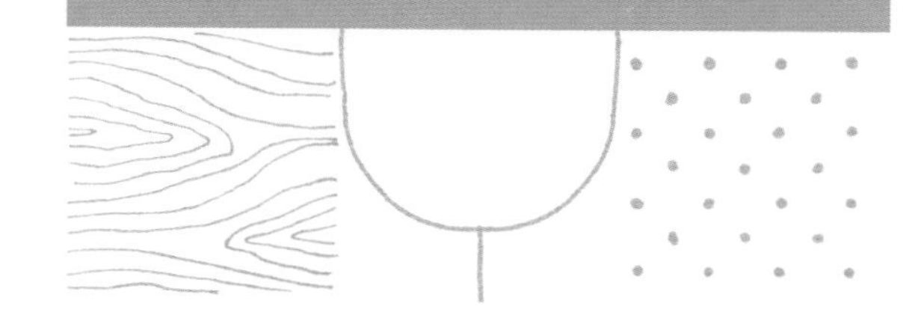

Wrenches

I have an 18-inch open-ended wrench displayed proudly on one of my bookshelves. It is covered in the signatures of ten of my students who gave it to me after we finished and opened the farmers' market structure we built together in their small town in North Carolina. I asked them why they chose a wrench and they said, "Because it symbolizes building and fixing things, and because we tightened SO many nuts and bolts!"

Screwdrivers and wrenches both work by rotating the head of a piece of hardware into place (a screw or bolt). But while screwdrivers usually require an internal slot or drive in the screw's head, a wrench works by grabbing the *outside* of a bolt's head and turning it using heavier torque (the force required to turn an object). While screwdrivers generally come at the screw in the same direction the screw is pointing, wrenches grab on to the bolt head and act as a lever, turning around the bolt's head to tighten it.

Tips for use

When using a wrench, here are a couple of tips to keep in mind:

- The longer your handle, the greater the turning power. Depending on the wrench type and size, your handle might be 5 inches to 2 feet long. The longer the handle, the greater the distance your hand is from the bolt or nut, and the more torque you can apply to turn it.

- Two sizing systems! You'll notice that wrenches come in both metric (mm) and standard (inches) sizes. If you see the abbreviation "SAE," that stands for "Society of Automotive Engineers" and refers to the American (standard inches) system. Regardless of which you select, make sure the size corresponds to the size of your bolt head or nut.

Types of wrenches

There are tons of wrenches designed for the most specific jobs (like fire-hydrant wrenches and lug-nut wrenches for car wheels). The following are a few you might find most helpful:

Open-ended wrench: You'll use an open-ended wrench to turn hex-head bolts and nuts, particularly in compact spaces where you only have access to one side of the bolt or nut. The open-ended wrench has one (or more commonly two) usable ends, each with a U-shaped opening (jaws) that fits perfectly around the same size nut or bolt. Open-ended wrenches with two usable ends have two different-size openings.

You'll notice that the jaws might not be perfectly in line with the wrench handle, but rather oriented at a 15-degree angle away from the axis of the handle. Some open-ended wrenches have one set of jaws pointing at a 90-degree angle (perpendicular) to the handle, too. Both these angles give you better access and movement.

Because open-ended wrenches are not adjustable, you'll want a standard set with sizes ranging from about ¼ inch to 1 inch. Make sure your wrench is snug around your bolt head or nut before turning, and don't use a wrench that is too big for your hardware! You will strip the sides of your bolt or nut, making it impossible to turn later. Turn the wrench and tighten the nut a little bit at a time. After each turn, remove your open-ended wrench and reposition it before turning again.

Box wrench: Unlike open-ended wrenches, box wrenches have a fully enclosed end that lets you grip the entire circumference of your bolt head or nut. They work great when you have limited space, like an open-ended wrench, and some come with a kink in their handle, which keeps your hands and knuckles farther from the surface and less likely to get scraped.

The circular openings of a box wrench have either six or twelve inside edges, or grip points. These edges give the box wrench a great grip on a hex-head bolt or nut. One box wrench will have two usable ends, each a different size.

Just like with an open-ended wrench, the box wrench isn't adjustable, so you need a set with various sizes. You can always add smaller or larger sizes as needed! Use the box wrench as you would an open-ended wrench, turning clockwise to tighten, and counterclockwise to loosen, repositioning after each turn.

Combination wrench: As the name implies, the combination wrench combines the open-ended wrench and the box wrench. A single combination wrench has an open end and a box end that fit the same-size bolt or nut.

The open end of a combination wrench usually has the same 15-degree orientation as a regular open-ended wrench. It's best to have a set around with sizes ranging from about ¼ inch to 1 inch.

Use either end of this wrench in the same way you use an open-ended or a box wrench. Just like with a box or open-ended wrench, you have to reposition it before each turn.

OPEN-ENDED WRENCH

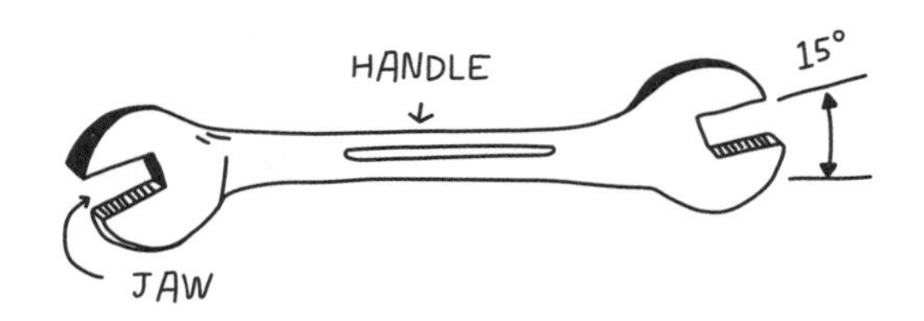

BOX WRENCH

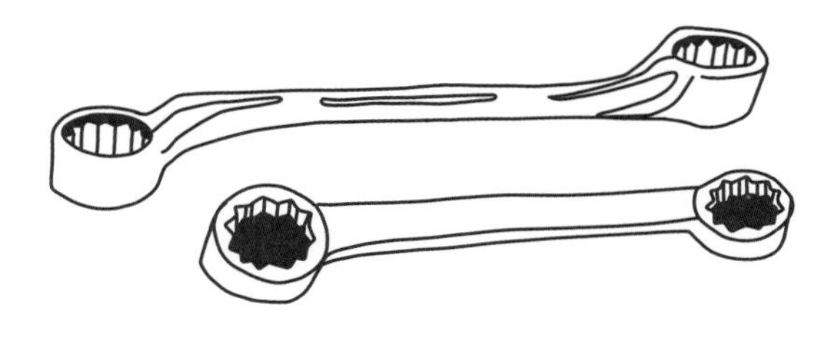

COMBINATION WRENCH

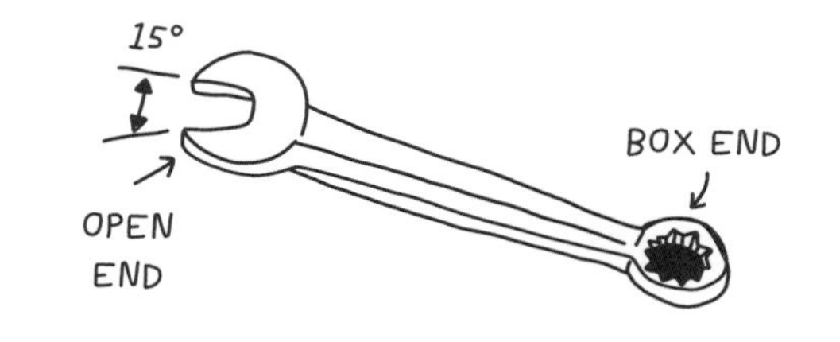

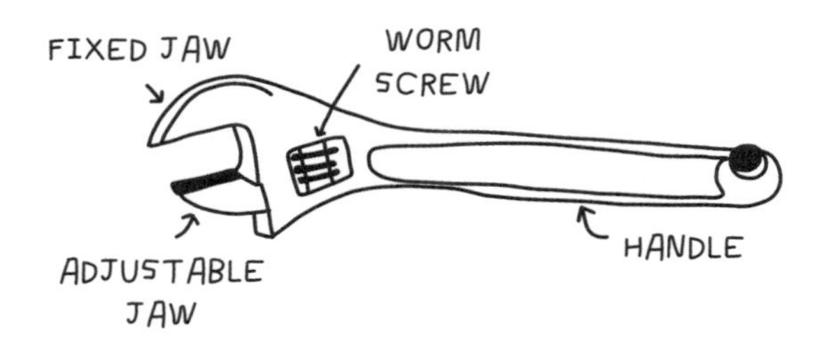

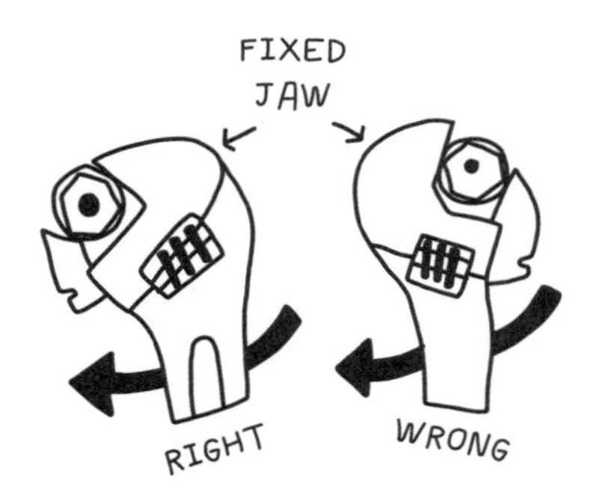

FUN FACT!

The first African American world heavyweight boxing champion, Jack Johnson, was also an inventor. While in prison (he was arrested for traveling with a white woman across state lines, which was illegal back in 1912!), Johnson designed an adjustable wrench that was later patented under U.S. Patent 1,413,121.

Adjustable wrench: The adjustable wrench is the jack-of-all-trades of wrenches, and if you can only have one wrench around, this is probably your best bet. An adjustable wrench has one fixed jaw and one adjustable jaw that can be narrowed or widened by rotating a small "worm screw." If you're working on a project with multiple sizes of hardware and don't want to constantly change tools, an adjustable wrench is a good choice.

You can buy adjustable wrenches in a range of sizes (usually measured in their handle length). Two good sizes to have around are an 8-inch wrench, with a maximum jaw opening of about 1¼ inches, and a 12-inch wrench, with a maximum jaw opening of 1½ inches.

Here's the tradeoff with adjustable wrenches: because of their moving parts and adjustability, these wrenches can slip more easily when you're using them. Some mechanics refer to an adjustable wrench as a "knuckle-buster" because your hand can easily slip and get scraped up.

To decrease the chances of slippage, and for a better grip on your bolt head or nut in general, always rotate an adjustable wrench toward the direction that applies pressure to the *fixed* jaw (not the adjustable one). A good way to make sure your jaws are tight around your hardware is to set the wrench in place around the bolt or nut, then tighten the worm screw in place until snug. Then tighten or loosen the nut or bolt.

Ratchet wrench: This is a great substitute for a combination or box wrench when you're short on time (or patience). The ratchet wrench looks nearly identical to a combination wrench, with one open-ended side and one box side. But the box end of a ratchet wrench has a built-in "grab-and-slip" action that grips the hardware to tighten in one direction but not the other. Essentially, you can move the handle back and forth, tightening and pulling back, without having to remove and reposition the wrench every time. A ratchet wrench is particularly helpful for jobs where you have many nuts or bolts to tighten all in a row.

A ratchet is a mechanism built into many helpful tools that allows movement in one direction but not the other. This is the mechanism in a ratchet wrench—you can tighten and pull back without removing the wrench from the hardware. It is the same mechanism you'll see in a socket wrench or in a ratchet strap. Like a combination wrench, you need a set with a range of sizes.

Because you can crank back and forth without removing the wrench, you don't need to turn the wrench that much before pulling back. Short strokes, back and forth, will do the trick.

Socket wrench: Owning a socket wrench is like having a whole set of ratchet wrenches in one tool. Unlike a ratchet wrench, where you need a separate size for each individual bolt or nut size, a socket wrench has interchangeable socket heads (shaped for hex bolts or hex nuts).

A socket wrench has a ratchet mechanism built into its handle, and a square drive where you attach sockets for whatever size hardware you're working with. On the back face of a socket wrench, there is also a lever that changes the direction of the ratchet, so you can tighten or

RATCHET WRENCH

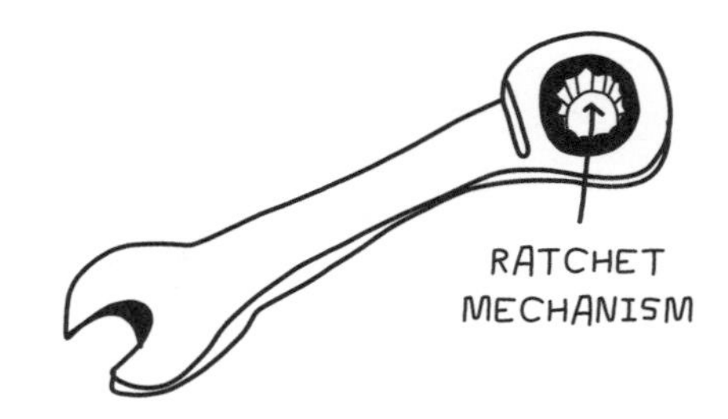

SOCKET WRENCH

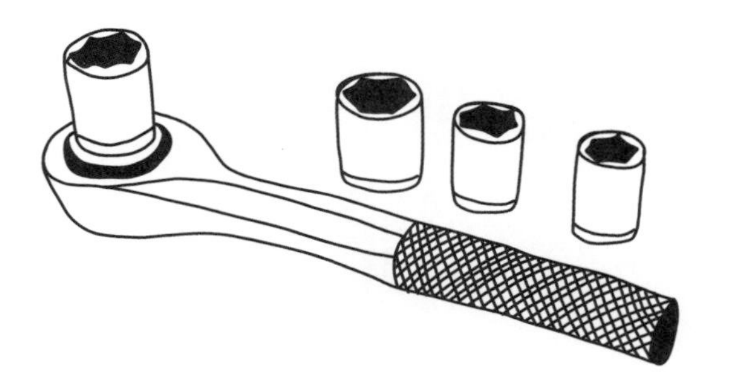

HEX-KEY (AKA ALLEN) WRENCH

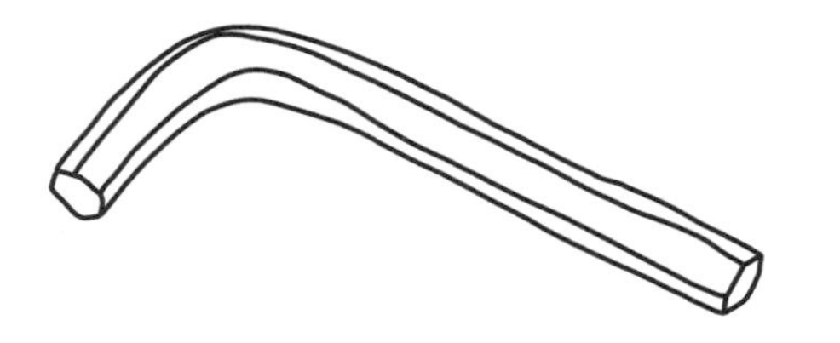

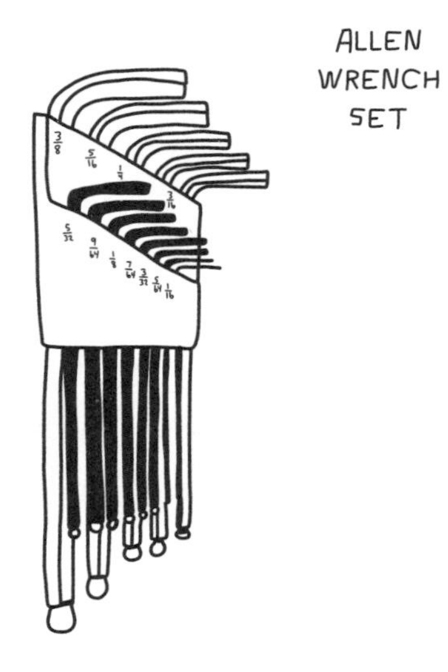

FUN FACT!

Hex-key wrenches are also often called Allen keys or Allen wrenches. Who's Allen, you ask? The first patent for the hex key was filed by W. G. Allen in 1909, who also founded his company of the same name, Allen Manufacturing Company. People now use the brand name "Allen wrench" interchangeably with "hex-key wrench."

loosen bolts. Socket wrenches are usually sold in sets, which include the handle with built-in ratchet, and a standard set of sockets ranging from approximately ¼-inch to ¾-inch sizes. When buying a set, pay attention to the drive size, which is usually ¼, ⅜, or ½ inch. The drive size is the size of the square "nose" on the handle where you attach your sockets.

To change sockets, use the release button on the back face of the socket wrench. When you press this button, your socket pops off the square drive so you can replace it with another one. Use the socket wrench the same way you use a ratchet wrench, turning back and forth to tighten your hardware without having to reposition at every turn.

Hex-key (aka Allen) wrench: Hex-key wrenches are usually L-shaped, with a hexagonal profile. Rather than wrapping around the outside of a bolt's head or a nut, *a hex key is inserted into the hexagonal hole,* or drive, in the top of some bolts or screws. Use them for hardware with this hexagonal drive, which is common in bikes, furniture assembly, and cabinets.

Hex-key wrenches can be purchased as full sets with sizes ranging from 1⁄16 inch to ⅜ inch.

The L-shaped hex key was designed that way for a reason! By inserting one "leg" of the L into your hardware, you can use the other leg as your wrench handle to rotate and tighten. Having one leg shorter than the other gives you more flexibility to get into cramped spaces.

KIA WEATHERSPOON

Interior Designer, Founder of Determined by Design

Washington, D.C.

I met Kia Weatherspoon at an event I spoke at in Washington, D.C., at the National Museum of Women in the Arts. After the event, I met so many incredible women who felt connected to my work, as I did to theirs. When Kia introduced herself to me and told me her story, I swear I almost said, "Can I quit my job and work for you?"

Kia is a dancer-turned-soldier-turned-designer—no doubt an unlikely path—but her determination as a builder is the common thread that has propelled her throughout life. As the founder of Determined by Design, she creates spaces for communities and groups who do not usually have access to beautiful, dignified spaces. She believes in social equity through design and walks the walk every day.

"It's funny—I never wanted to be a designer. But I was born with this innate sense of determination. I trained to be a professional ballerina from the ages of six to nineteen. But then, in my second year of studying dance in college, I did not receive any financial aid. So I decided to join the military to earn money to pay for school. This took me down a very nontraditional path that would eventually lead me to design!

"While serving in the United States Air Force, I was deployed in the Middle East shortly after September 11. I was stationed at a 'bare base,' which has only minimal housing and facilities. Our sleeping quarters were 15-foot-long tents that housed fourteen women. I wanted to cry, but there was no place to cry in private!

"Instead of putting my sheets on my cot, I hung them over some lengths of string to make three walls. That was the first space I ever created. Needless to say, I still cried like a baby, but creating that space made me feel something that affected me in a profound way.

"During four additional deployments between 2001 and 2004, I continued to create environments, even build furniture, in order to find comfort and solace. When I left active-duty military, I knew I wanted to create spaces, which is why I became an interior designer.

"Now I build, design, and create interior spaces for people who normally do not have access to well-designed spaces, including affordable housing, domestic violence shelters, and more.

"I read a quote once that said, 'Be the mentor you never had.' I constantly sought a mentor at various stages of my career, specifically when I was starting my own business, Determined by Design. I reached out to prominent women in the design and architecture fields but got no responses. So I decided to dig deep within myself and be my own mentor. I learned to trust my own judgment, make my own mistakes. While today I have people I value and admire, a mentor didn't come until much, much later. Now I am able to pay it forward to my many mentees.

"My role as a female builder of color is to let other young women know YOU CAN DO THIS, TOO! Every goal I have is not for personal accolades, but reassurance that any young girl can be where I am. They can build and craft their own environments, products, and, most important, their own lives.

"Building is about bringing your imagination to reality. It requires you to imagine, plan, and execute. If you look at the achievements of women in history, they demonstrate all the characteristics and skills required to build and make change. I think women have this innate ability to put pieces together and figure out how to make situations work. So learning how to build is in our genetic makeup!

"Learn to trust yourself first, always. Whatever you envision building is possible."

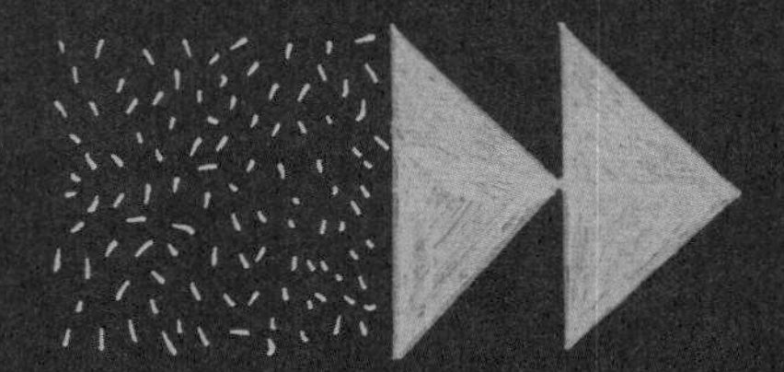

Pliers

Pliers are for grabbing! There are so many situations when you need to grab or hold something with enough strength that it won't slip, and pliers do the trick. Before we continue, a brief lesson in physics . . .

Pliers, in their most basic form, are made from two pieces of metal, crossed over each other in an X shape. The spot where the two pieces cross is the fulcrum, from which the two pieces rotate. The fulcrum isn't perfectly centered, though, so one side of the X has shorter lengths (these are the grabbing jaws of the pliers), and the other has longer lengths (this is your handle). This configuration gives your hand (and the handle side) the advantage to squeeze down and apply maximum force on the jaw side.

Tips for use

Pliers are especially helpful for plumbing, automotive repairs, and in situations where you're also using a wrench and need to hold a bolt head in place while turning a nut on the other end. Think of them as an extra hand for grabbing, holding, and squeezing.

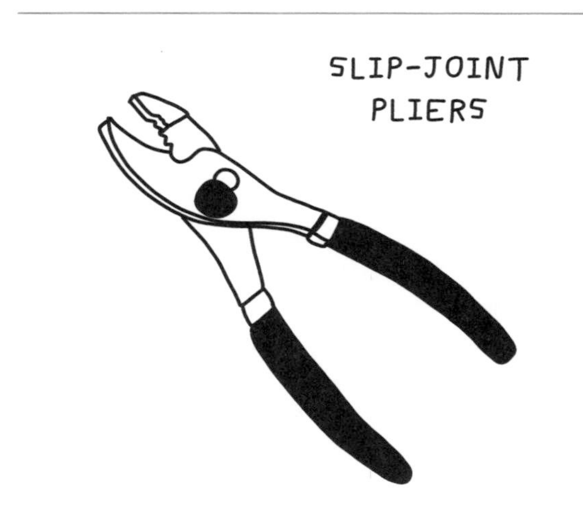

Types of pliers

Different types of pliers will help you work with various hardware, with different holding strength and maneuverability. Here are a few essential types of pliers to help you through most any project you decide to tackle:

Slip-joint pliers: Slip-joint pliers are probably the version of pliers you've seen most. I specifically remember a rusty pair of slip-joint pliers in my family's garage when I was a child, which I carried around like a MacGyver tool, ready to fix any problems that might arise. Slip-joint pliers are a great multi-tool, letting you grab, pinch, crimp, bend, and rotate a wide range of materials. They are characterized by an adjustable fulcrum (pivot point), which can widen the jaws of the pliers to accommodate wider objects.

Slip-joint pliers can be switched between two positions—one in which the jaws are completely closed and one in which the jaws remain slightly spaced apart, even when the handles are fully closed. This range is helpful because you can grip something as thin as sheet metal or as wide as a hex bolt—all with the same tool. The jaws of slip-joint pliers are serrated where the two jaws meet and have a wider, more open area toward the back of the jaws. This wider area lets you grip round objects like a pipe. You can find slip-joint pliers in many sizes: good go-to sizes are 6 or 8 inches long.

Play with the fulcrum adjustment to maximize your grip and look for pliers with a soft rubber handle, which will maximize your grip strength.

Diagonal cutting pliers: Diagonal pliers aren't exactly pliers, because they aren't meant for grabbing. Their diagonal-shaped jaws are actually intended to cut wire or thin metal by crimping the material between the angled jaws. Even though they've snuck their way into the pliers category, they're quite helpful and can be used for snipping or trimming extra lengths of metal wire, sheet metal, or other thin materials. Diagonal pliers are commonly used in electrical work, where wires frequently need to be cut and trimmed.

Unlike scissors, which use a shearing action (two parallel pieces of metal that slice material between them when they slide along each other), diagonal cutting pliers cut material by squeezing it between two diagonally angled jaws.

Diagonal pliers are fairly intuitive to use and don't require too much grip strength. Just place your material between the jaws and squeeze.

Combination (aka lineman's) pliers: You may have noticed by now that many tools come in some combination format, combining the best of two (or more) worlds. The combination pliers are no exception, integrating the grabbing action of slip-joint pliers with the cutting action of diagonal pliers. They are often referred to as lineman's pliers because electricians and linemen commonly use them to cut, grab, twist, and bend wire and cable.

The jaws of combination pliers, like slip-joint pliers, are serrated for added grip. The front section of the jaws has this serration, while the back part of the jaws has an integrated diagonal cutter. Most combination pliers have a rubber grip to help guard against electric shock (in reality, that thin layer of rubber won't do much, so never use pliers on live wires!). You can buy many sizes of combination pliers, but like other pliers, an 8-inch size is ideal for most jobs.

Combination pliers are what electricians use most, but they're a good tool for anyone to have around the house. As with other pliers, grab them firmly toward the end of their handle for the most gripping force.

COMBINATION (LINEMAN'S) PLIERS

FUN FACT!

Combination pliers are also sometimes called "engineer's pliers." Engineers use them for multiple cutting, stripping, pulling, and grabbing actions. While engineer's pliers are shaped the same way, they often have handles without a rubber grip.

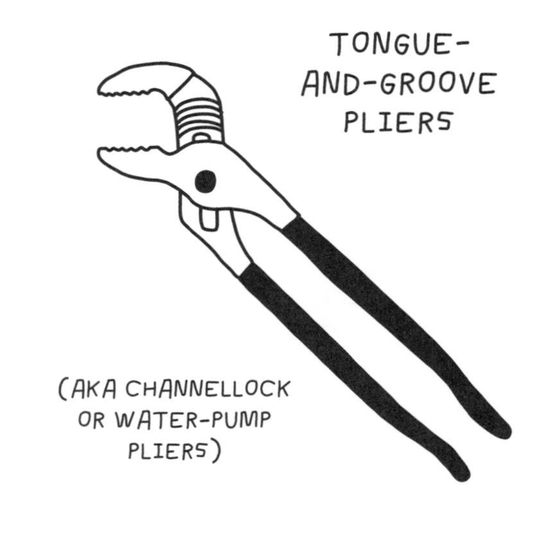

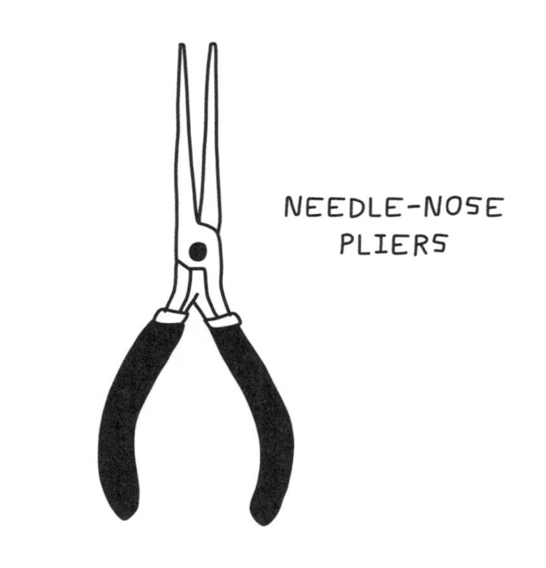

Tongue-and-groove pliers: A version of slip-joint pliers, tongue-and-groove pliers also have an adjustable fulcrum that lets you change the width of the jaws. Tongue-and-groove pliers are alternately known as water-pump pliers or Channellock pliers, a brand name. These pliers are generally larger than standard slip-joint pliers, with an angled set of jaws that help you access and firmly grip nuts, bolts, pipes, or oddly shaped objects.

The adjustability of tongue-and-groove pliers comes from an upper jaw piece with a built-in track. This track can be moved along a set of steps that widens or narrows the total jaw width.

Before using these pliers to grip, make sure they are locked in place at your desired position. Grip toward the end of the handle for maximum comfort and turning power.

Needle-nose pliers: As the name suggests, needle-nose pliers have slimmer jaws that are great for places your fingers can't reach. Because they are more delicate, they aren't the strongest member of the plier family, but they can help with more precise tasks, like placing or tightening small pieces of hardware.

Some needle-nose pliers also have an integrated diagonal cutter, making them helpful for jewelry work or cutting thin wire, too. There are various kinds of needle-nose pliers, some with an angled tip at 45 or 90 degrees, giving you even more dexterity for small tasks.

Because needle-nose pliers have smaller, more pointed tips, be careful not to bend them out of shape or poke or scrape yourself or your material.

Miscellaneous hand tools

The list of hand tools is long and comprehensive! These miscellaneous tools are all useful in their own right, even if they don't fit in a particular category.

Multi-tool

As the name implies, you can use a multi-tool multiple ways! A multi-tool is any tool that combines many functions in a single tool—a pocketknife is a great example! For most carpenters or builders, a Leatherman (brand name) multi-tool is a common choice.

Multi-tools like the Leatherman are usually between 3 and 6 inches long and come with a handy pouch you can proudly wear on your belt. They typically include pliers, many types of screw-drive bits, various knives or cutting tools, a can and bottle opener, and wire snips. You can also buy multi-tools optimized for particular work, like an electrician's multi-tool, which includes more wire-stripping features, for example. Most multi-tools are very intuitive and user-friendly.

When using a multi-tool, only pull out the tool you need, leaving all the others closed up in the case. When using the knives or screw-drive bits, lock the two arms together so you have a solid handle to grip.

Snips

Snips, also called tin snips, aviation snips, or side cutters, are basically scissors for heavy-duty use, especially for sheet metal. I've used snips to cut chicken wire and other sheets of material that are a little too thick for scissors.

There are three types of snips: right-cutting, left-cutting, and straight-cutting. Just like left-handed people prefer lefty scissors, snips come in different versions that optimize the direction of your cut. Use a right-cutting snip for cutting lines that curve to the right, and left-cutting snips for curves to the left. Especially when you

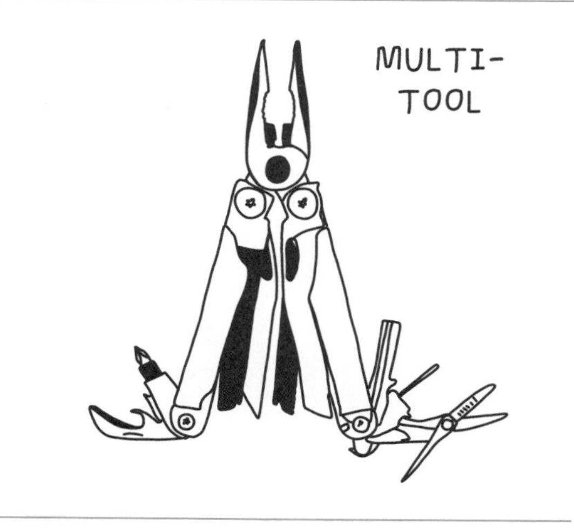

FUN FACT!

Snips are often sold in a color-coded set of three that includes a right-cutting (green), left-cutting (red), and straight-cutting (yellow) tool. If you find a lone pair of snips, you can immediately tell which type they are based on the color.

CHISELS

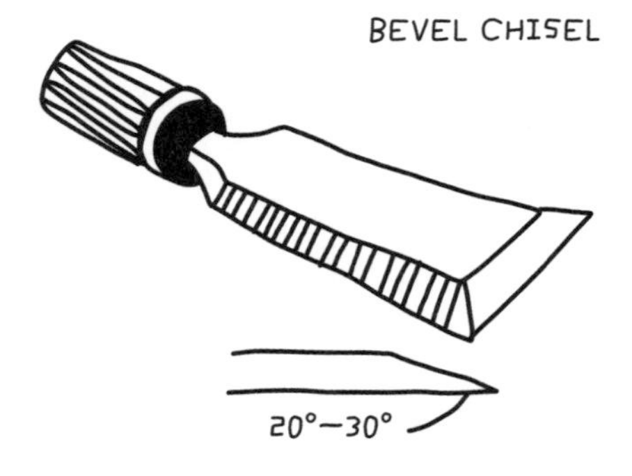

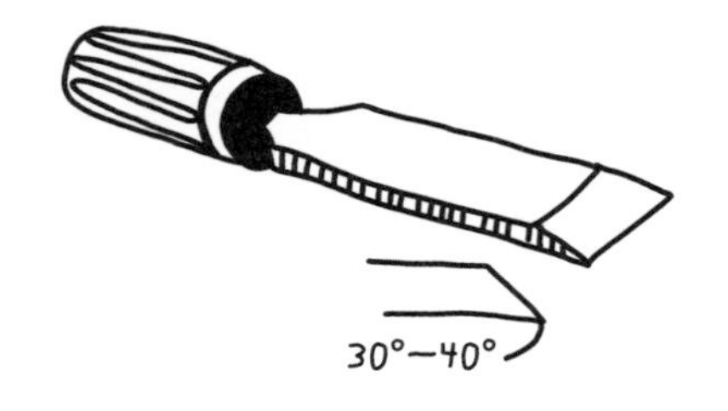

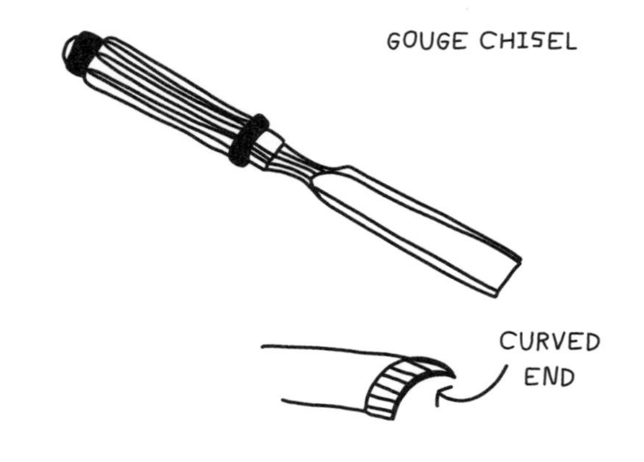

are cutting sheet metal, it can be extremely hard to cut curves without the correct snips.

Use snips just like scissors! It can help to draw a line on your surface before cutting. When cutting through sheet metal, chicken wire, or other sharp materials, always wear gloves.

Chisel

Use a chisel to carve, shape, and make slices or cuts in wood. A chisel has a wooden handle and a metal blade angled at the end to cut into wood. Most chisels are used in combination with a rubber mallet: place the tip of the chisel at an angle on the wood surface, holding it with one hand, and pound the end of the chisel with the mallet. Unless you're an avid woodworker, you don't need many chisels, but a couple go-to chisels to have around are a bevel chisel and a gouge.

The bevel chisel has a rectangular blade, with an angled edge on one face and a flat edge on the other. The bevel chisel is so versatile, you can carve shallow grooves with the angled side down, and deeper grooves with the angled side up. You can also use a bevel chisel to make vertical cuts into wood for detailed joint work. Mortise chisels are similar to bevel chisels, with steeper angles on their cutting edge, and are often used for more heavy-duty work. Gouge chisels have a scooped, rounded edge at the end of the blade (they look a little like a celery stick), which is helpful for scooping out round areas in wood. I use gouge chisels when making wooden spoons, to carve out the spoon's bowl.

Even though chisels appear to be easy to use, they are one of the hardest tools to master. Woodworkers spend years (even decades) perfecting their technique. **Chisels have very sharp blade ends and should be handled with extreme caution. Always point a chisel end away from your body, and treat the tool as you would a sharp knife.** To use a chisel for carving, it's all about the angle! Hold the chisel in your non-dominant hand, with the blade facing away from you (coming out of your clenched fist at the pinky side). Place the blade on the surface of the wood at the appropriate angle. The more parallel to your wood surface, the shallower your carve will be. A more vertical chisel will create a deeper cut. Use a rubber or wooden mallet to tap or knock on the butt of the chisel handle. You will see wooden chips or curls forming as your chisel does its work!

Hand plane

A hand plane is used to flatten, or "plane," the surface of wood. The hand plane is a sled-shaped tool with a cutting device on the bottom that slides along the surface of rough wood until its surface is flat.

There are bench planes and block planes, and planes designed for varying degrees of roughness: jack planes, jointer planes, smoothing planes, and more. Most planes, however, work using the same principle and parts. A plane has a flat sole (like a shoe), with a slot that runs across it. An angled blade (like a razor blade) fits into this slot and shaves off layers of wood a little bit at a time as you slide the plane across a piece of wood. Two wooden handles allow you to hold both the front and the back of the plane and slide it along the length of your wood, over and over until the surface is flat.

Like chisels, planes can be hard to master, **and have extremely sharp blades. Use with caution and always keep the sole of your plane flat on your table when not in use.** Start with a forgiving plane, like a jack plane, and get the feel for removing the top layer of rough wood. Keep a solid grip on both wooden handles. You can adjust the depth of your blade (and how deep your cuts are) by using the adjustment knob. Move along the top of the wood in the direction of the grain.

HOW TO USE A CHISEL

FUN FACT!

Chisels are also used for stone and metalwork. Many of the most famous Italian marble sculptures, including the famous statue *David* by Michelangelo, were carved using chisels. Michelangelo, who was only twenty-six years old when he began working on *David*, spent two years carefully shaping a block of marble, using chisels, into the iconic sculpture we know today.

HAND PLANE

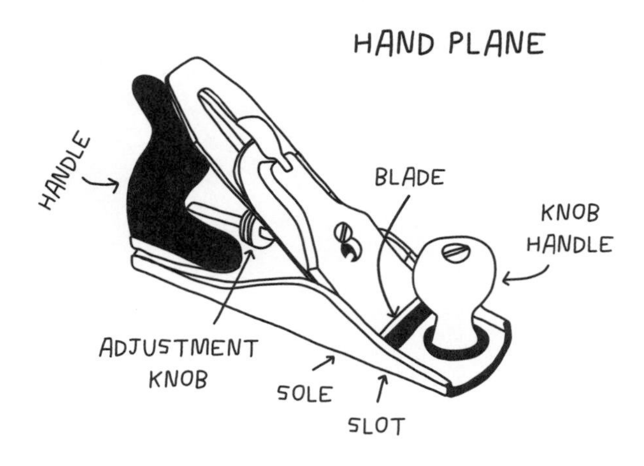

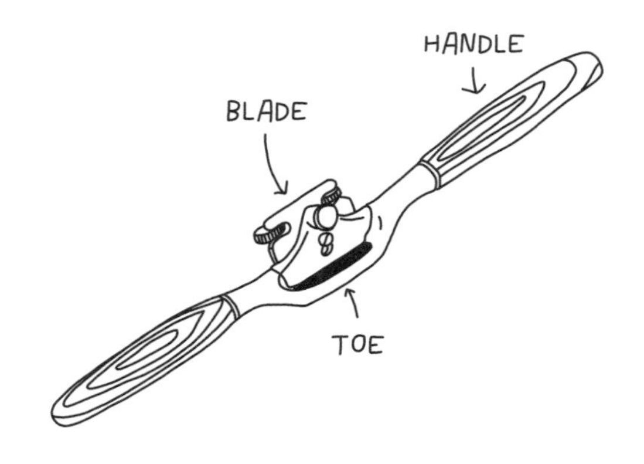

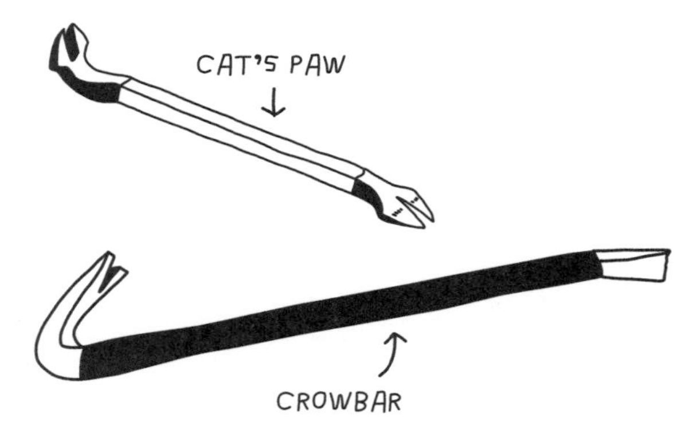

FUN FACT!

The crowbar has been around for a long time! In the days of William Shakespeare, it was called an iron crow. Shakespeare even mentions the iron crow in some of his most famous works, including in *Romeo and Juliet*, when Friar Laurence says to Friar John in Act V, Scene II: "Get me an iron crow, and bring it straight unto my cell."

Spokeshave

Think of a spokeshave as a carrot peeler for wood. Use it for slowly shaving off layers of the wood's surface, like if you had a square rod you wanted to turn into a round handle (see the Wooden Spoon project on page 266!). While a hand plane is ideal for flattening the surface of wide pieces of wood, a spokeshave is better for long and thin pieces, like a rod or chair leg.

The spokeshave looks a little like a small set of bicycle handles, with a blade in the middle. **Just like a hand plane, a spokeshave has an extremely sharp blade. Use with caution and set your spokeshave with the blade facing down when not in use.** To use it, hold both handles, one in each hand, and push or pull the tool in the direction of the wood's grain. The blade in the center shaves off thin layers as you go. Spokeshaves come in different blade configurations, with the blade either straight or with a slight curve in or out (concave or convex).

Pry bar

A pry bar, or crowbar, is used to pry apart two objects or to remove nails. While you might think of a crowbar as the stereotypical tool used by burglars (like the two hilarious bad guys in the movie *Home Alone*), a pry bar is a great multi-use tool to keep around. A crowbar is a long bar, usually steel, with a bent end. Most crowbars have a small notch or slot in the end that can be used to grab and pry up nails.

The standard crowbar works by wedging its bent end into a small crack and using the elbow bend of the bar as a rotation point (fulcrum) to pry objects apart. A smaller version of the crowbar is sometimes called a cat's paw, which is shorter, about 12 to 16 inches long, with a cupped end that works great for pulling nails.

Position the bent end of the crowbar in place (either wedged between two objects or with the slot around the nail head if pulling nails). Pull the bar back by pivoting it on the elbow bend in the bar. This will lift the end and pull apart your objects (or pull up your nail).

Hammer tacker and staple gun

Both hammer tackers and staple guns are quick-fire tools for attaching staples. They use a spring action to shoot heavy staples with greater force than an office stapler and are great for attaching thin materials to wood (think about all those fliers on telephone poles!).

A standard staple gun's spring action and squeezable trigger detaches a single staple from a strip of many and pushes it into the surface. A hammer tacker operates similarly but works by swinging the tool like a hammer. When it hits the surface, it automatically dispenses a staple. A hammer tacker is best to use when the precise location of your staple isn't that important, or when you need to fire off dozens of staples in succession.

Both tools accept a standard width of an industrial staple, but you can choose different lengths of staples depending on your job. Just like with an office stapler, you insert a whole strip of staples into the channel. **Both of these tools can be dangerous if not positioned and discharged correctly. Be sure to wear eye protection, load staples properly, and hold the tool firmly.** Position your staple gun as needed and squeeze the trigger. For a hammer tacker, aim and swing like a hammer, trying to hit the surface squarely so the staple enters at the correct angle.

HAMMER TACKER

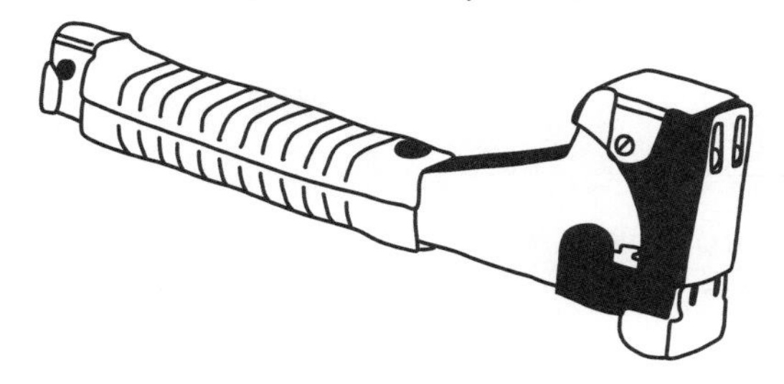

STAPLE GUN

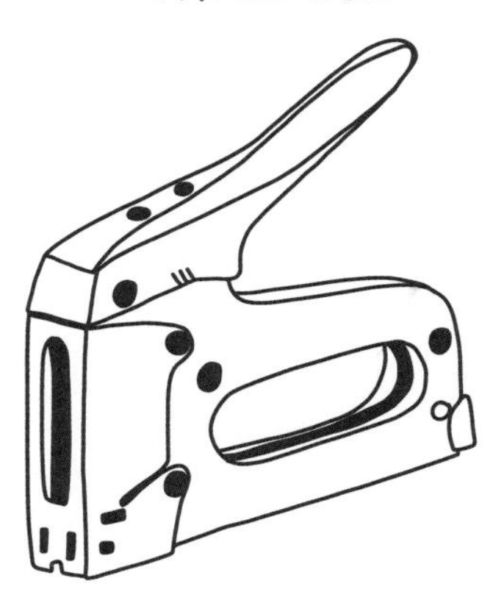

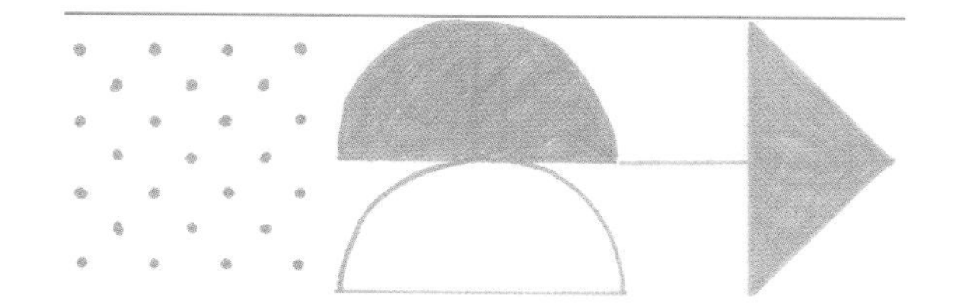

SAWS

Unless you plan to build using only whole tree stumps, very long 2×4s, or giant lengths of steel, you will most likely need a saw for any building project. In my opinion, cutting material with a saw is one of the most satisfying tasks, requiring precision and patience.

This section paints a picture of the wide landscape of saws, both handsaws and power saws. Like every other type of tool, you can geek out and track down dozens of other types, but the ones listed here are a great starter kit for most wood (and some metal) projects. Whether you're using a tiny keyhole saw or a burly miter saw, here are a few key terms and tips (and remember, before you work with any saw, make sure you have all the appropriate safety gear, and your (adult) buddy, too—see Safety and Gear, page 21).

TYPES OF CUTS

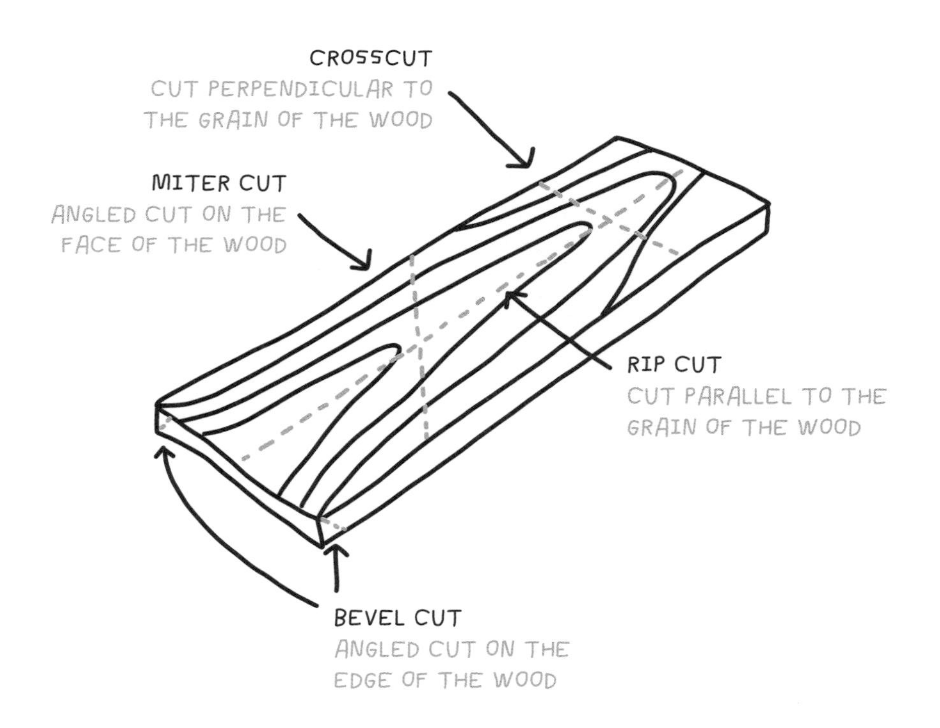

Tips for use and safety!

Do not use any saws without the supervision and hands-on support of a skilled and experienced adult.

Always wear safety glasses. Whether you're using a power saw or a hand saw, put on your safety glasses before you start working, and leave them on until you are done with all cuts. Safety glasses will protect your eyes from debris, sawdust, and in the worst case scenario, a flying piece of wood or broken sawblade (yikes!).

Read the manual. Because there are so many models and types of saws, a small detail or button on one can be different on another. It's crucial to read the instructions provided by a manufacturer to learn the specifics of your particular tool.

Know your grain and cut type. Wood has a "grain," which is the pattern of the tree's rings (learn more about lumber and grain in Building Materials, page 35!). The direction of your cut

in relationship to the grain determines the type of cut and also the type of saw you need.

- A **crosscut** is a cut across your lumber, perpendicular to the direction of the grain.
- A **rip cut** is a cut down the long length of your lumber, parallel to your wood's grain.
- There are also two types of angle cuts: a **miter cut** is an angled cut across the top face of your wood, while a **bevel cut** is a cut angled off the end of the wood.

Once you know your grain and cut type, it is easier to determine which saw you need.

WHICH SAW SHOULD I USE?

TYPE OF CUT	HANDSAWS	POWER SAWS
Crosscut	Crosscut handsaw	Miter saw, circular saw
Rip cut	Rip cut handsaw	Table saw, circular saw
Miter cut	Backsaw with miter box	Miter saw with miter gauge
Bevel cut	Backsaw with miter box	Miter saw with bevel gauge or table saw with bevel gauge, if bevel runs along the grain
Curved cuts	Coping saw	Jigsaw, band saw, or scroll saw for detail cuts
Crude cuts	Bow saw	Reciprocating saw, chain saw
Metal cuts	Hacksaw	Reciprocating saw, metal band saw, or abrasive saw

MAKE YOUR MARK

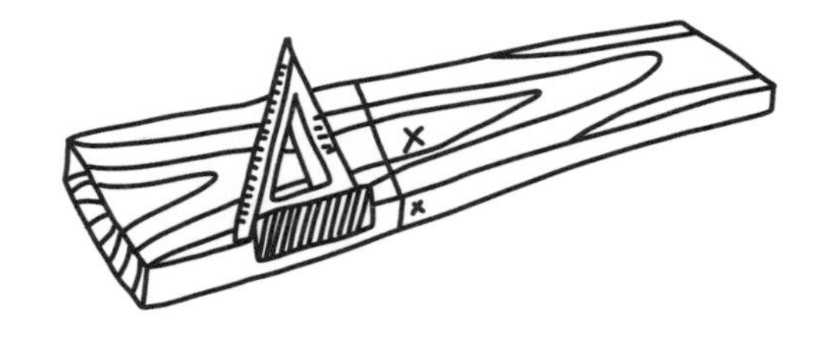

HOLD IT TIGHT

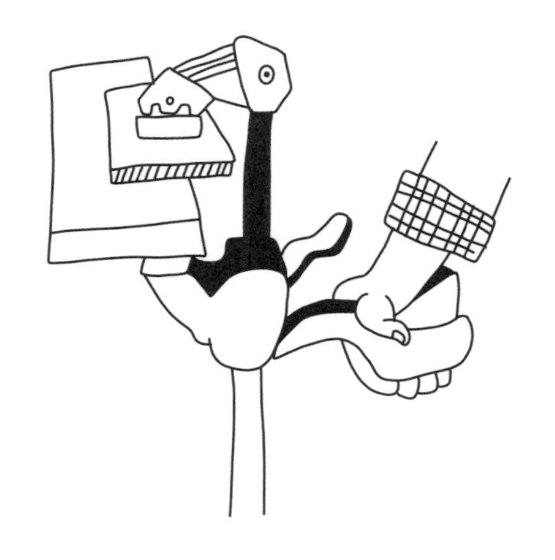

KERF (TOTAL SAW BLADE WIDTH)

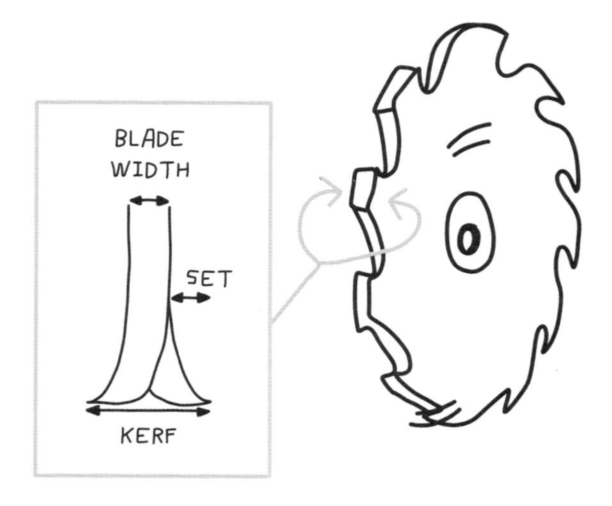

Make your mark. Whether you're using a scratch awl or a speed square and pencil, make an accurate mark before making your cut. Especially for curved cuts using a jigsaw, bow saw, coping saw, or band saw, this line is particularly helpful to follow. Now's a good time to refresh your memory on how to mark your measurement with a caret and use a speed square on pages 74 and 78!

Hold it tight. With some power saws, like a table saw or band saw, the blade stays put and you're in charge of moving the wood. But for others, like a miter saw or circular saw, the blade moves and your material must be totally still! For these, you will use clamps, vises, or the saw's fence, to help secure the wood when cutting. Refer to the Measure, Lay Out, and Secure section (page 88) to find the best clamping tool for each cutting job!

Mind the kerf! One of my favorite vocabulary words is "kerf," which comes from the Old English word "cyrf," meaning a cutting or cut. Kerf is the width of a saw blade, including the teeth, which are often angled slightly outward. Kerf matters because it is also the width of material you physically remove from your wood when you make your cut. You will need to remember and account for your kerf so you don't end up shortening your cuts.

Measure one cut at a time. Let's say you need four pieces of wood cut at 12 inches each. You might think you can make four marks on your wood, each 12 inches apart. But remember your kerf! If you are using a miter saw blade with a kerf of ⅛ inch, and if you make your cuts all at once, you will lose ⅛ inch of your 12-inch width from all subsequent cuts, and by the last cut you'll be ⅜ inch short! To avoid this, measure and lay out a cut, make the cut, then measure your next cut so you are always measuring from a fresh end.

The lower the teeth per inch (TPI), the rougher the cut. "TPI" stands for "teeth per inch" and corresponds to the density of teeth on your saw blade. The higher the TPI, the more teeth you have working on your wood, and the cleaner and finer your cut will be. Low-TPI blades chew up wood more crudely and are fine for rough lumber cuts. But if you are doing a fine wood-working project, use a high-TPI "finish blade."

Let the saw do the work. Yes, sometimes using a saw for hours can be a great arm workout, but actually when you're using a saw properly, the tool should do most of the work for you. With a coping saw, for example, which cuts as you pull the saw back toward yourself, you don't need to push too hard for it to do its work. In fact, with some saws, the harder you push into the wood, the more the saw will fight you or make for a dangerous situation.

MEASURE ONE CUT AT A TIME

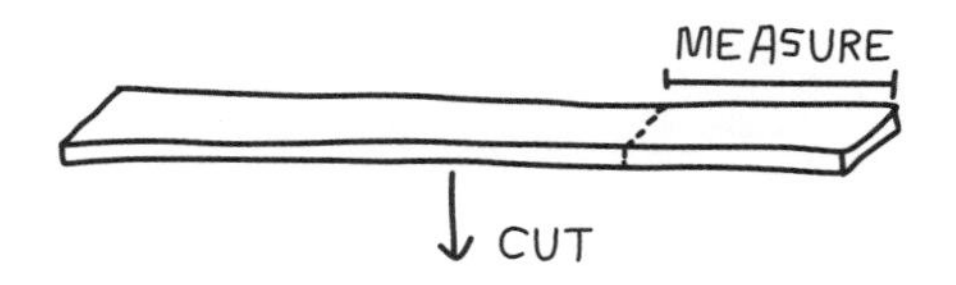

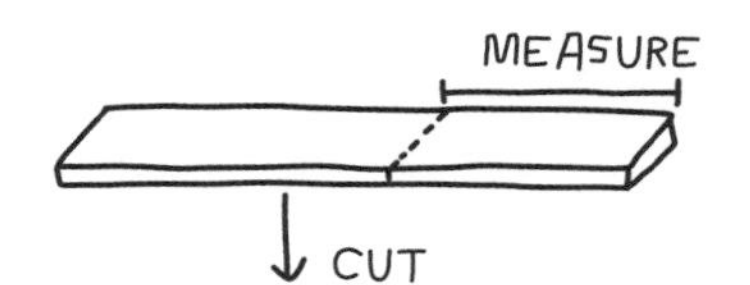

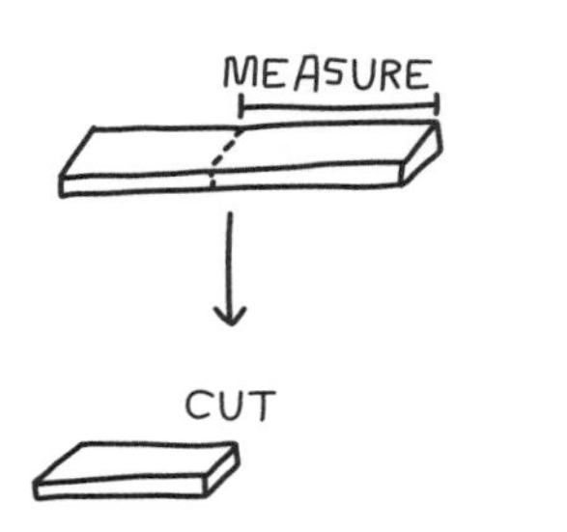

THE LOWER THE TEETH PER INCH (TPI), THE ROUGHER THE CUT

FEW TEETH (LOW TPI), ROUGH CUTS

MORE TEETH (MEDIUM TPI), MEDIUM CUTS

MOST TEETH (HIGH TPI), SMOOTH CUTS

CROSSCUT HANDSAW

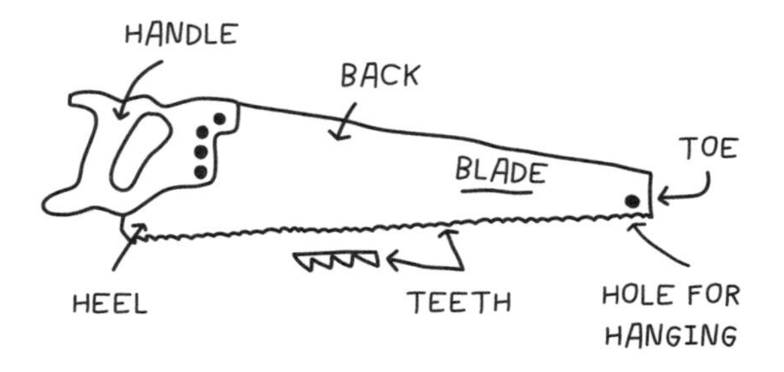

HOW TO HOLD A HANDSAW

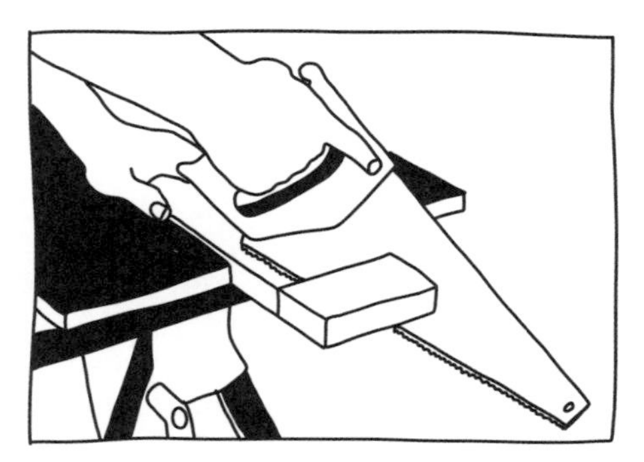

FUN FACT!

Most crosscut handsaws have two handy features: The end of the blade has a 90-degree corner, which can be used as a square to lay out right-angle cuts. Some also have a small hole cut through the end of the blade so you can hang it on a nail in your wall, which is far safer than hanging it from its handle.

Handsaws

Saws powered by your own muscle are a great option for rough cuts, precise woodworking cuts, and anytime you don't have access to electricity.

Parts of a handsaw

The handsaw is a simple tool that consists of a handle (usually wood) and a long metal blade. The bottom edge of the blade will have directional teeth to help you make an easy cut.

Types of handsaws

A handsaw's teeth are oriented in a particular direction—forward, backward, to one side, or the other. Because of this tooth direction, some saws cut on the push stroke (push-cut saws), and others on the pull stroke toward your body (pull-cut saws). The following saws are a solid collection for most cutting jobs.

Crosscut handsaw: When you picture a handsaw, this is likely the type you imagine. Crosscut handsaws are designed to manually make rough cuts across the grain of a piece of wood. You might use a crosscut saw when you don't have access to a miter saw or electricity but you need to cut some specific lengths of 2×4s or other lumber.

The crosscut handsaw has a handle and a blade about 2 feet long, with teeth that alternate in direction. You can buy crosscut handsaws of various lengths and tooth counts, but a 26-inch, 12-teeth-per-inch saw will do the trick for most projects.

The crosscut handsaw has a sister, the ripsaw, which has the same shape, with slightly different angles on its teeth to make cuts parallel to the grain of the wood like a series of mini chisels. In practice, though, a 12-teeth-per-inch crosscut saw is fine for rip cuts, too.

You can also buy two-person crosscuts (like the ones used to cut through giant tree trunks!), which have an extra handle on the other end for a friend to push and pull with you.

The crosscut handsaw can be either a push-cut or a push-and-pull-cut saw, which you can check by looking at the direction the teeth are pointing. If they are pointing forward, toward the end of the blade, you have a push saw that will do most of the cutting as you push it away from you. If the teeth are pointing straight downward, you have a push-and-pull-cut saw, which cuts on both the push and the pull strokes. In either case, it's easiest to tilt the saw downward, starting your cut at the farthest edge of your wood and working toward you. The saw should be at about a 45-degree angle, with its end lower than your hand and the handle. Make smooth, quick strokes back and forth until your cut is complete, and use sawhorses or a friend to hold the end that will fall off.

Backsaw: The backsaw has a rigid spine (or back) on the top edge of its blade that keeps the blade rigid for straight, fine cuts. Backsaws are ideal for making small, detailed cuts for joinery. Backsaws have a narrow blade with a small kerf, so they don't remove too much material from your wood.

Backsaws are usually pull-cut saws, meaning they make their cut as you pull the blade back toward you, with fine teeth and a high teeth-per-inch count. There are different types of backsaws, and a good starter backsaw is a 12- or 14-inch backsaw with a rectangular blade and wooden handle.

Unlike a larger crosscut handsaw, backsaws work best when the blade's teeth run flat atop the face of your wood (rather than at a downward angle, as with a crosscut handsaw). You can use a miter box to help guide these cuts, with the blade moving horizontally back and forth across your piece.

BACKSAW

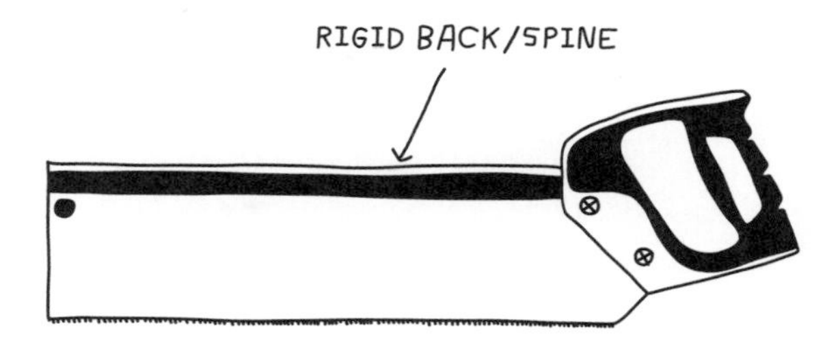

JAPANESE DOZUKI BACKSAW

FUN FACT!

The Japanese are known for a high-quality carpentry backsaw called a "dozuki." This pull-stroke saw is traditionally used for Japanese carpentry and woodworking. The dozuki often has a wrapped handle, which resembles the leather-wrapped handles of samurai swords.

BOW SAW

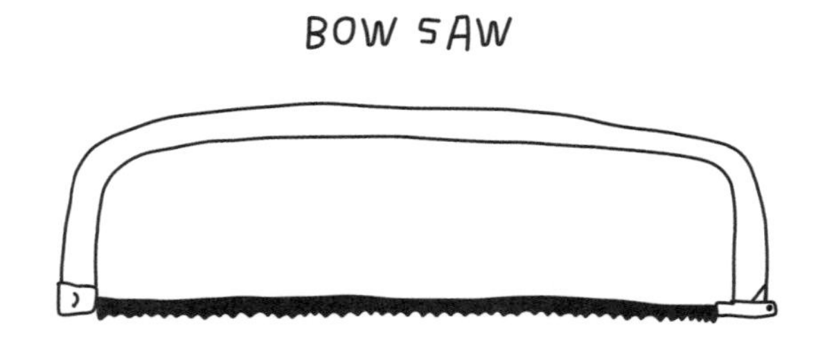

HOW TO USE A BOW SAW

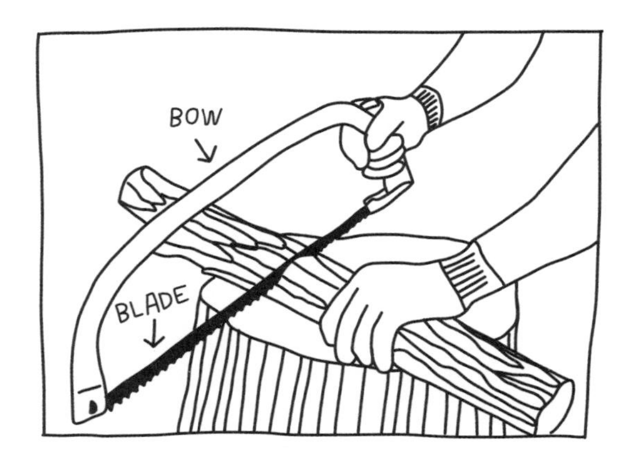

HACKSAW

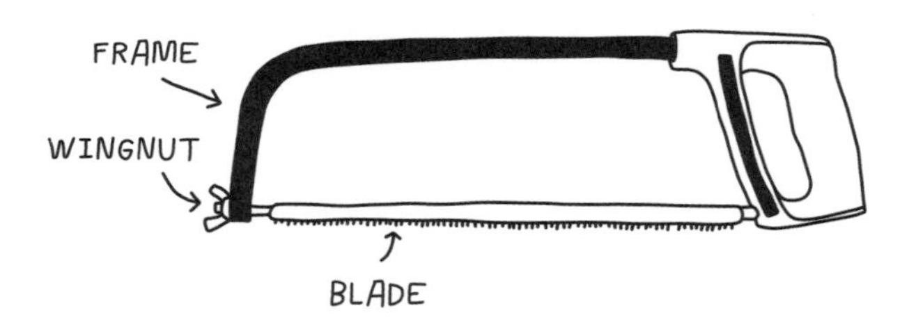

Bow saw: A bow saw has a thin blade and a large metal "bow" that holds it in place (picture an archery bow: the string of the bow is like the blade of a bow saw). Bow saws are lightweight and have been used traditionally for cutting branches or small pieces of firewood or kindling. Their thin blade is helpful when cutting through wood at odd angles or odd shapes, like a branch.

Bow saw blades are held in tension between the two ends of the bow, and the blade can be removed and replaced easily. Bow saw blades are usually fairly crude (only four to sixteen teeth per inch, so they're not the best choice for precision, but are very helpful for quick cuts out in the wild or in the shop. Because their teeth point downward (not out, back, or forward), bow saws cut on both the push and the pull stroke.

Some older bow saws do not have a separate handle, so you may simply have to grab the bow and go to work. Newer bow saws may have a built-in handle on the back end of the bow. Bow saws can also be used by two people, similar to a two-person crosscut saw, with one person holding each end of the bow saw and cutting back and forth.

Hacksaw: A hacksaw is similar to a bow saw, but it was originally designed for cutting metal. They can be used for wood or plastic (like plumbers, who employ them for cutting plastic PVC pipes), though their fine-tooth blades are ideal for cutting through metal stock.

Like a bow saw, a hacksaw has a replaceable blade that usually measures 12 inches long, with 18 to 32 teeth per inch. You can also buy a junior-size hacksaw, which measures 6 inches, but in my opinion, they are too small to be versatile enough for most jobs.

When cutting metal, angle the hacksaw blade for the most efficient cut. Blades can be brittle, and are likely to break if twisted, so keep your blade straight, not twisting to one side or the other.

Coping saw: Coping saws are easy to use and very versatile for making curved cuts. Use them for cutting curved shapes out of thin wood (like for the Whatever-Shape-You-Want Wall Clock project, page 262!), or thread their removable blade through a hole to cut from the hole outward.

Coping saws have a C-shaped wood frame ("C is for *coping*" is a great way to remember what this saw looks like) and a very thin blade. The blade's ends are held in the frame by tension, using a small pin at the ends of the blade that hook into slots in the wooden handle. These blades are inexpensive and easy to replace. Because of the large space between the blade and the back of the frame (the "throat"), coping saws can make deep cuts, far from the edge of your wood. You can also reverse the direction of the blade, with the teeth facing the frame, to make cuts at odd angles or from the underside of a piece of wood.

When working on a project where you need to cut a piece of wood starting from a small predrilled hole (picture trying to cut out the middle circle of the letter O). With any other saw, it might be very hard to get to that hole. No problem for the coping saw! Remove the blade, thread it through the hole, connect the ends to the saw frame, and start sawing. Coping saw blades work best at an angle to their surface and face in one particular direction. Point the teeth toward the handle (so they look like a lightning bolt with the zigzags pointing down), for pull cutting, which is most common and preferable. To replace your saw blade, compress the frame until you can pop out each end of the blade, then insert a new one by sliding the blade's pins into the pin slots in the frame.

COPING SAW

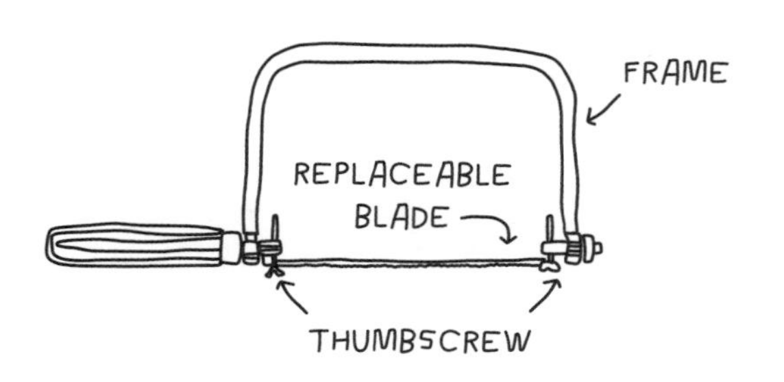

USING A COPING SAW TO CUT AN INTERIOR SHAPE

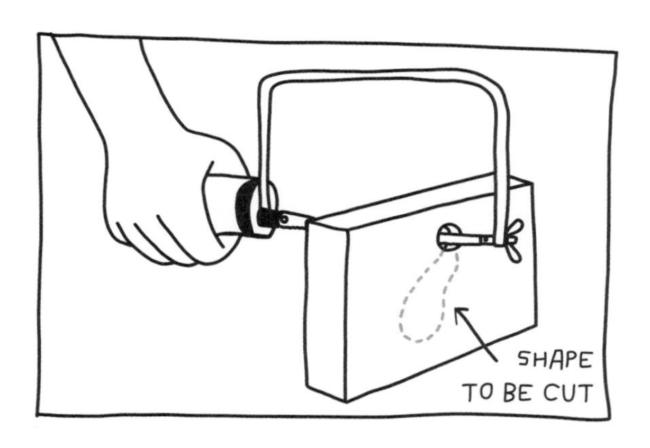

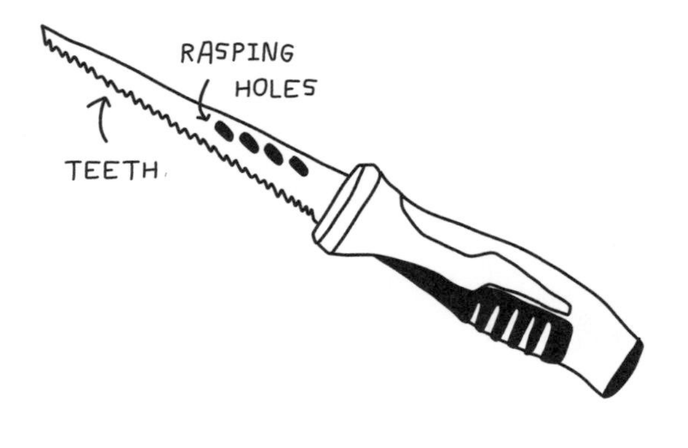

Keyhole saw: Also known as jab saws or alligator saws, keyhole saws are a particularly scrappy type of tool. This is the saw for the job that no other saw can reach. Use a keyhole to cut in small spaces, to jab the surface of a material, or to make cuts in holes. They are commonly used for cutting holes in drywall during wall construction.

Keyhole saws come in two versions: a fixed blade with a handle, and a folding version like a pocketknife, which can collapse and hide the blade. Some have an extra angled edge on the end of the blade, which can be used to pierce through materials. Rasping keyhole saws have small rasping holes (like a cheese grater) on the face of the blade, which help smooth surfaces as you cut through them.

When using a keyhole saw, don't use too much force, which might get your saw stuck and make it even harder to do the job in an already compact space.

BIBI AMINA

Master Carpenter, Member of Ciqam Management Committee

Altit Fort, Hunza Gilgit-Baltistan, Pakistan

Bibi Amina is one of the first female carpenters in Pakistan and part of a collective of tradeswomen called Ciqam. Her story is one of the most courageous I have ever heard, as she used building as a vehicle to improve her own life, family, and community. While I have not met Bibi in person, I often find myself thinking of her and the women she works with, working hard with handsaws on the other side of the world, and I hope every young builder girl finds the kind of grit and determination that Bibi models for us all.

"I was born in a small village in Pakistan and raised in a poor peasant family with a small amount of land. In 1996, when I was twelve, my father and uncle died in a road accident, which put my family in a very vulnerable financial position because my father was the only wage-earning member of the family. My mother, who came from a village near the Chinese border, had no relatives to help her. I was able to get my basic education while living with my maternal grandmother away from my own family and was admitted to college, but I eventually had to abandon my education because of the financial pressures in my family.

"To help my mother and family get out of poverty, I tried to find work and was offered a trainee's position in carpentry by the Aga Khan Cultural Service Pakistan. At the time, they were undertaking the restoration of Altit Fort, and I accepted the position eagerly. There were different professionals working on the Altit Fort restoration team, including engineers, architects, masons, carpenters, and electricians—but only a few girls were working as trainees. Initially, it was an extremely physically difficult job, but quickly my knowledge of historic timber framing improved, and it was fascinating! This experience afforded me access to technical knowledge that helped improve my confidence as a skilled person. After six months of training, I started repairing items around my house for my mother, which, in addition to my monthly income, was a great satisfaction and relief to her.

"In 2008, I officially started training in carpentry. At that time, there were very few women working outside of agriculture. I had to walk to work every day, which was dangerous and a source of tremendous fear and one of the biggest challenges I faced. At work, there were six male carpenters and two female trainees. I learned carpentry as an apprentice without following a curriculum.

"One of the major challenges at work was to maintain a strict cultural dress code, which posed a physical and emotional challenge. Outside of the job site, at the community level, we women were considered deviant individuals, acting against social norms to overcome poverty. I received many negative comments. But during that first year of training, I worked on a small structure located in a nearby town with great public visibility. We built it using traditional construction techniques in just six months, and it is what I am most proud of.

"My favorite tool I learned to use is a handsaw, as it allows me to shape timber, and with that timber, to shape spaces.

"For me, building is both a process and a tool with a tremendous potential to bring about changes that are necessary for our societies and physical environment. My profession as a female builder provides me with possibilities to do physical work in physical space, to use materials to create spaces and products for others. In doing so, I improve my own well-being while also giving others better physical conditions.

"Having experienced extreme poverty, I am interested in developing small energy-efficient houses to benefit vulnerable portions of the population. As a woman carpenter and builder, I would like to create greater opportunities for young women of rural Pakistan to become economically empowered and bring positive change to a patriarchal society."

Power saws

Even though I love handsaws because they demand focus and patience, sometimes you just need some extra oomph. Especially if you're working on a project that requires a lot of pieces to be cut, or when you're just looking for a more precise cut, power saws are a great option. Unlike a handsaw, where your primary job is to control each stroke of the saw's blade, a power saw spins or moves its own blade, so your main job is to safely and steadily control its motion in order to make a precise cut. Some saws do this with a spinning blade, others with a reciprocating blade, which moves back and forth like a sewing machine's needle. Others have large, thin blades on a loop that run around a track.

SAFETY CHECK!

For all power saws, make sure to take these safety precautions:

- **A skilled and experienced adult builder buddy**
- **Safety glasses (wear at all times!)**
- **Ear protection**
- **Dust mask**
- **Closed-toe shoes or work boots**
- **Long pants**
- **Hair tied back**
- **No loose clothing, hoodie strings, or jewelry**
- **Short sleeves, or sleeves rolled up to your elbows**

Types of power saws

Remember that the power saw you choose will largely be dictated by the type of cut you're making (crosscut, rip cut, etc.). You can reference the table on page 123 to help.

Regardless of which power saw you choose, get ready to make some sawdust!

Miter saw: A miter saw has a fixed base and a vertical disk blade that you lower in a chopping motion to cut narrow pieces of lumber like 2×4s. Use a miter saw to make crosscuts (across the wood's grain). The flat "bed" of a chop saw is usually only 6 to 12 inches deep, so it is an ideal tool for making quick cuts of long pieces of lumber. Keep in mind that a miter saw will only cut through pieces with a width that fits within the depth of the bed, so it is most often used to cut dimensional lumber like 2×4s and 2×6s. When houses are framed using 2×4s, a miter saw is an indispensable tool to have on the job site!

Miter saws are for wood-cutting only. **Don't try to cut metal, plastic, or other materials with your miter saw.**

A miter saw comes in a few different configurations.

- A standard miter saw has a blade and bed that rotate side to side so you can make diagonal cuts (miter cuts) across the face of your wood.
- A compound miter saw can make miter cuts and bevel cuts (diagonal cuts along the edge of the wood); the blade, normally vertical, tilts to the side up to 45 degrees to make an angled cut through the wood. See page 122 for a refresher on miter and bevel cuts!
- Some miter saws also have a slide, allowing you to pull the saw's arm toward you before making your cut. A slide miter saw gives you more depth to cut wider pieces of lumber, like a 2×12.

MITER SAW

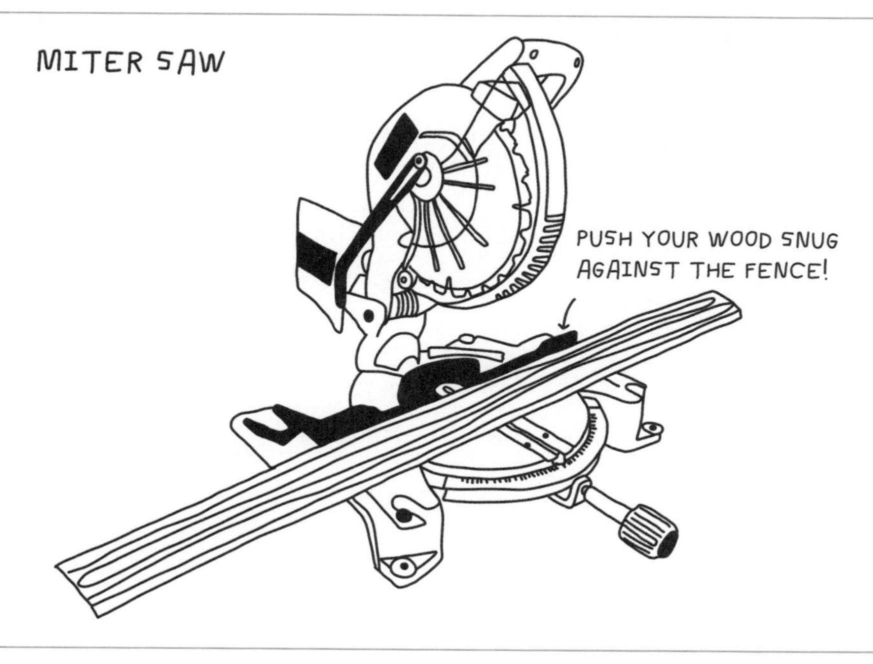

COMPOUND MITER SAW

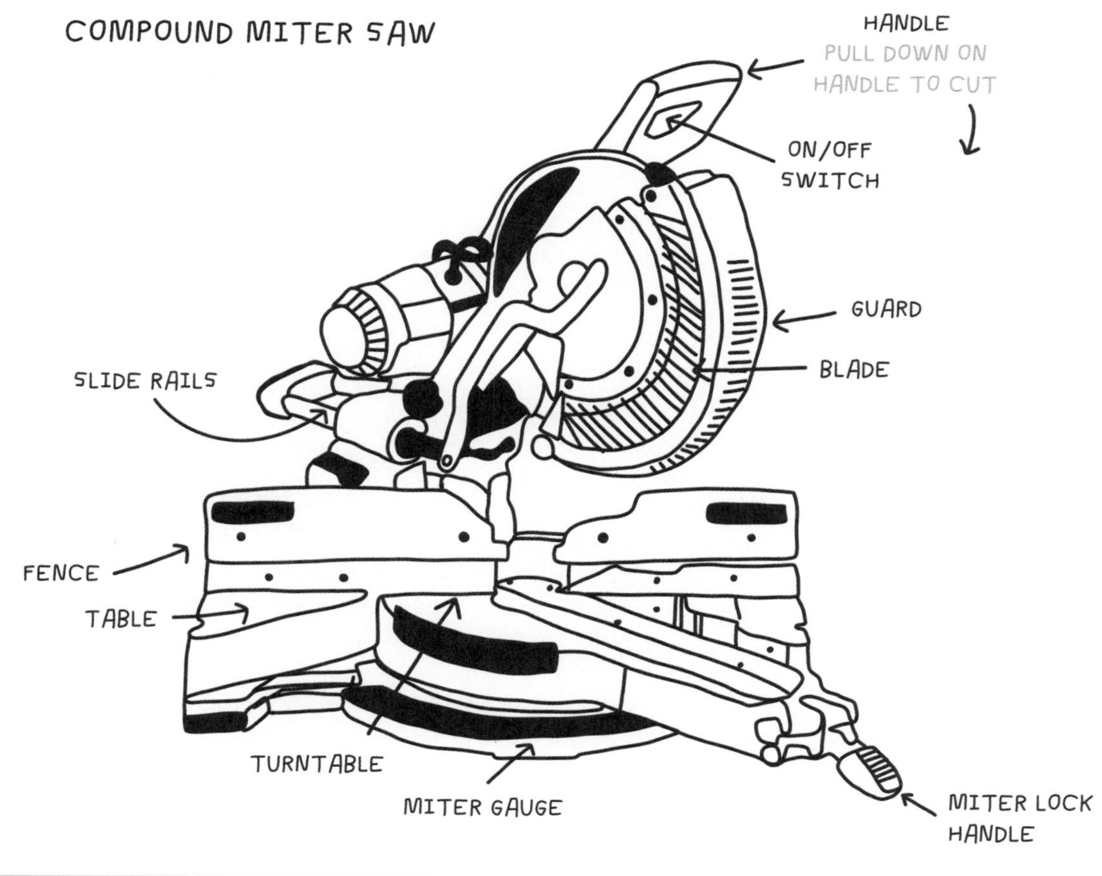

MITER CUT ON THE MITER SAW

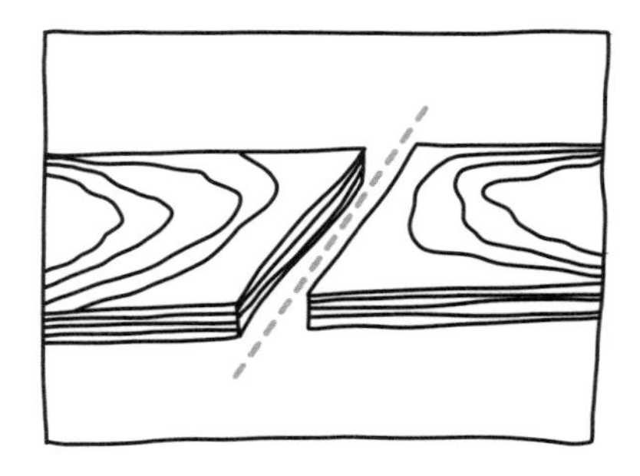

BEVEL CUT ON THE MITER SAW

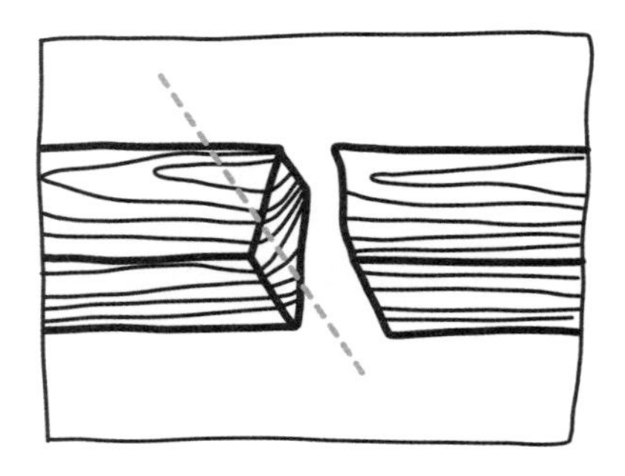

Two important safety features on a miter saw are the fence and the guard. Use the fence as the backboard for all your cuts! The guard is a plastic cover over your blade that shields you from the spinning blade. As you lower your blade to make your cut, the guard rotates slightly to allow the blade to make its cut. You can buy chop-saw blades in all kinds of varieties, though I prefer a wood-finish blade, which has more teeth than a standard blade and makes smoother cuts.

Do not operate a miter saw without the supervision and hands-on support of a skilled and experienced adult. Always read and follow the manual for your specific tool. Safety glasses should be worn at all times by both users and observers!

Once your wood is measured and marked, you're ready to cut it on your miter saw.

The cardinal rule of miter saws is: The fence is your best friend! When you place your wood on the saw bed, the first thing you should ALWAYS do is push it back against the fence, ensuring that the wood is tight against the fence AND flat on the bed of the saw. **Make sure that the entire length of your piece is in contact with the fence and bed.**

Use your nondominant hand to hold the wood in place, making sure that your hand is outside of the area marked with a "no hands" symbol on your saw. If your saw does not have this symbol, place your holding hand on the farthest end of the saw bed, as far from the blade as possible. Hold the wood firmly in place, pushing down onto the bed AND back into the fence. Tuck your thumb so that it is not protruding! This holding hand is incredibly important: Keep a firm hold against the fence for the duration of the cut.

Before you make your cut, double check that you're wearing your safety glasses! You will squeeze the trigger button with your dominant hand, which is usually located in the handle of the miter saw. Begin your cut by squeezing the trigger on the saw with the blade NOT in contact with the wood.

- For a standard miter saw, start your blade NOT in contact with the wood and allow it to come to full speed. Slowly lower it, cutting smoothly all the way through the thickness of the wood. When the cut is complete, release the trigger and wait for the blade to stop spinning before raising the blade arm back up.
- For a miter saw with a slide, pull the arm toward you so the blade is directly above the front edge of the wood. Begin your cut

NOT in contact with the wood, then lower the saw straight down first, and then back toward the fence (in an L-shaped motion). When the cut is complete, release the trigger and wait for the blade to stop spinning before raising the blade arm back up.

For any cut on a miter saw, **wait until the blade stops spinning completely before reaching for or moving your wood.**

Circular saw: A circular saw is a convenient, portable power saw for making long, straight cuts, most commonly in sheets of plywood. You may also hear a circular saw referred to as a Skilsaw, which is a common brand name. A circular saw has a disk-shaped blade held at a vertical angle within its casing. A flat base, or "shoe," sits flat on the wood, sliding along its surface as the blade spins and cuts along your cut line.

While a miter saw is better for cutting lumber like a 2×4, a circular saw's flat "shoe" base is awesome for sliding across wider pieces of material, like a sheet of plywood. If you don't have access to a table saw to make long cuts, a circular saw is a great alternative. I especially prefer cordless circular saws because there's no chance of getting caught or hung up on a cord during a cut.

Circular saw blades come in various tooth-per-inch counts and tooth patterns based on the type of material you need to cut. You can buy blades for wood cuts (rip cuts and crosscuts), metal, and even "nail cutting" blades designed to cut through wood that might have staples or other junk left in it. Buy a blade with a diameter that corresponds to the size of your circular saw: a common size is a 7¼-inch circular saw, which needs a 7¼-inch blade.

CIRCULAR SAW

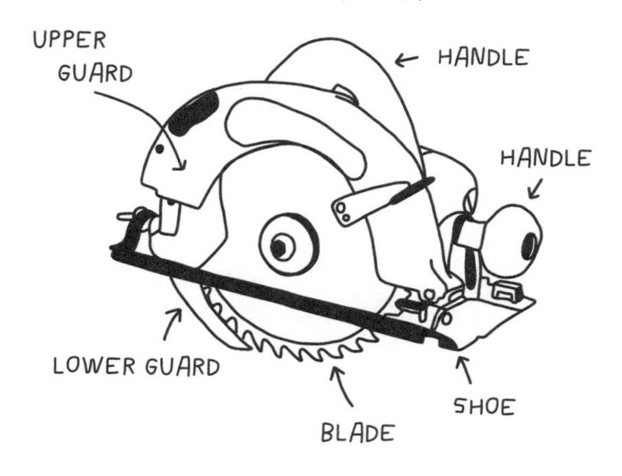

CIRCULAR SAW, MIDCUT

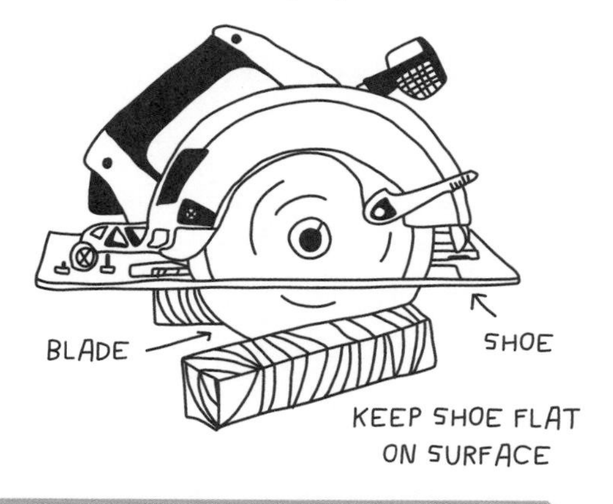

FUN FACT!

Miter saws are often referred to as "chop saws." If you hear the term "chop saw" in a woodshop, they are likely referring to a miter saw. However, a chop saw, technically, is the metal-cutting power saw with an abrasive disk that does not rotate to make angled miter cuts. In practice, they are used interchangeably: We often use the term "chop saw" at Girls Garage to refer to our wood-cutting miter saw!

Do not operate a circular saw without the supervision and hands-on support of a skilled and experienced adult. Always read and follow the manual for your specific tool. Safety glasses should be worn at all times by both users and observers!

The absolute number-one most important tip for using a circular saw is this: Keep the base ("shoe") of your saw flat on the material's surface throughout the entire cut. Secure your material using clamps on a worktable, or use sawhorses for larger pieces. Mark your entire cut line using a long straightedge or chalk line. Start the saw with the blade OFF your material, not in contact with your starting point. Hold the saw with both hands at all times. Most circular saws have a large grip handle for your dominant hand and an additional handle up front to hold with your nondominant hand for extra stability. Let the saw do the work; no need to force it. A helpful habit is to use another piece of wood, like a 2×4, as a makeshift fence or rail. To do this, clamp the 2×4 parallel to your cut line at the correct distance to accommodate the width of the saw's shoe, and make your cut easily by sliding the saw along the 2×4 fence.

Table saw: Use a table saw to make long cuts in plywood or lumber. For lumber, these cuts are called rip cuts, which are cuts that run parallel to the grain of your wood. For plywood, which has layers of grain running in different directions, use a table saw to make cuts along the long dimension of your piece of plywood. A table saw is ideal for making these long cuts to trim the width of a sheet of plywood or lumber, with the blade running parallel to the wood's grain. A table saw has a spinning vertical disk blade like a miter saw, but the blade is fixed in one place within a large table surface. Instead of moving the blade (like with a miter saw), your job is to move the material along the surface of the table using the fence to support the longest side of your wood, sliding it over and past the blade to make your cut.

Industrial woodshops usually have a table

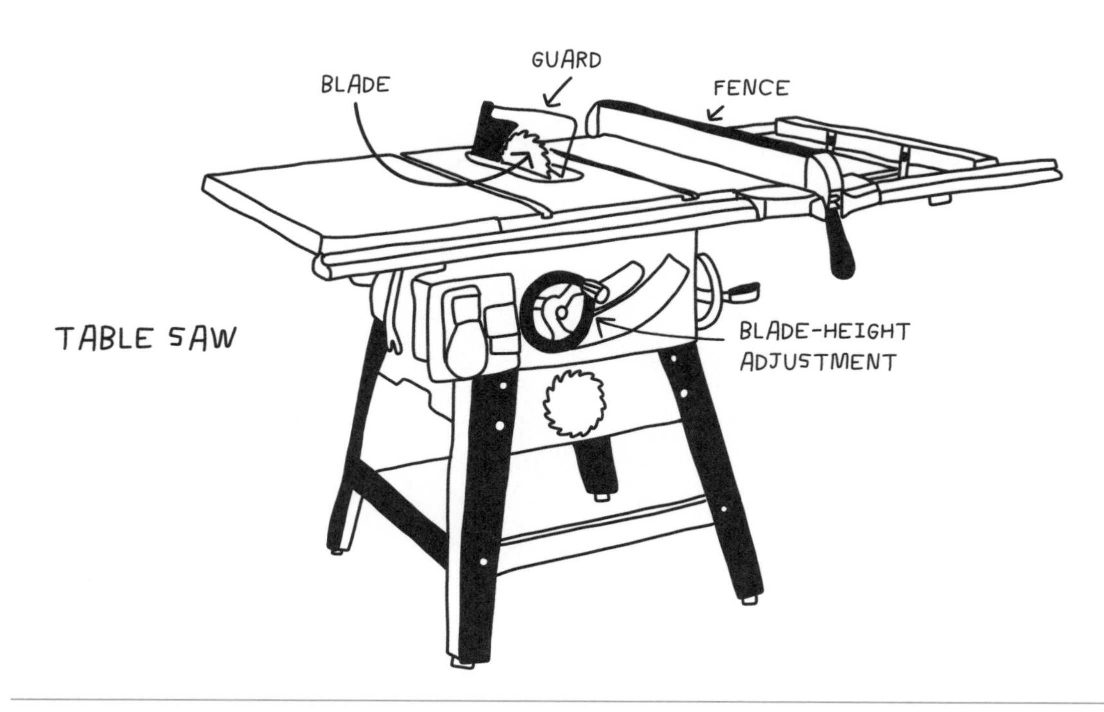

saw as their "centerpiece." They take up a ton of room and require a lot of clearance around them because of the size of the sheets they can cut. But there are also great options for portable table saws, which you can fold up like a folding chair and easily transport from the garage to your backyard.

Table saws have a fence and guard over the blade, which are the most important safety features. For any table saw cut, keep the edge of your material tight against the fence to ensure a straight, safe cut. **Never leave material unsupported in between the fence and blade.** Using the fence helps you make a straight cut and lowers the possibility of kickback—when a table saw grabs and hurls a piece of wood. Kickback is scary and tends to happen when cuts are not straight and there is a funny sideways or twisting force on the blade. Kickback is not common, especially when table saws are used properly and safely, but definitely good to know about.

The guard is a plastic cover that protects your hands from the blade and sometimes has a built-in dust collector that sucks up sawdust as you cut. Table saw blades come in a few varieties that can help you make different cuts: a standard cutting blade, or a dado or rabbet set, which are multiple parallel blades that can cut channels and notches in wood. Table saws also let you raise and lower the height of your blade, and some have a bevel adjustment that tilts the blade to make angled bevel cuts.

Of all power saws, the table saw is the most powerful and robust, requiring the most knowledge, safety, and confidence. **The table saw is for experienced adult users only. Do not use this tool if you are under the age of 18. Always read the manual for your specific tool (the following tips are general guidelines). Safety glasses should be worn at all times by both users and observers! Do not operate a table saw alone: A table saw is a two-person machine! Always work with another experienced adult**

CUTTING ON THE TABLE SAW

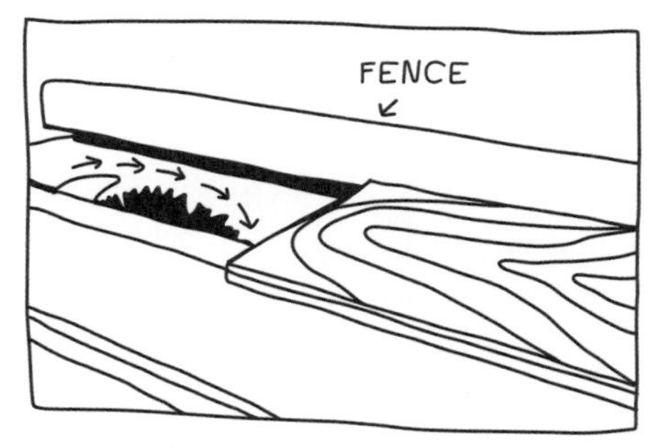

ALWAYS KEEP WOOD SNUG AGAINST THE FENCE

FUN FACT!

One table-saw manufacturer, SawStop, is extremely popular in industrial and school woodshops for its brilliant safety features (we use a SawStop at Girls Garage). The SawStop table saw has a mechanism built into its blade system that detects any material that conducts electricity (like a metal staple, or, more importantly, a human finger). Wood does not conduct electricity. When the blade detects an electrical current, within milliseconds it drops below the surface of the table to protect the user from potential finger loss. SawStop has an excellent video showing this dropout safety feature, using a hot dog. Do NOT try this at home.

who can help guide the material, especially large sheets of plywood, by holding one side and "catching" as the wood comes off the other end of the table. Your builder buddy is also an extra set of eyes to check that you are doing everything safely.

When making a cut, set your material on the table with the edge up against the fence, with the appropriate measurement and layout. Your blade should be raised above the table so the lowest part of its teeth (the "gullet") sits about ¼ inch above the surface of your material.

Remember that table saws are optimal for making long rip cuts along the grain of wood. You can make crosscuts with the use of a miter gauge or a "sled" that works as a horizontal fence to help you safely push through a piece that is wider than it is long. **Don't make crosscuts on a table saw without one of these devices.**

You always want the portion of wood you intend to keep between the fence and the blade, with the "off cut" on the other side. Your "keep" section should always be wider than your off-cut portion, giving you more stability when sliding the material across the table. For narrow pieces of wood, a "push stick" should be used to push wood through the table saw.

Remember that the table saw is extremely dangerous and should only be used by experienced adults. Do not operate a table saw without reading and following your own machine's manual! To make your cut, start the table saw

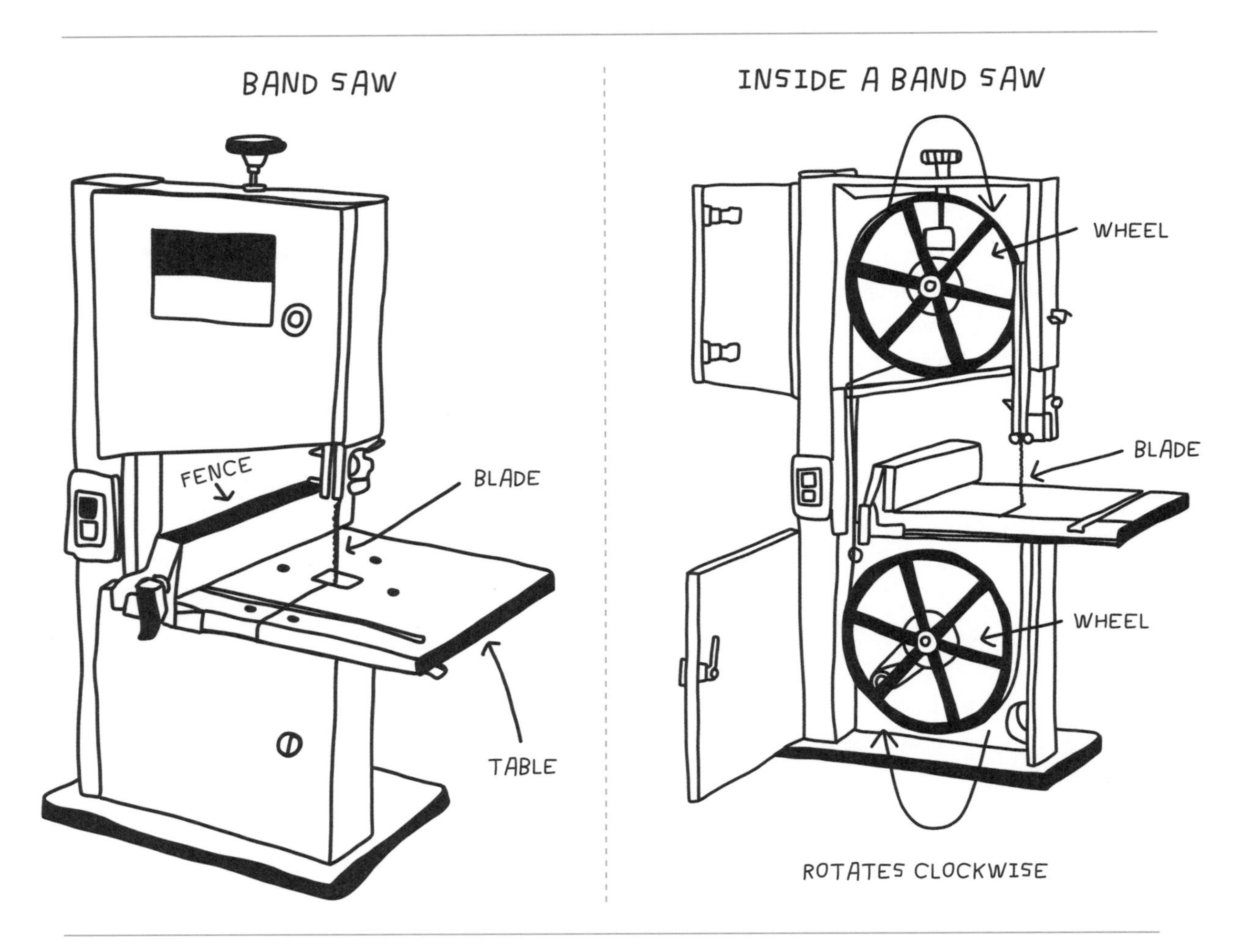

with your wood NOT in contact with the blade and with its long edge pushed up against the fence. Make sure that your builder buddy is ready to "catch" your cut pieces on the other side of the saw. Start the saw, and slowly and steadily push the material straight forward across the table surface, using both hands to hold the rear edge of the wood and ensure that the long edge remains in contact with the fence as you progress. To finish the cut and push your wood past the blade, use your right hand (closest to the fence) to push it through, using a push stick instead of your hand for pieces narrower than 12 inches. Never lean your body over the blade or reach across the blade.

Complete the cut, turn off the machine, and wait until the blade stops spinning before reaching for your cut material.

Position your body outside of the "kickback zone" in case a kickback occurs. If your fence is on the right side of the blade, stand to the left of the blade. If your fence is on the left side of the blade, stand to the right side of the blade.

Band saw: A band saw uses a long, continuous steel band looped around two wheels to make precise cuts. The band itself has teeth on one side; as the wheels turn quickly, the teeth move downward toward the table to make cuts in the wood. Use a standing band saw to make more precise cuts in thin material, especially curved cuts.

A standing band saw has a vertical band blade that loops around an upper wheel and a lower wheel. The total length of a typical band-saw blade can be 6 to 12 feet long, welded into a loop. About halfway between the upper and lower wheels, the blade is exposed and runs through the center of a table where your material sits. Like a table saw, the band saw has a fixed blade, and you are in charge of moving the material along the blade to make your cut. Unlike a table saw, you can make cuts without the support of a fence or miter gauge, and also cut curves.

The blade of a band saw rotates downward toward the table surface, which means it is a safe option that is unlikely to kick back and hurl material, like a table saw sometimes can. While a standing band saw is the most common (and one of my favorite go-to tools for many jobs), there are also portable handheld band saws.

Do not operate a band saw without the supervision and hands-on support of a skilled and experienced adult. Always read and follow the manual for your specific tool. Safety glasses should be worn at all times by both users and observers!

A band saw has an exposed blade, so before making any cuts, it's important to raise or lower the band-saw head to limit the length of blade that is actually exposed. You only need enough blade to cut through the height of your material, and any additional height is unnecessary and an added risk. Use the wheel or dial on your band saw to raise or lower the height of the head to just an inch or so above the top of your material so you have enough visual clearance to see the material and the blade.

FUN FACT!

In the long evolution of the band saw, it was a woman who made it all work. Though the band saw was first patented by a British man in 1809, it did not have much success at first because the metal band blades could not be produced with accuracy, or without snapping easily. Then in 1846, a Frenchwoman named Anne Paulin Crepin discovered a far superior way to weld the bands seamlessly so they would run more smoothly. It was this innovation, plus new steel alloys developed at the same time, that led to the band saw's widespread use throughout Europe and beyond.

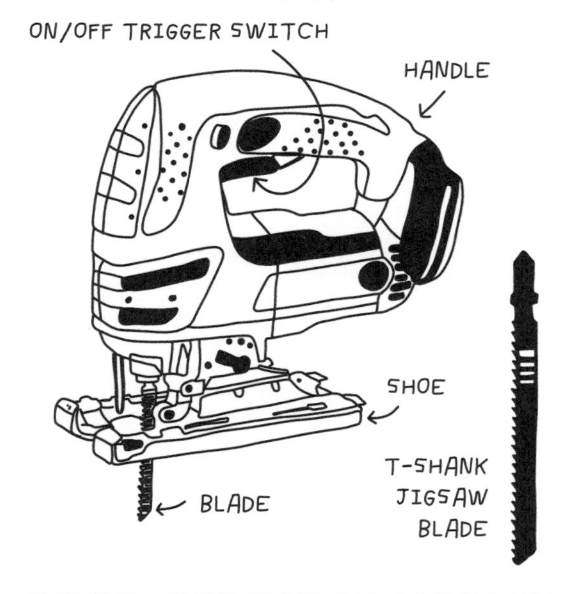

FUN FACT!

If a jigsaw reminds you of a sewing machine, that's no accident! The jigsaw was first invented in the late 1940s by Albert Kaufmann, who tinkered with his wife's sewing machine, replacing the needle with a short saw blade.

With any band-saw cuts, use two hands at all times, with one hand on either side (left and right) of the blade and your thumbs tucked in. Your fingers should never be directly in front of or in line with the blade! Keep your fingers at least 4 inches from the blade at all times, moving them back incrementally as the material advances through the blade.

- Turn the machine on with your material NOT in contact with the blade, and let the band rev up to full speed.
- For straight cuts, use the adjustable fence as a guide, pushing your wood straight through.
- For curved cuts, make slow turns, being careful not to put too much awkward twisting tension on the blade. Listen to the sound of the blade: if it makes a squeaking or squealing sound, it's a sign you're putting too much twisting pressure on the blade, and it might break.

Jigsaw: A jigsaw is a portable tool with a thin, vertical "reciprocating" blade, meaning it moves back and forth quickly. When you don't have access to a band saw, a jigsaw is a great option for making curved or diagonal cuts through thin pieces of material.

Nowadays, most jigsaws use a "T-shank" blade, which has a T-shaped top that locks into the saw mechanism without having to use a tool to install it. These blades come in various tooth-per-inch counts and tooth directions for different materials and smoothness. As with a circular saw, I prefer a cordless jigsaw for its portability. Most jigsaws have a "variable speed" trigger, so the harder you squeeze the trigger, the faster the blade will reciprocate. Do not operate a jigsaw without the supervision and hands-on support of a skilled and experienced adult. Always read and follow the manual for your specific tool. Safety glasses should be worn at all times by both users and observers!

Jigsaws are relatively easy to use, compared to other saws, though also less precise. Always clamp your material securely to a table surface with two or more clamps. Position your piece so that your cut lines are NOT on the table surface (otherwise you will end up with gashes and slices in your table edge!). Before making your cut, position your jigsaw with the blade NOT in contact with your wood, lined up with the start of your cut line. Your number one goal is to keep the base ("shoe") of your jigsaw flat and steady on the surface of your material without tilting to the sides, front, or back. Use your nondominant hand to hold the top of the tool steady, pushing downward into the wood to help stabilize the cut. Use your dominant hand to squeeze the trigger. Always hold the jigsaw with two hands while operating. With the trigger squeezed, advance the jigsaw along your cut line slowly, letting the saw do the work. It can be hard to follow lines with a jigsaw and this requires some practice, especially as sawdust can hide your line as you cut. I try to blow away any sawdust as I'm cutting, to keep my eye on the line. Even though a jigsaw is a great tool for making curved cuts, make turns slowly so as not to break your blade.

One cool trick: If you need to cut out an interior shape (like the hole of a capital letter P), you can drill a hole wide enough to fit your jigsaw blade through, and then start your cut from the hole, working your way outward until you have cut out the entire shape.

Reciprocating saw: A reciprocating saw, which you might hear referred to by the trade name Sawzall, has an exposed blade that pushes and pulls, moving back and forth quickly to make rough cuts through basically any material. Reciprocating saws are used for demolition jobs and by policemen or firefighters who need to quickly cut through walls or debris.

RECIPROCATING SAW (SAWZALL)

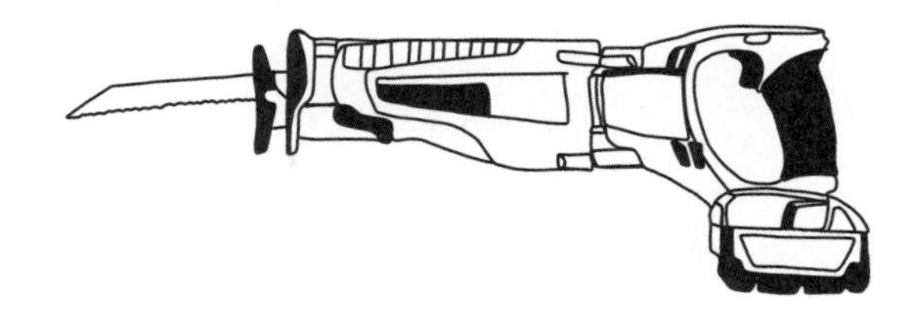

Reciprocating saws might remind you of an electric turkey carver! You can buy various blades for heavy-duty or lighter-duty jobs, and they have a trigger that's similar to a jigsaw, with variable speed, depending on how hard you squeeze it.

Reciprocating saws are definitely not for precision work, but what they lack in precision, they easily make up for in versatility and power. You can use a reciprocating saw in a horizontal orientation to cut though materials like drywall or wood, and most also have a small "shoe" like on a jigsaw, so you can use it vertically with the sole resting on your material. Do not operate a reciprocating saw without the supervision and hands-on support of a skilled and experienced adult. Always read and follow the manual for your specific tool. Safety glasses should be worn at all times by both users and observers!

Because of the exposed blade, be extra careful when using a reciprocating saw, and use a trigger lock if your tool has one so the saw does not accidentally get turned on before you're ready.

Scroll saw: A scroll saw is like a baby band saw, but with a reciprocating blade that moves up and down like a sewing machine, instead of in a complete loop. Scroll saws have tiny blades, making them ideal for cutting tight curves or

SCROLL SAW

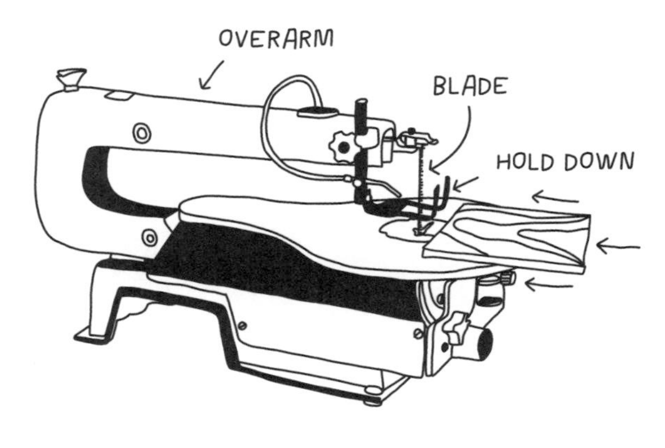

CHAIN SAW

FUN FACT!

Despite how burly and crude you may think a chain saw is, it is also a tool used by artists! If you have ever seen an ice sculpture or a giant tree trunk carved into an adorable bear, you can thank the chain saw. Chain-saw artists and ice sculptors use a chain saw to make the majority of their carvings.

small details. They are a good alternative to the coping saw, as they are more precise.

The scroll saw's small blade is suspended between two points, just like a coping saw. And also like a coping saw, this blade can be removed and threaded through a hole in your material, to start a cut from inside a hole.

Do not operate a scroll saw without the supervision and hands-on support of a skilled and experienced adult. Always read and follow the manual for your specific tool. Safety glasses should be worn at all times by both users and observers!

You'll use a scroll saw in a similar way to a band saw, by placing your material on the table and moving it into the blade. And as with a band saw, keep your fingers at a safe distance from the blade, moving them back as the wood moves forward.

Chain saw: While your first association with a chain saw may be from a horror movie, it is a common tool used for cutting down trees, cutting firewood, and more. Chain saws use a set of sharp teeth attached to a chain that loops around a metal plate to cut through lumber and trees.

Like a bike chain, the chain is looped around a metal guide bar and rotates around it at high speeds. Most chain saws have a small engine that requires gas and has a pull-string to start it, similar to how you'd start a lawn mower. **The chain saw is for experienced adult users only. Do not use this tool if you are under the age of 18. Always read the manual for your specific tool (the following tips are general guidelines). Safety glasses should be worn at all times by both users and observers!**

A chain saw is a hefty tool to hold, and **should be held with two hands at all times,** one on the handle at the rear, and one on top of the body of the saw. When making a cut, your motion should always be downward, allowing gravity to do most of the work. Like the table saw, a chain saw is one of the most dangerous saws and is for adult use only.

QUETZALLI FERIA GALICIA

Girls Garage student

East Bay Area, California

Quetzalli has been a student of mine at Girls Garage for a few years now, and she is truly a ray of hope and sunshine. As a participant of our Protest + Print class, Quetzalli, along with eleven other teen girl colleagues, hand-drew and screen printed a collection of activist posters that expressed anger, love, curiosity, and optimism about the pressing social and political issues of our time. Since that first class, Quetzalli has also joined a cohort of teen girls to build projects for our community, like a hexagonal sandbox for a local preschool. In every Girls Garage class, she is a calm and considerate presence and a fierce builder through and through.

"My name is Quetzalli, and I *really* like animals. I really love dogs, and making their lives better just makes me happy.

"I'm also a Chicana; my family is from Mexico, but I was born in Illinois. To me, being a Chicana means that I have experienced things that other people of any other classification will never experience, and also that I have not experienced things that other people have. Being Chicana gives me a sense of community: I have people in my life who have this label, too, who identify with this label, and who relate to me in this way.

"From a young age, as a young girl, I knew how hard it was to be loud, to have your opinions be heard. So I ended up being quiet for a really long time. Then I went to an all-girls middle school, which was such a safe environment to speak your mind. For the first time, I was surrounded by people who had similar hopes as me, even if we had differing opinions. Being in an all-female environment gave me another sense of community. It made me feel like, 'we are all girls, we are all a minority in some way, and we get to defy the stereotypes together, talk about these things together, and be understood.'

"I found Girls Garage through my school! I took the Protest + Print class and got hooked. It was a physical place just for girls, and it's hard to find places like Girls Garage in the world. Even though I'm not sure if I want building or making to be my career, it is something I want to do in my life, because it gives me yet another community of people who are all doing something real together.

"The first big building project I did at Girls Garage was building a hexagonal sandbox for a nearby preschool with other teen girls from the Bay Area. None of us knew each other, and we all had such different experience levels—actually, most of us had *no* experience using tools or building anything! But we all helped each other, and that's how we overcame the challenges and the fear—by asking each other for help when we needed it.

"When we built the sandbox, we first cut all the wood pieces, which required so much math to figure out the angles and dimensions for the miter cuts. I love the miter saw! It makes such sleek, smooth cuts, and I like the jigsaw. I was scared of making even a tiny mistake, but then I got over it and got better every time I did it, just like riding a bike. Then we assembled all the sandbox wall pieces on site with drills and drivers, connecting each side into the hexagon shape. We also wrote little notes of hope or encouragement on wooden shims and hid them inside the sandbox for someone to find in the future. And then we opened the sandbox for the preschool kids, and they all jumped in! Getting to build something physical, to have my little mark on the world, was really special to me.

"For girls who want to try building, I would say just do it! You don't have to second-guess yourself. And if people tell you that you can't do it, then do it because of that, to prove them wrong. For my own future, I hope that people will be more empathetic and listen to each other, and really see each other. I think empathy and kindness would help us all have hope for the future."

POWER TOOLS

One of my unofficial slogans is "Girl Power and Power Tools!" Now that you've got all your pieces cut, thanks to a wide array of saws, you'll need some other tools to help assemble. While wrenches and screwdrivers and other hand tools might also be helpful, there are some awesome power tools that can amplify your human power and make the work of building more efficient and enjoyable.

Even though so much of the imagery we see makes power tools seem like accessories for the manliest men, I can assure you that power tools have no idea whether a man or a woman is using them. In my experience, my young girls (and their mothers, sisters, and friends) are skillful, precise, and brave around power tools. The ability to hold a machine that automatically makes your physical ability more powerful can be an awesome and transformative experience. So put on your safety glasses and flip that on/off switch!

SAFETY CHECK!

For all power tools, make sure to take these safety precautions:

- A skilled and experienced adult builder buddy
- Safety glasses (wear at all times!)
- Ear protection
- Dust mask
- Closed-toe shoes or work boots
- Long pants
- Hair tied back
- No loose clothing, hoodie strings, or jewelry
- Short sleeves, or sleeves rolled up to your elbows

Types of power tools

In addition to all the power saws you may have used to cut your materials (see page 134), here are the go-to power tools you might find or use in any shop to assemble and finish your project.

Drill

A drill is arguably the most common power tool in any house or woodshop. However, if there's one lesson to be learned from this book (okay, there are many), it is this: **a drill is first and foremost for drilling holes in wood, metal, masonry, or other materials.** You can also use a drill to install screws, but an impact driver is a much safer tool for that job because of its internal hammering mechanism that delivers greater turning force (torque), so you don't strip your screw heads or tweak your arm (more on impact drivers next!). A good rule is to use a drill for drilling holes and a driver to drive screws. Most drills are advertised as drill/driver combo tools, but if you can, it's best to have both a drill and a driver around, which are often sold as a convenient set.

A power drill is called a pistol-grip tool because of its shape and its trigger-style on/off switch (though it should go without saying that a drill should never be used as a gun or toy). The drill has a handle, replaceable and rechargeable battery, trigger, and chuck. The chuck is like the jaws of the drill, gripping the shank of a drill bit. You can manually adjust the torque using the torque collar, which changes how much strength the drill exerts to rotate its bit. Drills also have a forward/reverse switch to change the direction of the rotation of the bit (forward to drill a hole). The forward button is usually on the right side of the handle, and the reverse on the left. When the forward button is pushed in on the right, the reverse button pops out on the left, so you can always feel with your fingers which direction the drill is turning. These forward/reverse buttons are usually labeled with arrows—one pointing forward, toward the chuck of the drill, and the other pointing backward.

This button is also a locking mechanism—when the button is in its neutral or central position (not pushed forward *or* backward), the drill is locked and you cannot squeeze the trigger. When drills are not in use, it's a good practice to lock them in this position.

Do not operate a drill without the supervision and hands-on support of a skilled and experienced adult. Always read and follow the manual for your specific tool. Safety glasses should be worn at all times by both users and observers!

There are a few important skills to practice in order to use drills safely and efficiently:

- **Insert and take out bits:** Your drill's chuck has two parts: metal jaws that hold the bit and a rotating plastic chuck (usually black). While looking directly at the front of the chuck, loosen the metal jaws by turning the chuck *counterclockwise* (righty tighty is clockwise, lefty loosey is counterclockwise!)

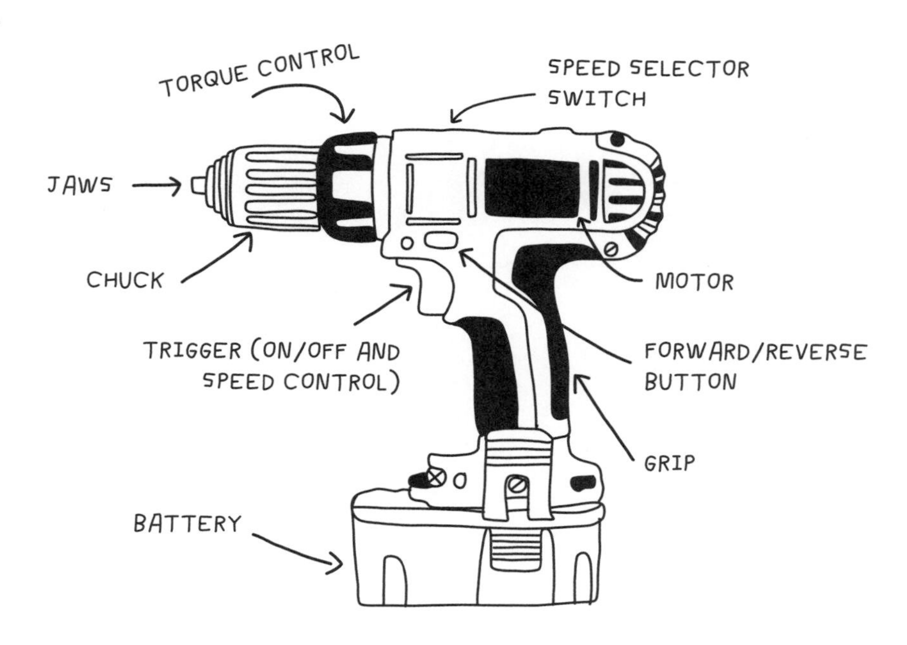

and watch as the metal jaws widen. Now insert the shank (smooth) end of a drill bit.

- **To tighten, you have two options:** Hold the bit in place with one hand while tightening the chuck by turning *clockwise* (righty tighty!) until the metal jaws squeeze firmly around your bit. Or (this technique requires more practice), pointing the drill away from you, hold the bit in place with your thumb and index finger, and, with the rest of your fingers and palm, hold the chuck firmly so it doesn't move. Then with your other hand, squeeze the trigger slowly, which rotates the metal jaws and tightens them around your bit. Practice both ways and see which feels more comfortable.

- **Hold and drill:** To use the drill, your number-one priority is to keep your drill bit perpendicular to your work surface (for example, if you're drilling through a piece of wood lying horizontally, you want your drill bit to be perfectly vertical). Grab the handle with your dominant hand. Your index finger should easily reach the trigger, and you can also use your index finger to check that the forward switch is pushed in. Squeeze the trigger slowly, then faster (the harder you squeeze, the faster the drill will spin). Use your nondominant hand to stabilize the drill; I like to place this hand on the back end (the butt) of the drill motor to push into the drill direction. Once your hole is drilled, you might think you need to reverse the direction of the bit's rotation. However, it's easiest to maintain the forward-spinning direction, squeeze the trigger, and simply squeeze the bit back out.

- **Adjust your torque:** The torque collar is useful if you use your drill to install screws (but remember that a driver is better!). Because you don't want to overtighten screws, having the ability to adjust the torque (*how much force your drill exerts when rotating*) is good. The torque collar has a range of numbers (1 to 20, etc.), with the lowest number being the lowest amount of torque. Set this on the lowest number, and when the screw is tightened sufficiently, the drill will make a clicking noise and not turn any further.

- **Blowout:** If you're drilling all the way through a piece of wood, place it on top of another piece of wood. This will minimize "blowout"—when the drill bit chews up the underside of a piece of wood when the bit exits. It will also keep you from drilling into your own table!

FUN FACT!

In the early 1990s, rock star Eddie Van Halen started using a drill as a musical instrument. He would turn the drill on and hold its spinning chuck against the strings of his electric guitar, making a squealing "revving-up" sound that matched the melody of the song. He even painted the drill to match his guitar's graphic pattern. His live performance of the song "Poundcake" is his most famous with the drill and guitar.

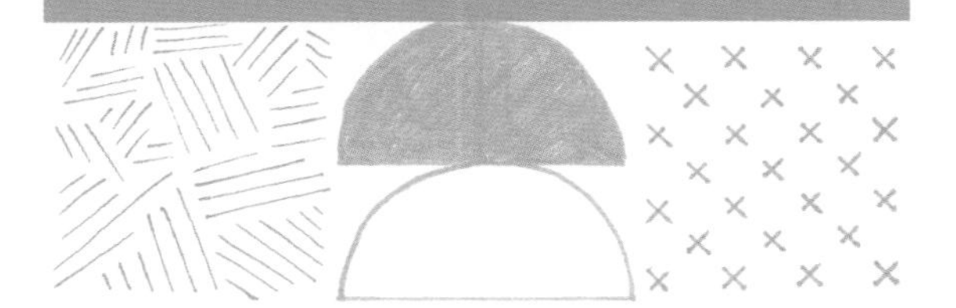

Drill bits

As with all the types of hardware, there are specific drill bits for any hole you might need to drill in any type of material. As the size of holes and hardness of materials might vary, you'll find all these types of drill bits to be very handy.

Twist bit—HSS—High Speed Steel: This is the most common type of bit and the kind found in most drill-bit sets. Use this bit for drilling holes through wood. Sets of twist bits come in a range of sizes, from about 1/16 inch up to 1/2 inch, but you can also purchase single bits in the exact size you need. Use twist bits to make pilot holes (reference the table on page 49 to figure out the right size).

Spade bit: This bit has a point and a flat-square shape for drilling wider holes. The pointed end will help you locate the starting point and center of your hole, and the paddle shape will follow, removing the wider diameter of your hole. Only use spade bits with wood.

Wood-spur bit: These are sometimes called "brad-point bits" or "lip and spur bits" and are like a hybrid of a spade bit and a twist bit. Wood-spur bits are shaped like twist bits but have a point on the end like a spade bit that makes them easier to guide and start your hole.

Masonry bit: Masonry bits are made for drilling small holes in bricks, stone, and concrete blocks. They have a wide tip, usually made of tungsten carbide, which is very strong and can drill through hard masonry without breaking the bit. Use masonry bits at a slow speed and pull the bit out every 5 to 10 seconds to remove dust and debris and to prevent overheating.

Countersink bit: Countersink bits are made for screws designed to be countersunk, with a regular twist bit and an angled collar about halfway up the shank. The shape of the collar matches the shape of this specific screw's head. Use a countersink bit to predrill holes for screws that need to be countersunk to sit flat on the wood's surface.

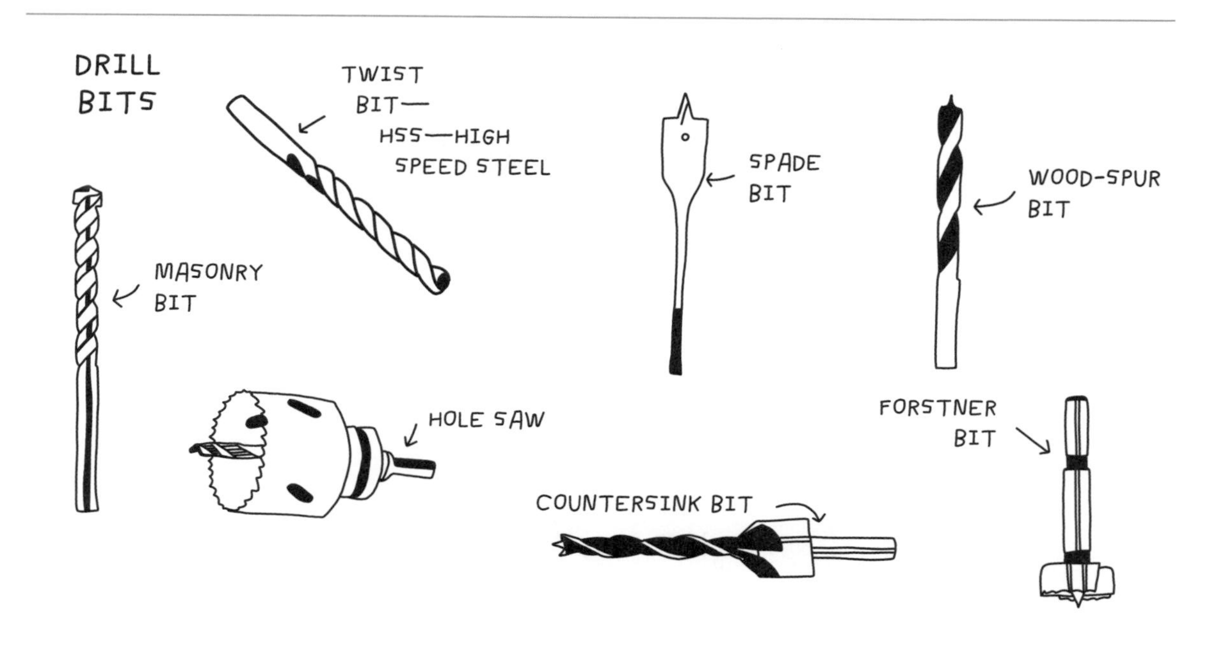

Forstner bit: Like a spade bit but burlier, a Forstner bit helps you drill wide holes with flat bottoms. The Forstner bit has cutting edges around the circumference of its circle, so it cuts efficiently through thick material. Forstner bits work best with a drill press (more about those on page 154!), rather than a handheld drill.

Hole saw: Use a hole saw to cut a large hole out of a piece of wood, plastic, or other material. The hole saw has saw teeth along its circumference, with a drill bit in the center to help start your cut. Like Forstner bits, hole saws work best with drill presses.

Impact driver

This is a drill's best friend, as it does the work to install screws safely and efficiently. Use an impact driver to drive screws (and some bolts!) of any size and with any drive type (Phillips, square, star, etc.). A drill and driver work best as a pair: make the pilot hole with your drill, then install the screws with your driver.

A driver is easily confused with a regular drill, but it has some benefits all its own. While a drill's chuck only rotates, an impact driver rotates *and* exerts an additional impact on the screw head to help drive it in. An impact driver makes a distinct hammering sound (like the power tool version of a woodpecker) as it drives screws. This hammering action turns screws with more force, prevents stripping your screw's head, and also protects your arm from violent motion.

The biggest visual difference between a drill and a driver is the chuck. A driver's chuck does not have metal jaws that widen and narrow to hold the bit (like a drill). Instead, a driver has a small metal sleeve in the shape of a cylinder (you may also hear a driver's chuck referred to as a "collet"). This sleeve pulls out and retracts, letting you release and insert your drill bits more easily than a drill.

Do not operate an impact driver without the supervision and hands-on support of a

IMPACT DRIVER

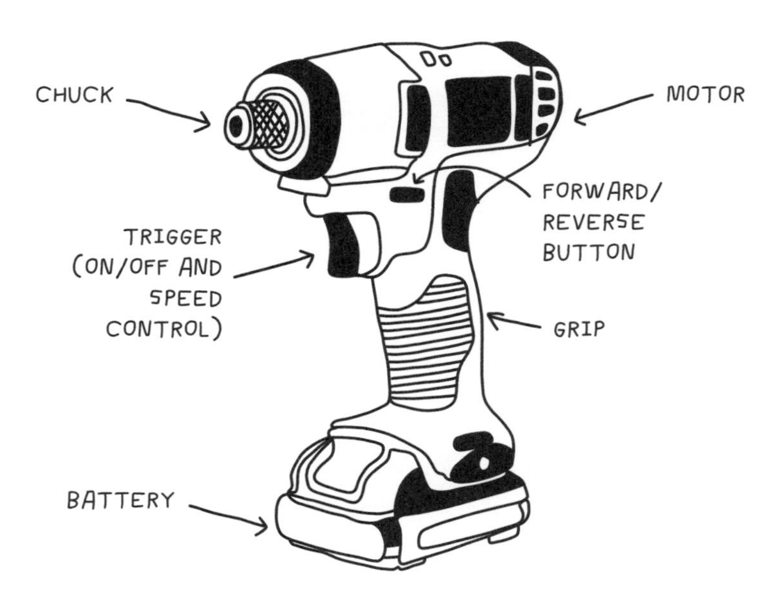

skilled and experienced adult. Always read and follow the manual for your specific tool. Safety glasses should be worn at all times by both users and observers!

When driving screws, hold and use a driver as you would a drill, with your dominant hand on the handle and trigger, and your nondominant hand on the butt of the driver for stability.

You can hold the screw with your nondominant hand and start the driver, driving the screw just enough so it bites into the wood. Then let go of the screw and move your nondominant hand to the driver to help push as you drive the screw the rest of the way. Remember that the driver has a variable-speed trigger just like a drill, so the harder you squeeze it, the faster the bit rotates. As the screw is almost totally inserted, you'll hear that distinct impact driver "hammer" sound (which is a good sign!). If you need to remove screws, reverse the rotation direction using the forward/reverse switch. It may sound counterintuitive, but you'll need to apply pressure downward, into the head of the screw, while squeezing the trigger, in order to pull the screw out. This pressure is necessary to maintain a good connection between the bit and the screw head.

Driver bits

Unlike round-shank drill bits, driver bits are hexagonal in profile. Almost all drivers only accept ¼-inch hexagonal shank bits. If you recall from the screw section, there are many drive types: Phillips, flat, star, and more. You'll need a bit for your driver that matches both the shape and the size of your screw drive: a Phillips #2 screw needs a Phillips #2 (PH2) bit, and a star/Torx size 25 screw needs a T25 bit. In most sets of driver bits, the bits are marked on their sides with the type and size.

You'll also notice that some driver bits are long, while others are little stubby things. The longer bits can be installed directly into the chuck, but the short ones require a magnetic bit holder, which fits into the driver like any other bit. This can be helpful if you're working with

DRIVER BITS

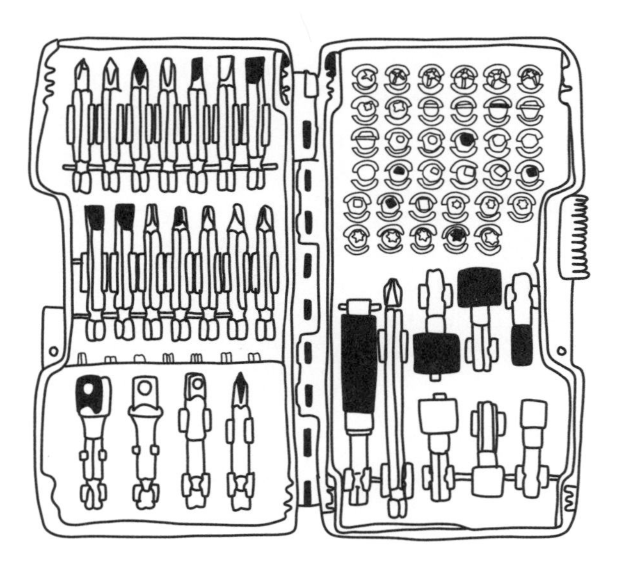

multiple types of screws and need to swap out bits frequently; you can leave the bit holder in the driver and just pop the shorter bits in and out as needed.

To remove or replace your driver bits, simply pull the metal sleeve on the chuck outward, which releases the grip on the bit, and you can easily remove it. With the metal sleeve still pulled outward, put in your new bit, then release the sleeve. The bit should now be locked in place! If you're using a bit holder and the shorter bits, just pull out the short bits and replace.

Drill press

A drill press is merely a stationary drill, meant for drilling holes very precisely, safely, and vertically. Unlike using a handheld drill, where you manually control the direction and motion of the bit, a drill press has a stationary chuck that holds the bit perfectly vertical for you. You then lower and raise it without worrying about its alignment. A drill press is also handy for drilling many holes in a row.

The drill press has a stationary chuck, a table with a crank you use to raise and lower the table, depending on the height of your material, and a handle on the side that usually has three arms and knobs that you rotate like a wheel. The chuck of a drill press takes a bit just like a handheld drill, with metal jaws that hold it in place. Most drill presses, though, come with a "chuck key," which is a small metal key device that loosens and tightens your chuck, rather than doing this by hand like with a drill. The chuck key inserts into a hole and lines up with the horizontal gearing of the chuck, rotating it to loosen and tighten. Remove the chuck key and store it safely before operating the drill press.

Do not operate a drill press without the supervision and hands-on support of a skilled and experienced adult. Always read and follow the manual for your specific tool. Safety glasses should be worn at all times by both users and observers!

Clamp your material securely to the table surface before starting the drill press. This lets you focus on lowering the bit and prevents your material from wobbling or spinning. You also need to adjust your table height based on your material, so that the drill bit starts above the surface, and can bore through the entire material. If you're drilling all the way through your material, you'll also need to center your table so that the drill bit can pass through the hole in the center of the table. You should also use a piece of scrap wood as a "backing board," which will ensure a clean hole without the underside tearing out. If you are drilling a hole that is wider than the hole in the table, you will also need a backing board underneath your work material, so that the bit does not hit the table once you drill through your piece. Insert and tighten your selected bit in the chuck using the chuck key. Turn the drill press on and let the bit work up to full speed. Using the knobs on the handle, rotate the handle toward you and lower the bit to drill your hole. Once your hole is complete, raise the bit back up by reversing the rotation of the handles!

DRILL PRESS

ON/OFF SWITCH
HANDLE
CHUCK
TABLE HEIGHT ADJUSTMENT
TABLE
BASE

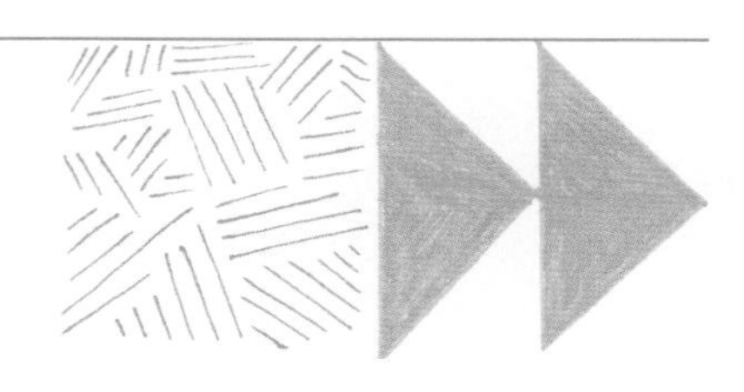

ALLISON OROPALLO

Engineering Technology Teacher

Petaluma, California

Allison is a dear friend (and hero!) of mine, a woodworking and carpentry instructor at Girls Garage, and a high school shop teacher. She is also hilarious, has a dog named Aubin, and played ice hockey in college. Allison is always up for a crazy challenge, from play structures for goats to her own impromptu kitchen renovations, and loves teaching young people how to tackle the seemingly impossible.

"When I was six years old, I built my first handmade creation. It was a wooden piggy bank that I made with my dad's help. I liked it so much that my dad immediately built me my very own workbench in the garage so I could work side by side with him, no matter how dangerous, difficult, or complicated the task. My dad never treated me differently because I was a girl, and, as a result, I always knew I could do and build anything.

"I decided to become a high school engineering, construction, and woodshop teacher because of my middle-school technology teacher, Jeff Sova. He taught me how to use big machinery and introduced me to drafting and woodworking; after I graduated from college, I worked on his construction crew.

"Now my purpose as an educator and builder is to demonstrate that girls can build anything. My approach is to lead by example. It is also important for young boys to learn how to build from a female builder because it changes their mind-set.

"One summer, I built a hexagon tree house in Northern California with a former student. It was 30 feet aboveground and was the most difficult structure I've built. We had to carry all our tools and materials manually up to the tree house every day. It felt so tremendous to build in such a physically and mentally challenging way.

"My advice to girls who want to build? Try to do everything, and no matter how hard something is, finish it. Knowing how to build, realizing the independence it takes to make decisions, and following through with ideas can and will forever help you and the communities in which you live.

"And PS: My favorite tool is my impact driver. I use it for every job, and it never lets me down. Everyone should have one!"

TS3000
GRAVITY RISE

Router

A router has a spinning bit, like a drill, but the bit acts like a saw to cut a particular shape profile out of wood. A router can be used to cut ("rout out") channels in wood or, more commonly, to shape the edges of a table or cabinet in a rounded or specific shape or profile. Router "spindle" bits come in different shapes and profiles; when they spin, they cut that same profile out of the wood. Lots of table edges, molding, and cabinetry are edged using a router.

Do not operate a router without the supervision and hands-on support of a skilled and experienced adult. Always read and follow the manual for your specific tool. Safety glasses should be worn at all times by both users and observers!

A router can be a dangerous tool because of the sharp spinning bit that sticks out. When using a router, treat the spinning bit like a saw blade, and start your router cut with the bit NOT in contact with your wood. While making your router cut, maintain a straight path and keep the base of the router flat on the surface. When finished routing, keep the base in contact with the material until the tool has stopped spinning.

The most common router for woodworking is a handheld spindle router, with a flat base that sits on the surface of your material and a spindle bit that sticks out farther (lower) to make your cuts. Routers usually have two knob handles on either side so you can hold the tool steady.

Router bits

Router bits come in many shapes for different jobs: round, straight, or V-groove bits cut channels in wood. A "round-over" bit in various diameters rounds out the edge of a piece of wood; other fancy bits, like the Roman ogee, carve decorative edges for cabinets and crown molding.

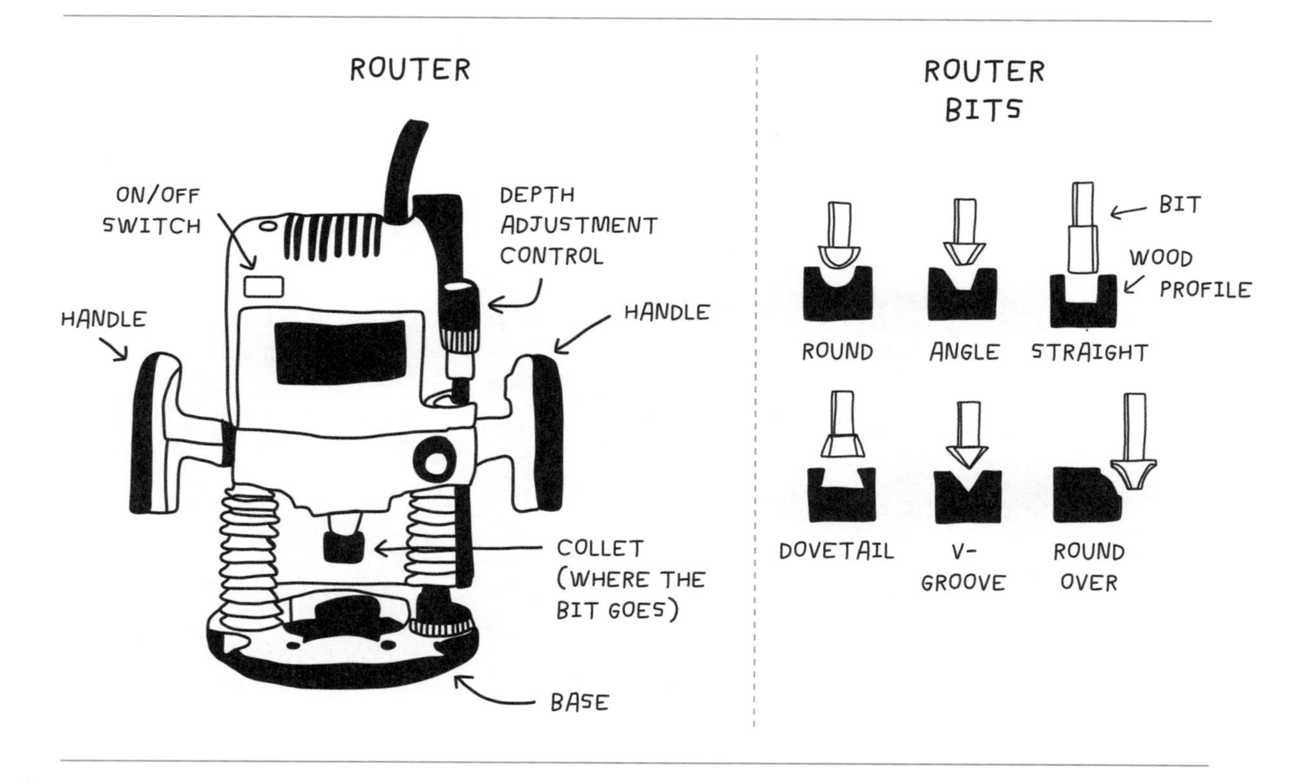

Rotary tool

Most people refer to this tool by its most common trade name, Dremel. A rotary tool has so many functions because it comes with an entire set of bits, ranging from sanding bits to buffing bits, engraving bits, and more. This small-but-mighty handheld tool will spin a bit at speeds up to 35,000 rpm (revolutions per minute).

The rotary tool spins faster, but has less rotational force, than a drill. The small chuck works similarly to a drill or a driver chuck, with a wrench key that lets you remove and replace bits easily.

One of my favorite uses for a rotary tool is engraving, using a diamond-tip bit. Many people use a rotary tool with sanding, grinding, and buffing bits to finish metal. A disk-cutting bit can also cut through thin-sheet metal. **Make sure you're wearing safety glasses when using a rotary tool, and secure your work material to a surface. Hold the tool like a thick pen or pencil so you can control it well.**

Nail gun and brad gun

For jobs where you have hundreds (or thousands!) of nails or brads to install, use a nail gun or a brad gun! Most nail or brad guns you'll encounter are pneumatic, meaning powered by quick blasts of compressed air from an air compressor that pushes the brad or nail out of the gun and into the material. Nail and brad guns work quickly, rapid-fire style, like a staple gun, so they're quick and efficient.

There are nail guns for framing nails, which are often used for building the framed walls of houses and attaching a bunch of 2×4s. Nail guns take a strip of attached nails and fire them one at a time. Brad guns work the same way, but with much smaller brads; usually 1 inch or less in length. You might use a brad gun to quickly attach thin materials to another structure.

ROTARY TOOL

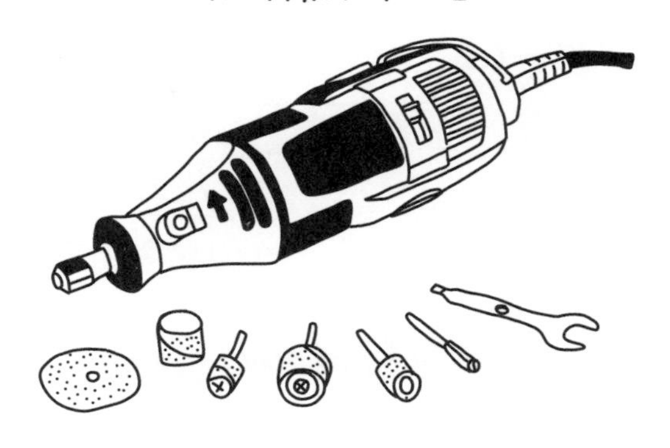

FUN FACT!

Dremel now also makes specialty rotary bits, including pumpkin-carving and pet-grooming sets.

NAIL GUN

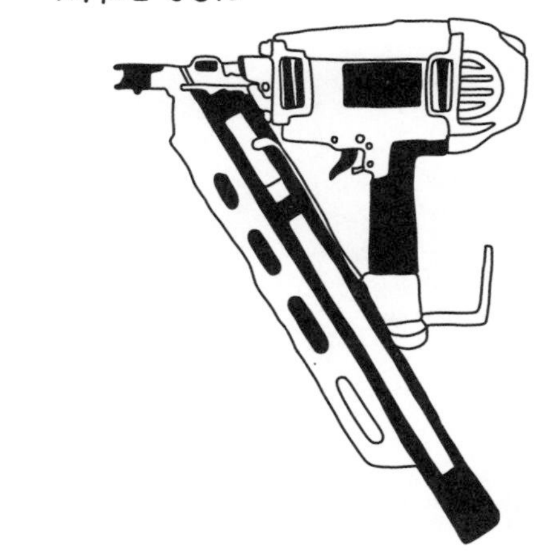

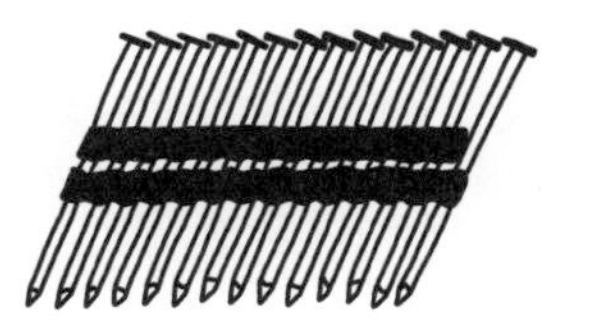

NAILS FOR NAIL GUN

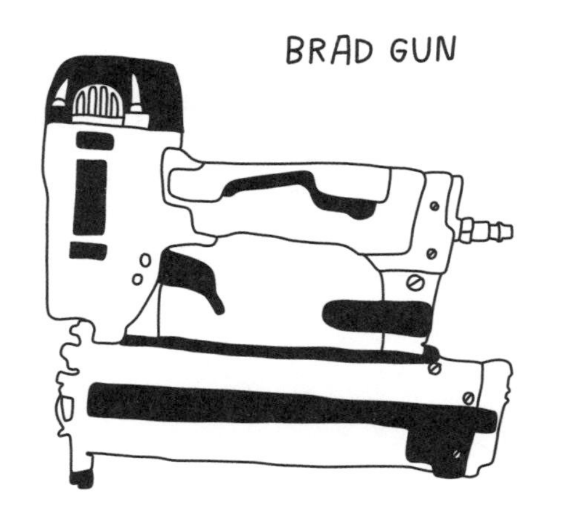

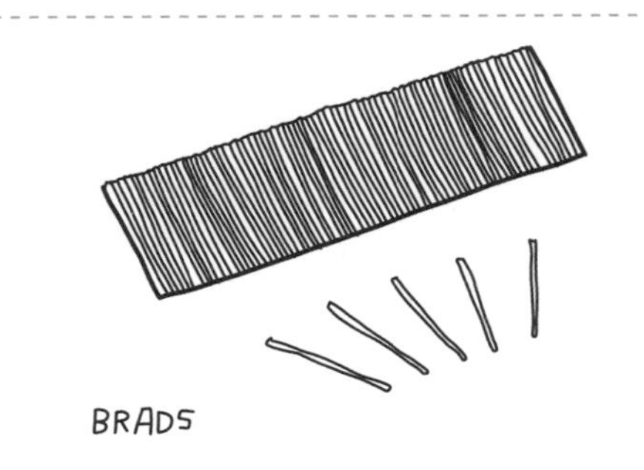

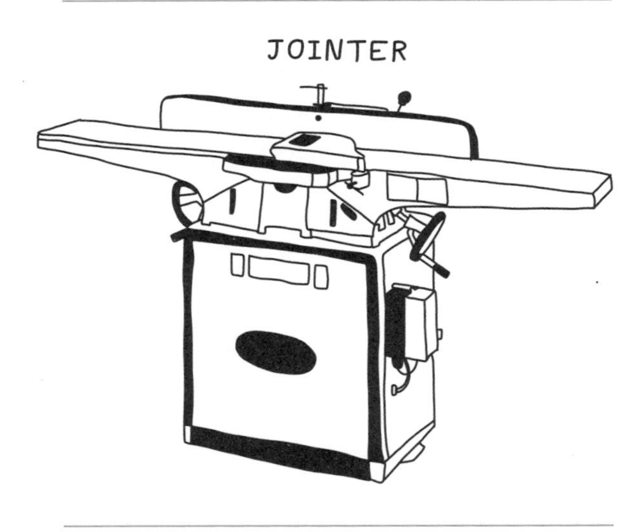

Nail and brad guns can both be very dangerous. Most do have a safety tip that must be pushed in against your wood surface in order for the brad or nail to come out. **Take extreme caution not to bump or drop the gun, which can trigger this safety tip and fire your nails or brads. Brad guns and nail guns are for experienced adult users only. Do not use this tool if you are under the age of 18. Always read the manual for your specific tool (the following tips are general guidelines). Safety glasses should be worn at all times by both users and observers!**

Before use, make sure you have a clear work space and your nail or brad gun is connected to its air-compressor hose. Check the suggested air pressure for your tool and match it on your air compressor.

When using your nail or brad gun, align the center arrow on the tip to the spot you want to nail, and keep the square tip of the gun flat against the surface; this ensures that the nail or brad goes into the material perfectly straight and not at a funky angle. Despite the efficiency of a nail or brad gun, take your time and set up each nail carefully to avoid a misfire or a poorly angled nail. Make sure your non-dominant hand is clear from the work area, as well as the hands of your builder buddy!

Jointer

Use a jointer to flatten the long face of a piece of lumber or to square its edge for precise applications like furniture projects. Sometimes lumber can "cup" or warp, so a jointer will take those curves away by trimming them to a flat surface. A jointer should only be used for pieces of lumber that are 12 inches or longer and never for plywood or other materials.

The jointer is for experienced adult users only. Do not use this tool if you are under the age of 18. Always read the manual for your specific tool. Safety glasses should be worn at all times by both users and observers!

A jointer has a long, flat table and a fence like a table saw. You use the fence to slide your wood along and over a spinning "cutter head." That cutter head is set at a certain height to remove a specific amount of material from your wood.

Like a table saw, a jointer has an exposed spinning cutter head. Always use a push stick or push block to push material over and past the cutting head, and never your bare hands. As with a table saw, it is imperative that your wood remain in contact with the fence along its edge for the entire duration of operation.

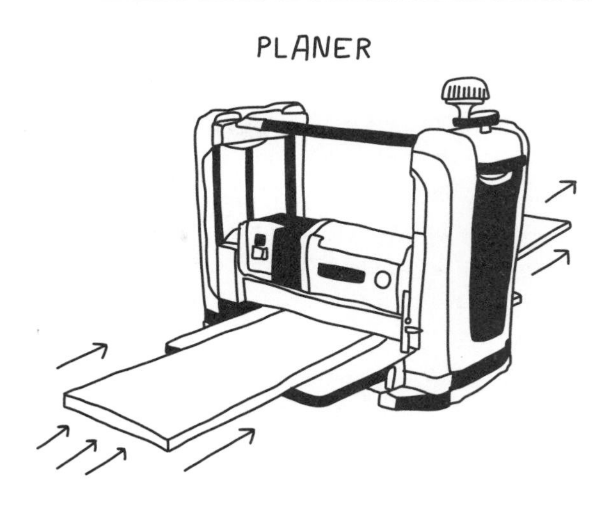

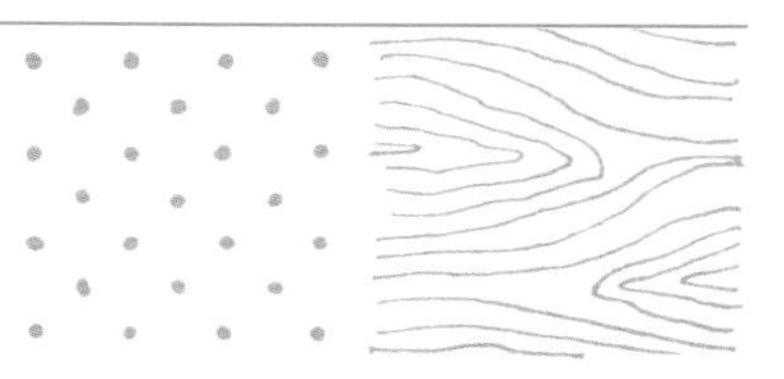

Planer

In general, when preparing wood for a furniture project, use a jointer first to flatten the faces of your wood, and then use a planer to trim the entire thickness of your wood to a specific dimension. For example, maybe you're building a bookshelf using a lovely hardwood, and you've designed it to have ¾-inch-thick walls and shelves. You might buy 1-inch-thick wood boards, flatten them using your jointer, and then run them through your planer, set exactly to ¾-inch thickness. This ensures that all your wood is flat, level, and the same dimension.

The planer, sometimes called a "thickness planer," has a flat table, and an overhead cutting head that you set to whatever height you want your wood thickness to be. As you slide your wood through the planer, the cutting head shaves off that material from the top. Rollers help move the wood through the machine.

The planer is for experienced adult users only. Do not use this tool if you are under the age of 18. Always read the manual for your specific tool. Safety glasses should be worn at all times by both users and observers! As when using a jointer, a planer is a hefty tool, so it's good to have an experienced builder buddy helping you. Move your wood through the planer in a perfect line, in the direction the cutting head is cutting, and with the wood's grain.

Lathe

A lathe holds a piece of material from both ends, and spins it so it can be carved, shaped, or sanded. There are wood and metal lathes, but they both work the same way. Wooden bowls, candlesticks, and baseball bats can all be made on lathes by shaping the wood as the lathe turns, using a carving tool.

Lathes have two sides: the headstock and the tailstock. Your wood or other work material is held in place between these two points and spins along its long axis. A lathe also comes with a tool rest, where you can rest your carving tool handle while making contact with the spinning wood to slowly carve it away. If you have ever used a pottery wheel to shape clay, a lathe works sort of like that, but on its side, allowing you to carve and shape your material as it spins.

The lathe is for experienced adult users only. Do not use this tool if you are under the age of 18. Always read the manual for your specific tool (the following tips are general guidelines). Safety glasses should be worn at all times by both users and observers!

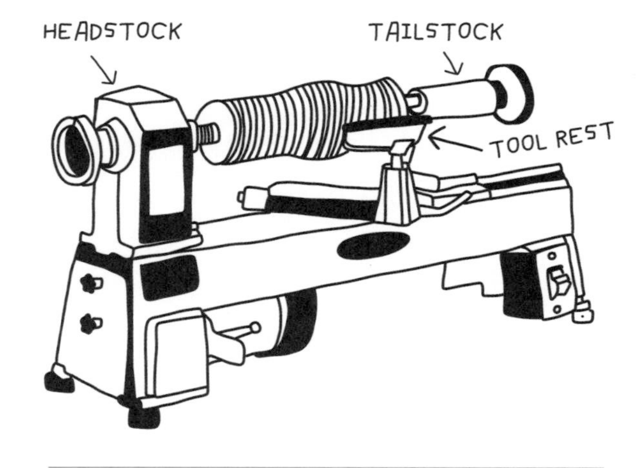

Because a lathe is an exposed spinning tool, **do not have any dangling sweatshirt strings, hair, or other hanging objects that could get caught in it.** You will always work on the side of the lathe where your material is rotating downward toward the ground (and also toward you).

Once you have your wood attached to both the headstock and the tailstock (you'll have to hammer it onto the pointed tip of each), start the machine, place your carving tool on the tool rest, and slowly make contact with the wood, carving away a bit of material each time into the profile you want. Spend some time reading about the features and recommended use for your particular machine before using it. And **wear a face shield** when using a lathe because curlicues of wood will fly everywhere!

Biscuit joiner

If you're working on a wood project where you need to join pieces of wood without visible hardware, you might use a biscuit and biscuit joiner. The biscuit itself is a small wooden connecter, shaped like a football flattened into a

FUN FACT!

The lathe is sometimes called the "mother of machine tools" because before and during the Industrial Revolution, it was the machine tool that allowed people to make other tools.

wafer, that connects two pieces of wood internally. Use a biscuit joiner to cut out the half-football shapes in each piece of wood so you can insert the biscuit and connect the two pieces.

A biscuit joiner has a flat front plate and a small circular saw blade that sticks out of that plate when in use. When joining two pieces of wood, this blade cuts a crescent-shaped slot into each piece. Then the biscuit is glued and inserted, with half of its football shape going into each piece, joining the two together.

Do not operate a biscuit joiner without the supervision and hands-on support of a skilled and experienced adult. Always read and follow the manual for your specific tool. Safety glasses should be worn at all times by both users and observers!

The key to biscuit joining is alignment! Your half-football shape has to be in the same location on each piece of wood so they line up perfectly. First, place the two pieces of wood together, in the position you want to connect them. Make a small pencil mark across the seam where they meet to note the location of the center of the biscuit. Now, grab your biscuit joiner. Line up this pencil mark with the center mark on the machine (the widest part of the biscuit). Adjust the thickness plate on your biscuit joiner to match the thickness of your material (this ensures that the biscuit hole is cut in the middle of the wood's width). Press the face of the biscuit joiner against the edge of your wood and turn the machine on. The circular saw comes out of the face, makes its cut, and then goes back into its hidey-hole. Cut the same slot in your other piece of wood, then glue and insert your biscuit. Clamp to ensure a solid connection (added bonus: the wet glue expands the biscuit, making it an even tighter fit!).

BISCUIT JOINER

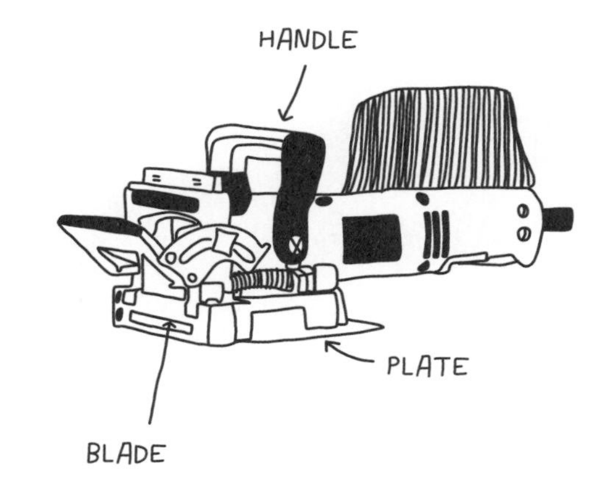

BISCUIT-JOINED PICTURE FRAME CORNER

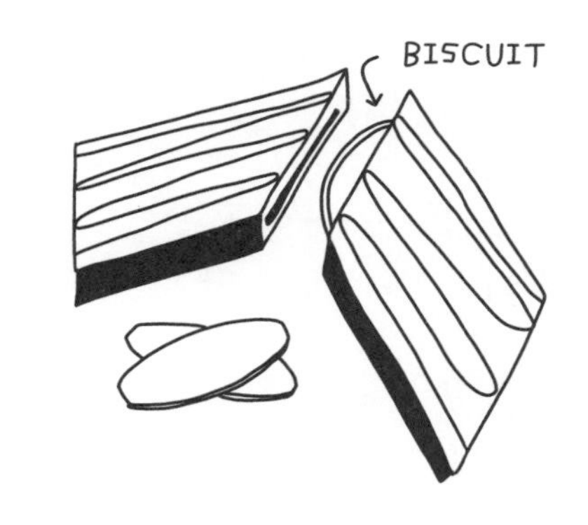

Heat gun

Basically a high-powered hair dryer that blows slower but hotter, a heat gun is a useful tool for stripping old paint or drying certain sealants.

Most heat guns blow air from about 200 to 1,000 degrees Fahrenheit. They are shaped like a drill, with a trigger switch, and though air does not blow as fiercely as a hair dryer, it works on the same principle.

FUN FACT!

One of our favorite projects at Girls Garage, created by my co-instructor Allison Oropallo (page 156) is designing, shaping, and racing air-powered dragster cars. We shape small wedges of wood into aerodynamic vehicles, decorate them and give them axles and wheels, and then hook them up to an air compressor and track. The air compressor fires air into the back of the car and sends them speeding across the room! Our favorite version of this activity is a mother-daughter dragster car race.

Be super careful not to touch the metal ring around the front of the heat gun, which (duh!) gets very hot. Use a heat gun to remove old paint from a door, furniture, or other surface so you can refinish or repaint. With the heat gun on, move it slowly back and forth across the surface, not staying in one place too long, at least 12 inches above the surface. The paint will release from the surface and you can easily scrape it off with a metal spatula or scraper. You can also use a heat gun to quickly dry certain paint or sealants (like screen-printing ink!).

Air compressor

This compact tool stores compressed air and shoots it out of a hose to power tools like a nail or brad gun. An air compressor can also help you clean dust or debris off a surface. Anything powered by air is called "pneumatic," and the air compressor is the quintessential pneumatic tool!

Air compressors for power tools are most commonly used in the 90 to 100 psi range, which is relatively low, compared to some larger industrial-scale compressors. There are portable electric compressors and gas-powered compressors, both of which are (relatively) compact and can easily be hooked up to your air-powered tools with a hose.

An air compressor usually has two dials: one showing the tool pressure and one showing the tank pressure, so you can set your psi based on the requirements of your tool. Connect the air-compressor hose to your tool (such as a nail gun), make sure the valve on the air compressor is open, and go to work. Read the specifications for both your tool and the air compressor, as some require you to drain the air from the hose to prevent rust.

SANDING AND FINISHING

While making and building as a category is pretty darn satisfying, sanding has a particular kind of gratification. While some sanding and filing applies to metal and other materials, I'm mostly talking about wood for our purposes here (which is also why the metal tools section comes next). For as rough and burly as wood might feel, using the proper order of grits of sandpaper and a little love, you can achieve a bunny-rabbit-smooth finish on wood projects. I once picked up a wooden baby rattle in a fancy store that had been sanded to perfection; I swear it could have been made of velvet. You might also find yourself sanding to remove a layer of paint or other sealant so you can refinish a surface, clean it, or otherwise work on it.

It's also worth noting that while we think about sanding as the last step in a building process, there might be other moments during your project when sanding is useful. For example, you might want to lightly sand the edge faces of two pieces of wood before using a drill and driver to attach them. Or you might need to sand down the rough edges of a drilled hole prior to assembly.

Tips for use and safety!

Here are the cardinal rules of sanding:

Always sand with the grain. Just like when cutting wood, it's important to know which way your grain runs. Sanding in the direction of your grain helps smooth your piece consistently and without scratches or marks. Think of this as swimming with a river's current versus trying to swim straight across. You're so much better off just floating along with the current.

The lower the grit, the rougher the paper. Sandpaper grit ranges from about 40 to 600 (though most are in the 80 to 300 range). Think about those numbers like actual particles on the paper. With only a few, you'll really feel the scratch. With lots and lots of tiny particles, the entire paper feels smoother overall. You can also think about grit like pixels or resolution, if you're a digital photographer. The higher the resolution, or pixel count, of an image, the smoother and cleaner the image is. With fewer pixels, the image becomes grainy and rough.

Start rough and work your way to smooth. Your best bet is to work in a progression, starting with a rough sandpaper, then medium, and finishing with a smooth (high-grit) sandpaper. If your material is particularly rough or uneven, a file or rasp will be helpful for a first pass. Either way, start with a low-numbered grit and work your way up to the 200- and 300-grit papers.

Wear a dust mask and safety glasses. Even though most power sanders come with a dust bag attached, all sanding produces massive amounts of dust, so give yourself some respiratory protection and save your sneezing for allergy season. **Wear safety glasses to protect your eyes, too, both from dust particles and any other wood pieces that could go flying while sanding.**

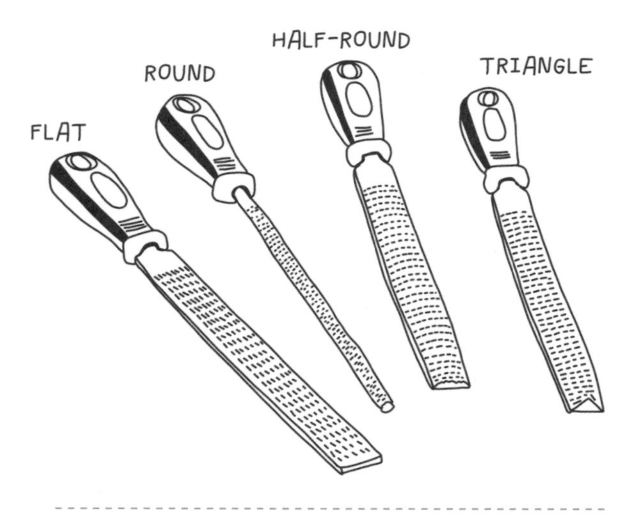

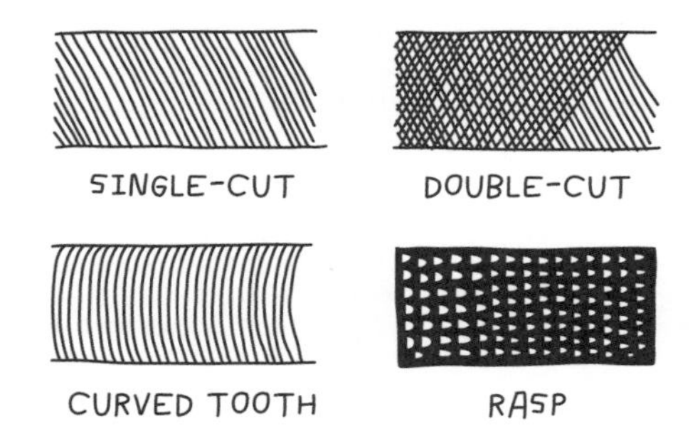

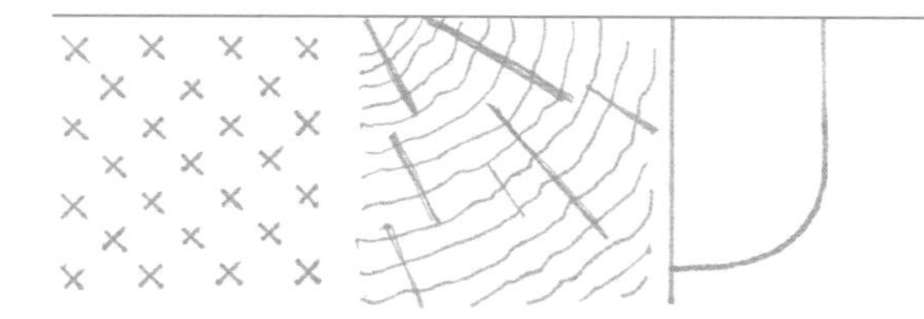

Keep it moving! With any sanding job, especially with power sanders, don't stay in one place too long. Remember, you're removing material from your surface, and you don't want to end up taking off too much unevenly (just like with a haircut, you can always take more off, but once it's gone, it's gone—so take it slow!). With power sanders, staying in one place can also leave distinct sanding marks that might be hard to undo.

Let the sander do the work! You shouldn't have to put too much muscle into sanding, whether you're using a file or sandpaper or a power sander. In fact, if you apply too much pressure, you can sand off too much or end up with bumps and scratches in your wood. If you feel like your sanding tool just isn't doing its job, it might be time to replace the sanding surface (such as the disk on a random orbital sander or the belt on a belt sander).

Types of sanding tools

To achieve those perfectly smooth finishes on your lovingly built projects, you'll use some of these most common sanding tools (both hand and power tools).

Files and rasps

If you need a handheld tool for your first pass of sanding, especially along an edge or corner, a file or rasp is a great option. You can use files for metal projects, too, to file down or smooth out any sharp or jagged edges. Made from hardened steel, these long metal bars usually have a plastic or wood handle. They come in different shapes—flat, round (or "rat tail"), half-round, triangular—as well as different file patterns, depending on the level of abrasion you need.

The rasp is a particular type of file with holes in it, sort of like a cheese grater, and is generally the roughest of files.

Having a set of multiple files around is super helpful, and you can change shapes or file patterns, depending on your wood or desired smoothness.

Sandpaper (sheets)

Spoiler alert: Sandpaper isn't made from sand! Sandpaper, sold in sheets of various sizes, is paper with small abrasive particles (usually aluminum oxide or silicon carbide) glued to it. The size and density of these particles determine the sandpaper's grit.

For wood projects, a grit range of about 80 (rough) up to 220 (fine) works great, and remember to *work with the grain*, starting with the rougher sandpaper and working your way up to the finer ones. You can use the super-fine grits, like 320 and up, between layers of some paints, and the super-rough grits, like 80 and lower, for the toughest jobs. Sheets of sandpaper are particularly helpful because you can tear off tiny pieces to get into small spots, and also because the standard 9-inch-by-11-inch sheets fold and tear into four equal pieces that fit a rubber sanding block perfectly!

Rubber sanding block

We have about twenty-five rubber sanding blocks at Girls Garage because they are so satisfying to hold in your hand and can take any grit of sandpaper you put in it. The block itself is flat on one side (the sanding side) and curved on the other (where your hand holds it). At each end is a slot in which you slide the end of your sandpaper sheet. Inside the slots are small nails that act like teeth to grip the paper.

To load your sanding block, fold and tear a 9-inch-by-11-inch sheet of sandpaper into fourths, so you end up with four pieces that measure 9 inches long by 2¾ inches wide. Pry open the jaws of one side of your sanding block and insert one end of the sandpaper, then clamp the jaws shut so the nails bite into the paper. Wrap the paper around the bottom of the block tightly, pry open the other side, and insert the other end of the sandpaper, letting the nails bite down into it as you close it. When one piece

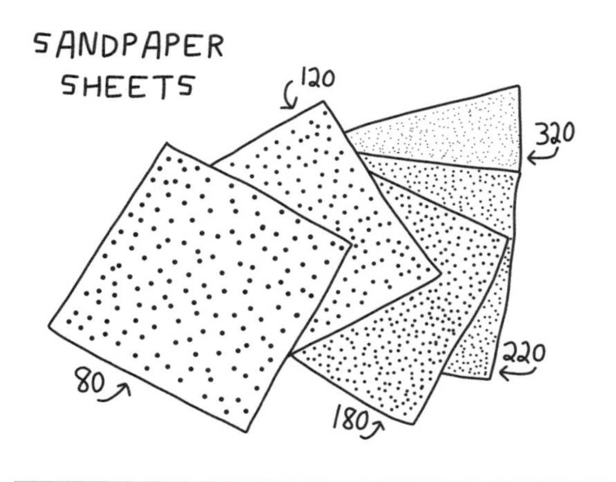

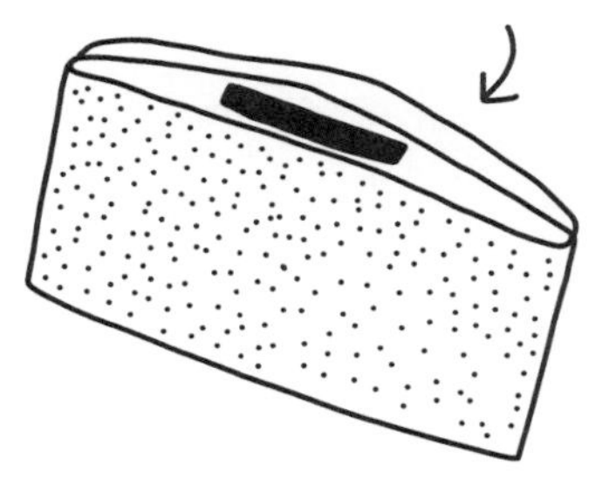

HOW TO FOLD AND TEAR A SANDPAPER SHEET SO IT FITS YOUR RUBBER SANDING BLOCK TO PERFECTION

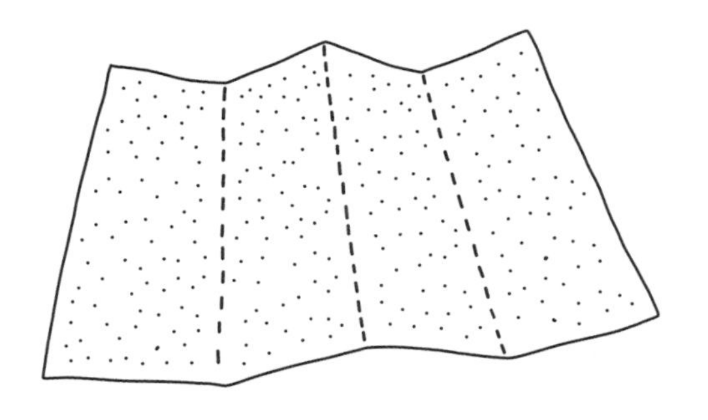

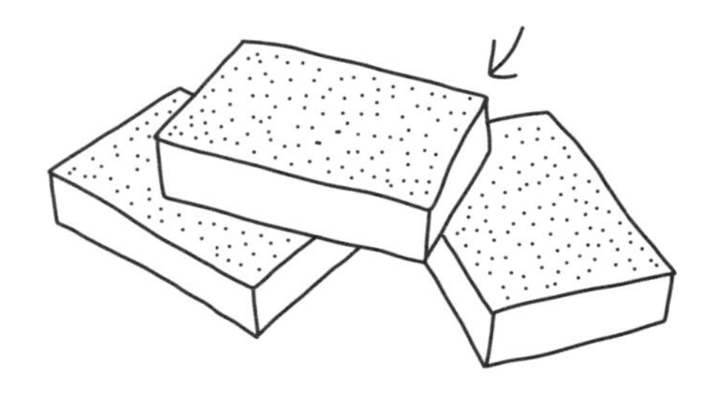

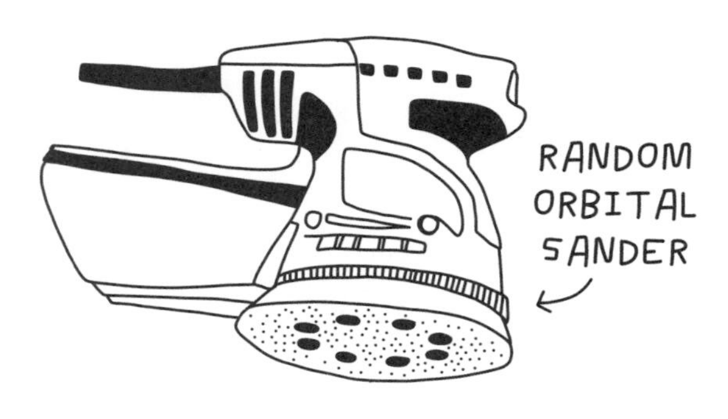

of sandpaper is torn to shreds, just pull it out and put in a new one!

Because these sanding blocks have a flat sanding surface, they work best to sand other flat surfaces, though you can also use them to round out edges and corners.

Sanding sponge

If you're working on a project with tight spaces or any wet sealants, a sanding sponge is a great alternative to the sanding block. Because it's a sponge, it can smoosh into small spaces, conform to oddly shaped surfaces, and you can actually rinse it and clear out all your dust before reuse. Unlike a sanding block, you can't replace a sponge's sandpaper, so once its surface is kaput, you'll need a new one. Just like sandpaper sheets, you can buy sanding sponges in various grits and some even come with an angled edge to get into sneaky corners.

Random orbital sander

What makes a random orbital sander *random*? Unlike a square-shaped palm sander, a random orbital sander has a circular sanding disk that vibrates in tiny circular motions (aka random orbits) while also rotating. This randomized motion makes for a more even sanding job than something more directional, like a belt sander or a disk sander.

And the random orbital sander is compact and portable. It's a great tool for finishing wood surfaces as well as taking off paint or finish. Most random orbital sanders take replaceable 5-inch or 6-inch sanding disks that attach with Velcro.

The sanding disks have holes in them, which help suck dust up into the attached dust bag. A random orbital sander can also be hooked up to a shop vacuum for even better dust collection!

Don't forget your dust masks, safety glasses, and adult builder buddy.

Belt sander

Belt sanders come in handheld, standing, compact, and gigantic versions, but they all have one part in common—a continuous belt of sandpaper. The stationary belt sander is the most common version you'll see in shops. The sanding belt makes a loop that runs over two parallel cylinders, very quickly, over and over again, like a running treadmill. Belts for belt sanders are easy to replace, fairly inexpensive, and come in various grits.

The advantage of the belt sander is that it gives you a long, flat sanding surface, with the sanding action happening in one direction, so it's great for sanding the long edges of a piece of wood. Often these stationary belt sanders come with a small disk sander, too.

Do not operate a belt sander without the supervision and hands-on support of a skilled and experienced adult. Always read and follow the manual for your specific tool. Safety glasses should be worn at all times by both users and observers!

When using a stationary belt sander, take advantage of the fence and place your material against it to keep it from following the speed of the belt and flying across the room. Remember to let the belt do the work; you shouldn't have to push too hard!

Disk sander

A disk sander has a large circular sanding surface that rotates with a table surface just below the equator of the circle (picture a halfway-setting sun). While the random orbital sander is great for sanding large surfaces of wood, a disk sander is totally superior when it comes to sanding the ends of smaller cut pieces (the end grain).

Do not operate a disk sander without the supervision and hands-on support of a skilled and experienced adult. Always read and follow the manual for your specific tool. Safety glasses should be worn at all times by both users and observers!

Let's say you cut a small length of a 1×4 piece of wood for a picture frame using a miter saw. The edge you just cut is likely a little rough and splintery. Take that piece to the disk sander, hold the cut end against the spinning disk, and you'll have a buttery-smooth edge. You can also

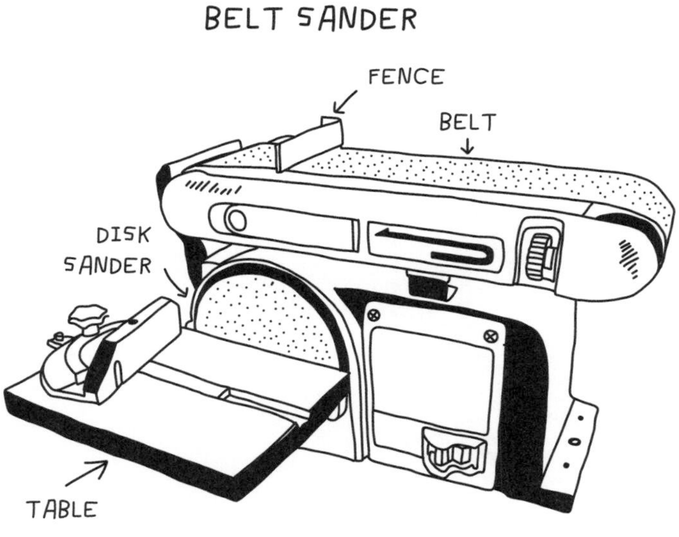

BELT SANDER

FENCE

BELT

DISK SANDER

TABLE

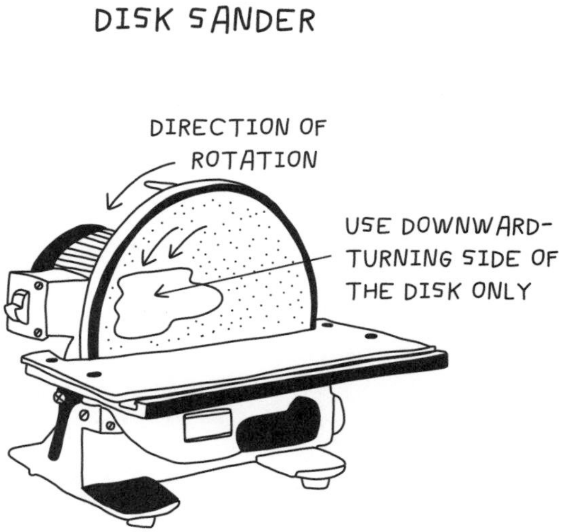

DISK SANDER

DIRECTION OF ROTATION

USE DOWNWARD-TURNING SIDE OF THE DISK ONLY

use the miter gauge built into the table to line up and sand angled miter cuts.

A word to the wise: Because the disk sander rotates in one direction (clockwise or counter-clockwise), **only sand on the "down turn" side of the disk.** This is the side of the disk where the disk is rotating downward toward the table, instead of up to the sky. If you sand on the upward-rotating side, the sander can grab your piece and fling it in the air. On the "down turn" side, the direction of the spinning disk works with you to keep your piece pressed down against the table. For a clockwise-spinning disk, this means you should sand using the right side of the disk. For a counterclockwise-spinning disk, use the left.

Spindle sander

Let's say you cut a letter *C* out of plywood using a jigsaw. Your cuts are precise but pretty chewed up from the rough jigsaw cut. A belt sander will be no help in getting to that inside curve. So for curves, arcs, and otherwise uncommon shapes, a spindle sander is an awesome choice. The cylindrical spindle rotates and also oscillates up and down.

Do not operate a spindle sander without the supervision and hands-on support of a skilled and experienced adult. Always read and follow the manual for your specific tool. Safety glasses should be worn at all times by both users and observers!

Always sand with the work piece flat on the table. Sand against the rotation, because if you are moving your work piece in the direction of rotation, there's a chance the spindle might fling the piece out of your hand. Place your weirdly curved cuts against the spindle and move smoothly along the spindle surface for perfectly finished curved edges. Most spindle sanders come with drums of multiple diameters that fit on the spindle so you can sand curves and arcs of various sizes and geometries.

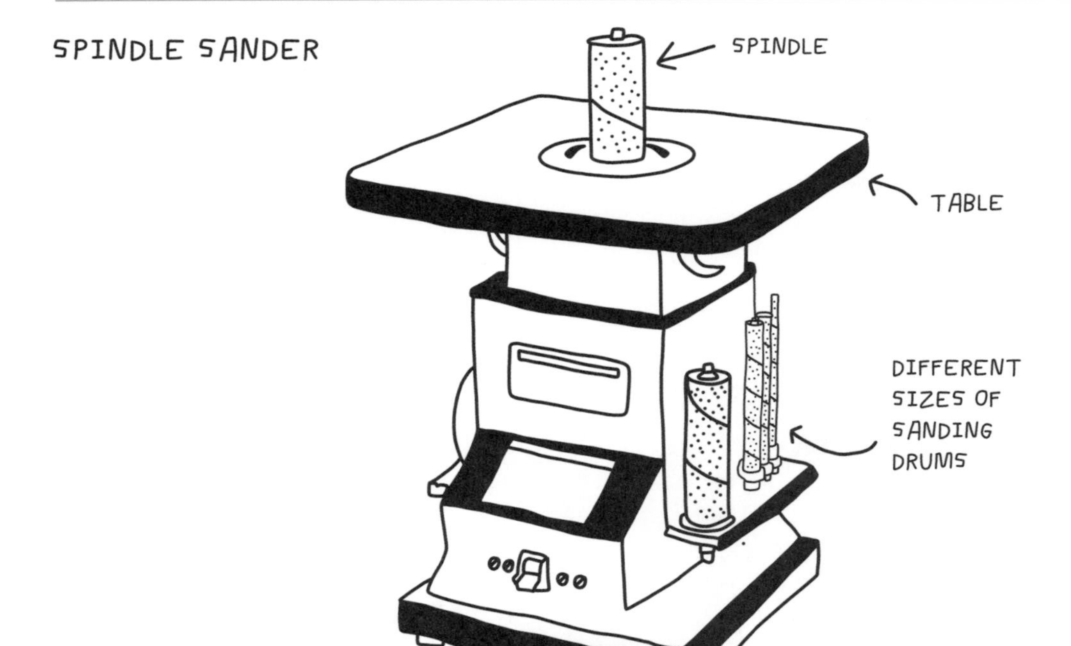

METAL TOOLS

Metal is an incredible material that might seem impenetrable, but it is remarkably easy to work with. Though metal and metal tools might be used in a project in combination with wood, they warrant their own section because the physical properties of metal are quite different than wood.

Welding

Welding is like a superpower and will make you feel like Wonder Woman (you can fuse metal!). I learned how to weld in graduate school at the School of the Art Institute of Chicago, and it was a game changer. It's also worth noting that welding instruction is the subject of many *whole* books and multiple years of study. I'm including a primer on how one type of welding works and some basic concepts, but it's hardly a step-by-step welding lesson. I hope it's enough to entice you to investigate further and give it a try!

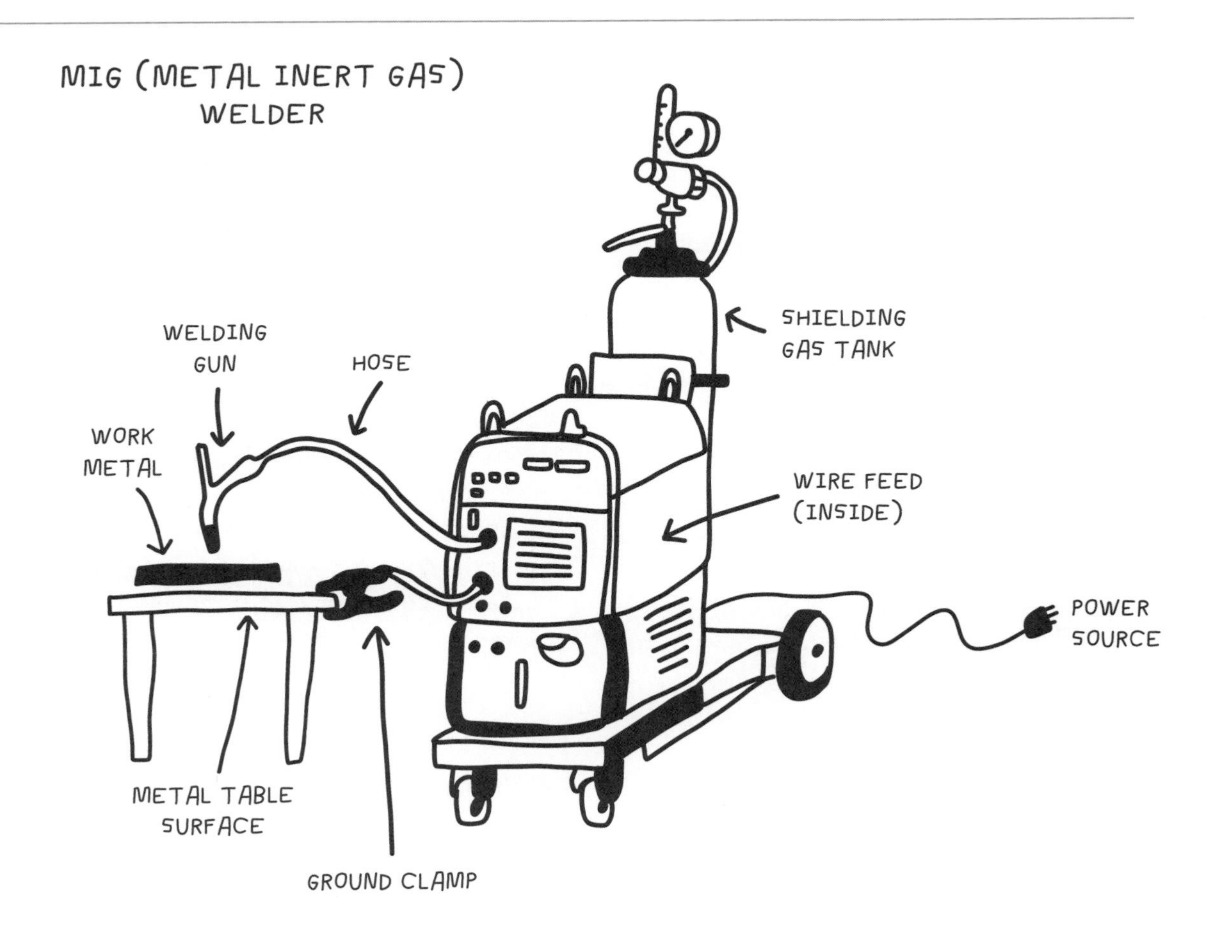

What is welding, and how does it work?

Welding is a way of melting and joining metal using very high heat. The heat can come from different sources, but the most common method is to use an electrical current. This is called "arc welding." The three must-have ingredients for arc welding are an electrical current to create the heat, a "filler" material that helps to join your two pieces, and a shielding gas to protect the weld as it forms and cools.

Arc welding works by using an electrical circuit (think of a circuit as a circular path through which electricity flows). To create this completed circuit, the welder has two different "arms" that must both be connected to a metal work table. The first arm is a ground clamp that gets clipped to your metal work table, which makes the first circuit connection. The second arm is the welding gun, which you as the welder hold and control to dispense the filler material. When you start your weld, the filler material touches your work metal (which is sitting on your metal work table) and the circuit is complete! This creates the electrical *zap* and high heat that melts and fuses your metal. You might be thinking, *Am I going to get electrocuted?* Not to worry; the electrical current flowing is looking for the "path of least resistance." In other words, electricity is far more interested in flowing through highly conductive metal than into you. That being said, it's important to remember that you are in close proximity to electrical currents, so you should work in a well-ventilated area, keep anything flammable away from your welding space, and wear the right gear (more on that next!).

Some types of arc welding use a stick of filler metal that you have to hold in your other hand, while others use a spool of wire that is fed out of the tip of the gun.

Either way, the filler metal is what melts into the "bead" of metal you see left on the surface. This bead has done a very important job, letting the electrical current run through itself to melt, mix with, and join the two pieces of metal underneath it. Unlike soldering, where the melted metal is acting like an adhesive, a welded bead is merely the evidence of a much stronger structural fusing. In fact, you could grind off that bead entirely, and the two pieces underneath would still be fused together! You can see this on some high-end welded metal furniture, where an angle grinder was used to grind off the bead, showcasing the seamless welded joint underneath. So while soldering uses melted metal as glue, welding creates an actual fuse, joining two pieces into one.

Arc welding also requires a shielding gas that protects the hot molten metal so it can cool quickly and solidly. This shielding gas creates a little protective bubble that helps your work metal fuse properly, protecting the welded area from exposure to oxygen, nitrogen, and hydrogen that might make it a bubbly or corroded weld when it cools. Some welders get this gas from a connected gas tank, while others use a hollow wire that has the gas built in ("flux core").

KAY MORRISON

World War II Journeyman Welder, from 1943–1945

Richmond, California

Every Friday, in a small building on the San Francisco Bay in Richmond, California, a group of women meet. They are all in their mid to late nineties, have children, grandchildren, and great-grandchildren. They all wear matching vests adorned with many pins marking their many honors. And they are all living Rosie the Riveters, who worked as welders and electricians and draftswomen and laborers during World War II. It was here, on a random Friday, that I met Kay Morrison, a pistol of a woman who sat and talked to me about her experience as a welder during the war.

She was born in 1923 in Chico, California. After her older brother went off to fight in the war, she too felt compelled to join the ranks. The first time she tried to get a job as a Rosie, working on the shipyards in Richmond, the hiring hall displayed a large sign that read NO WOMEN OR BLACKS WANTED. She came back two years later, determined, and was assigned as a welder. She took the government's navy welding test and practically aced it and was assigned to the graveyard shift—from 11 p.m. to 7 a.m. every day—welding on the ships being built in the Richmond shipyard. She welded flat, vertical, and overhead joints, sometimes on her back or in tight spaces. She came home with scars on her chest from where sparks flew down her shirt between her bandana and collar. Kay and her counterpart Rosies are a national treasure and without question laid down the legacy that many of us female builders follow today.

"During the war, they needed all the help they could get from the women, doing work here at home while the men were away fighting. So I went to the union hiring hall in San Francisco and I was told, 'You're going to be a welder.' And I said, 'What the heck is a welder?' I didn't know what it was, but I knew I could do it. Because women can do anything when they want to.

"I felt like a nitwit when I went to welding training, which lasted ten days. It was entirely different from anything I had ever experienced. When I was going to high school, I worked in a soda fountain. When I graduated, I worked for J.C. Penney. Welding was totally different! The clothes we had to wear were so different. The cowhide jacket, the coveralls, the heavy boots with the steel toes. And it was strange working with fire. But I liked it! It made me feel like I was going to get out there and win the war myself if I had to. I think all of us Rosies felt that way.

"In my era, there were people who made it challenging. People were afraid of change. But I had a husband who was all for the rights of women. Equal rights, equal pay—and that gave me support and incentive. But I was always the type of person who knew I could do what I wanted to do, and I wouldn't let people stop me or deter me.

"I would tell young girls today that you can do anything you want to. Keep your eye on the goal, and do not let anyone deter you. Just do what you want to do. You have to really want something bad enough, and then just go and do it."

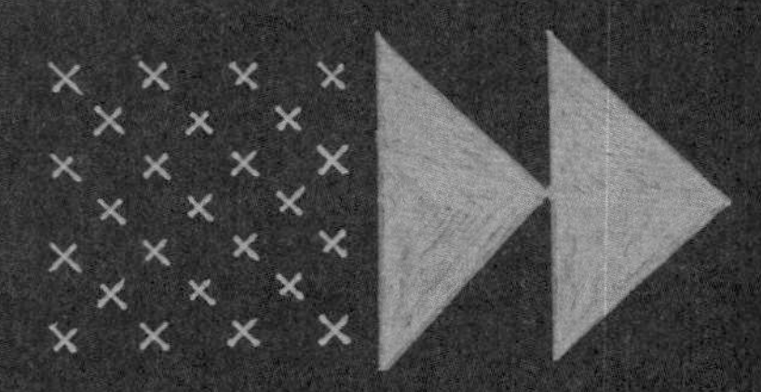

MORRISON
ROSIE THE RIVETER
WORLD WAR II HOME FRONT
NATIONAL HISTORICAL PARK

Types of arc welding

There are many kinds of arc welding, each with specific advantages and purposes. MIG welding is probably your best starting place, because it is relatively forgiving and used for so many applications. Flux-core welding is also great for beginners, especially because it can be set up easily without a big gas tank to tote around. TIG welding is more precise and requires a little more hand-eye coordination (and patience!), and stick welding is used the most for industrial applications. Once you learn one type of welding, you'll probably want to learn them all! The main differences between these welding types is how they deliver the three ingredients: the electrical current (and heat), the filler metal, and the shielding gas.

MIG (metal inert gas) **welding**, also referred to as "GMAW" (gas metal arc welding), is one of the easiest and fastest welding techniques to learn. A MIG welder has a welding gun and a "wire feed," which is a spool of filler wire that feeds continuously out of the tip of the gun. The machine also gets connected to a gas tank that dispenses the shielding gas (which is usually CO_2, argon, or a mix of both). MIG welding is ideal for welding steel (though you can also MIG-weld aluminum), is relatively quick and strong, and can weld thick pieces of metal.

TIG (tungsten inert gas) **welding**, also referred to as "GTAW" (gas tungsten arc welding), is ideal for smaller, more delicate welds on thinner materials. All the welds on your bike are probably TIG-welded. A TIG welder requires you to hold a filler rod in one hand while using the welding torch in the other. The welding torch creates the electrical circuit that melts both the work metal and the filler rod into a bead. TIG welding also requires a shielding gas tank, typically 100 percent argon. TIG welds are very clean, without much slag to clean up afterward. TIG welding can be used for steel, aluminum, and even titanium!

Flux-core welding is the most plug-and-play type of welding because it doesn't require a gas tank. Flux-core welding still uses an electrical current and a filler metal, but instead of a gas tank, it uses a special flux-core welding wire that releases its own shielding gas as it melts! The most accessible hobby welders use a small diameter flux-core wire, which you can use in a MIG welder machine. Flux-core welding is ideal for welding outdoors, because as the wire melts, its shielding gas hovers closely around the bead weld to protect it. If you were using a gas tank, the shielding gas would more easily blow away with even the tiniest outdoor breeze. The downside of flux-core welding is that the welds are dirtier and will have a lot of slag and dust that you'll have to clean off with a wire brush.

With **stick welding**, also referred to as "SMAW" (stick metal arc welding), a filler rod (called an "electrode") is attached to the electrode holder, which looks and functions like a clip. This electrode serves many purposes: it is the filler that melts onto the work metal and creates the bead, it completes the circuit, and it also has the shielding gas built into it (similar to flux-core welding, stick welding does not require a gas tank). You can buy different kinds of electrodes for stick welding, but in general, you want one that is the same type of metal as the metal you are welding. Stick welding is fast and burly, so it is used for large-scale construction projects like bridges and skyscrapers. Stick welding is also the type of welding used by the female "Rosie the Riveter" welders during World War II, like Kay Morrison (page 173).

A closer look at a MIG welder

My educated guess is that if you take a welding lesson or find someone to teach you, you'll be working on a MIG welder. A MIG welder is an ideal entry-level welding machine, as it is forgiving, quick, and you can weld large and long enough welds to learn quickly from mistakes. So let's take a look inside a MIG welder!

Spool and feed: A MIG welder looks like a giant toaster, and will often have a door on the side that you can open and look inside. Behind this door you'll see a spindle for a spool of wire. That wire runs along a track and out the front of the welder, into the hose. When the welder is in use, this wire travels all the way through the welder hose and out the tip. As you just learned, the wire is what carries and completes the electrical circuit when it touches your work metal.

Gun: The welder gun that you hold in your hand has a bend in it, so you can hold it at the correct angle against your work metal. It also has a trigger, which dispenses your filler wire and your shielding gas at the same time when you hold it down. The welding gun also has a tip with an outside covering (plastic or metal) and a smaller copper tip the wire comes out of. This copper tip has a hole that needs to be matched to the size of your wire (.030 inch or .035 inch, most likely). It is important to keep this tip clean from debris and replace it every so often.

Voltage and speed settings: On the front of your welder, there are two important dials—speed and voltage. The speed dial controls how fast the wire feeds out of your gun, and the voltage dial controls how much of an electrical charge it carries. The two settings work together to create the best weld conditions based on the thickness of the wire you're using and the thickness of the metal you're welding.

Luckily, you don't have to figure out these settings for yourself. On the inside of your welder door, the manufacturer usually includes a setting chart so you can quickly look up the correct settings for your material and wire!

Welding safety

First of all, do not operate a welding machine without the supervision and hands-on support of a skilled and experienced adult. Always read and follow the manual for your specific machine. Safety glasses and nonflammable head-to-toe protective clothing should be worn at all times by users and observers!

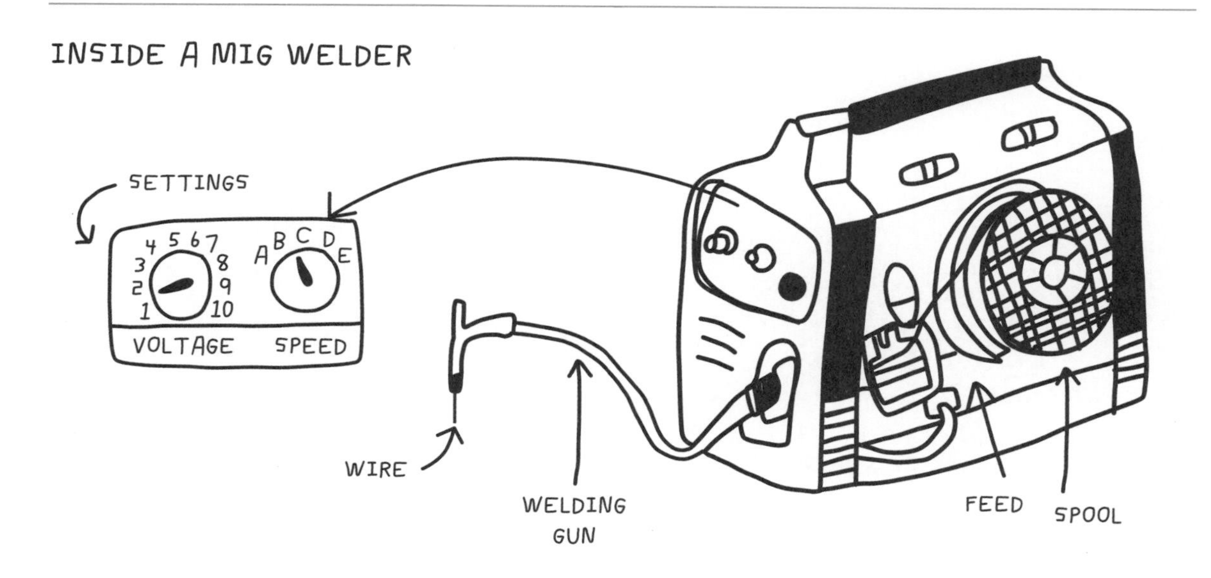

Welding for the first time, those sparks can be scary. Oh, and the heat, and the UV light! Yes, welding can be dangerous and gives off multiple by-products that can be harmful, but with the recommended protective clothing and eyewear, you'll be good to go.

Welding gives off HUGE amounts of UV light (like staring into an eclipse, which you never do, right?). You'll wear safety glasses and an auto-darkening welding mask or hood (more about this on page 180) that protects you against this light but still lets you see the weld. You also need to cover your whole body, so wear long pants and closed-toe boots, leather welding gloves, and a welding jacket to protect you from UV light and flying sparks. **Don't wear any synthetic materials, like fleece or polyester!** Denim, cotton, and natural fibers are best. If a spark does hit your skin, don't worry, it will feel like a little insect bite and then it's over. Lastly, you'll want to protect your hands from the heat of the molten metal. Wear leather welding gloves that are thick enough to protect you (and go up your forearm enough to cover your jacket cuffs), but still flexible enough to allow you a full range of motion. And be very careful with just-welded metal; it can be tempting to touch your beautiful welded bead! Let it cool sufficiently before handling, because that weld can reach temperatures in the thousands of degrees Fahrenheit!

SAFETY CHECK!

- A skilled and experienced adult welder buddy
- Safety glasses (wear at all times!)
- Welding mask (auto-darkening preferred)
- Closed-toe shoes or work boots
- Long pants and nonsynthetic clothing
- Welding jacket
- Welding gloves
- Hair tied back
- No loose clothing, hoodie strings, or jewelry

WELDING TECHNIQUE

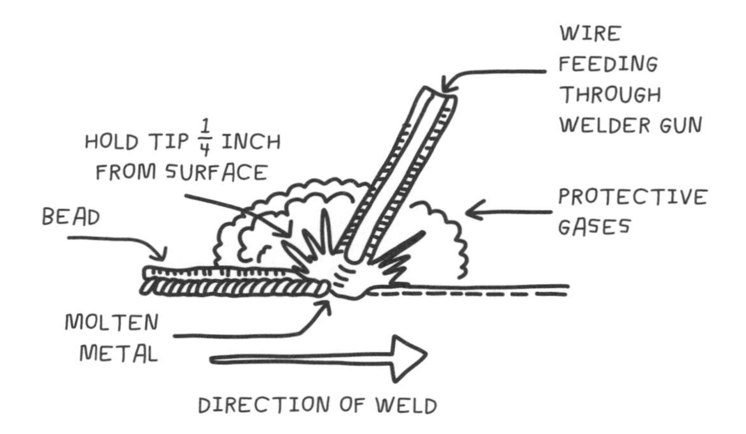

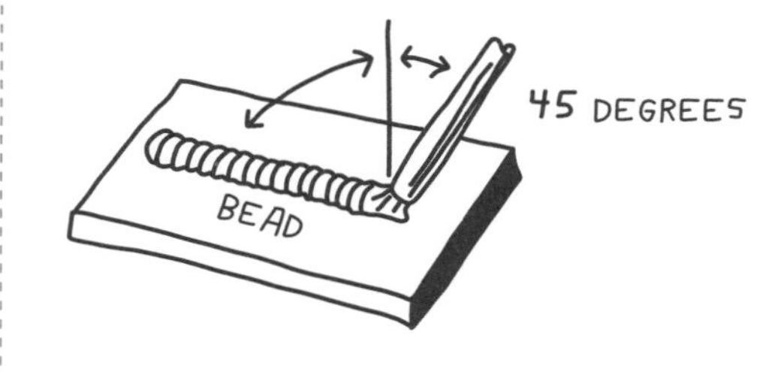

BEAD PATTERN: PIGTAIL OR HORSESHOE

Welding technique

Every welding machine is different, and, even with the same welder, you will likely have to alter your technique for different materials, thicknesses, and wires. However, there are a few important tips you must consider for any weld:

- **Distance:** If the tip of your gun is too far from your work metal, you won't be close enough to create a consistent flow of electricity. If you're too close, your copper tip can get stuck to the work surface and your wire might jam. With our MIG welder, I shoot for about **¼-inch** distance between the surface of my work metal and the end of my copper tip. A good way to ensure this distance is to clip your wire to a ¼-inch length before you start your weld. With the end of the wire touching the surface, that's the distance you should maintain for your entire weld. The wire will feed out, so you won't feel it in contact with the surface (it's melting!); just hover and keep that distance. I sometimes describe this to girls as similar to a weed trimmer, when you're cutting grass to a certain height. You're in charge of maintaining the right distance from the surface, even though you won't feel physical contact with your metal as the wire is melting.

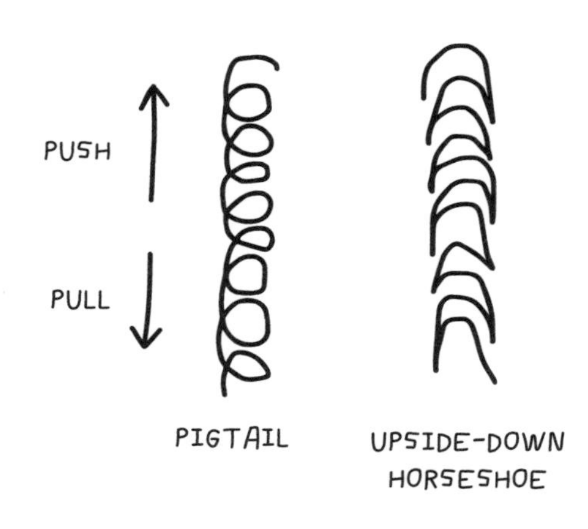

- **Angle:** I aim for a **45-degree angle** between my work surface and the angle of my gun. You don't want to be perfectly vertical because the debris and sparks can muck up your gun quickly and block the shielding gas from doing its job, and too horizontal means you run a greater risk of touching your contact tip to the surface and getting stuck to the surface.

- **Motion:** Everyone has their preference when it comes to welding motion, called a "weave pattern." Your weave pattern is like your handwriting. Sometimes at Girls Garage I can tell which project belongs to which girl because of the distinct "signature" of their weld's shape and pattern! I prefer a pigtail motion (little curlicue circles), made up of small circles that partially overlap the last loop you made. The pigtail motion creates a nice consistent bead, and if you did a good job, you can actually see each of the loops of your pigtail, kind of like a fingerprint. Other welders prefer an upside-down horseshoe, back and forth, with each horseshoe semi-overlapping the previous one. During the very first year of Girls Garage, I taught the horseshoe motion, and then switched to the pigtail for years after that. I still have girls from that first year (like Erica Chu on page 103) who are horseshoe-loyal! With either motion, the trick is to create a solid bead by overlapping your previous motion. Try both and see which you prefer! There are other variations in weave patterns, such as a triangle pattern that you might learn as a professional welder for specific uses or position of the weld. You can also weld in

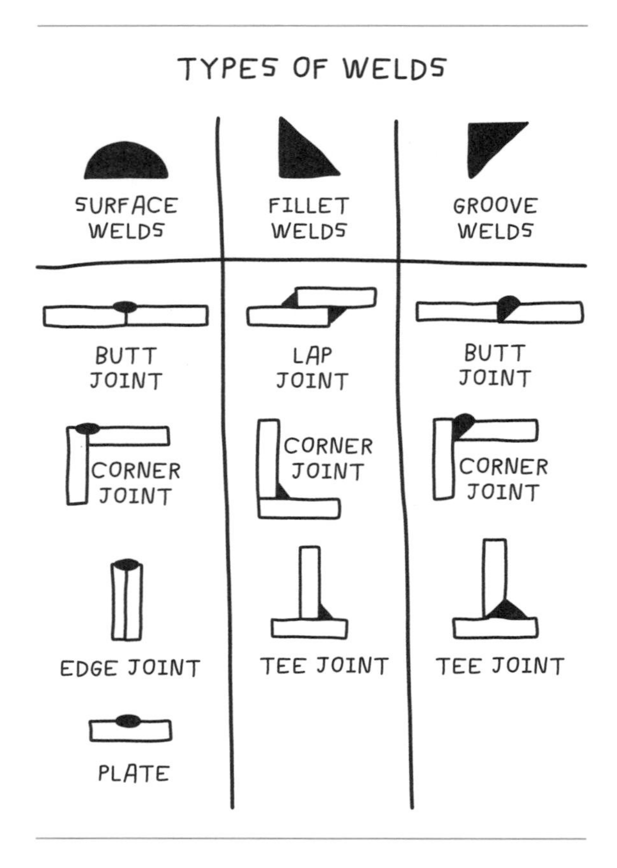

different directions: in the "push" direction (forward, in the direction your gun is pointing), or in the "pull" direction (backward, in the direction of your grip hand). "Pull" welds are generally easier to see as you do them, but "push" welds are usually stronger and more penetrating.

- **Speed:** The wire will be set to feed at a specific speed, but you also need to move your hand (and pigtail or horseshoe motion) at the correct speed. Welding speed is a bit like *Goldilocks and the Three Bears*: not too slow, not too fast, but *just right*. Your speed should be determined by how well you can maintain a good "puddle" of melted filler metal as you go. You have to move slowly enough to have a consistent bead for a solid weld, but quickly enough that you don't burn a hole in thin material. You'll have to play with this a bit, depending on your wire speed and voltage settings, but most of my welds that are set to medium-low speeds require a "1 second per pigtail" count. For each loop of my pigtail, I count "one one-thousand."

Types of welds

Depending on how your pieces need to be joined, you'll need a different type of weld! There are three basic categories of welds: a **surface weld**, which sits on top of a surface; a **fillet weld**, which sits in a corner to join pieces at right angles; and a **groove weld**, which fills in a precut groove in one of the pieces to be joined. In each category, there are different kinds of surface, fillet, and groove welds.

For example, a fillet tee weld joins a vertical piece to a horizontal plate. Or a plate surface weld is just a bead that sits on top of a flat piece (utility companies often weld words or numbers on top of manhole covers using plate surface welds). Groove welds are used more for thicker pieces of metal. With these thicker pieces, it can be helpful to cut this angled groove in the end of one piece, which gives the weld more surface area to connect both pieces.

Welding tools

The tools in this section are all specifically designed for working safely with, cutting, fusing, and finishing metal and will be a great addition to your own toolbox!

Welding helmet with auto-darkening lens: Arc welding creates a bright flash of UV light, so an auto-darkening UV protective welding helmet is a must-have! Plus, you'll feel like Darth Vader wearing it. Many welders pride themselves on their welding helmet stickers, which are commonly skulls and flames and such. An auto-darkening welding helmet has an adjustable head strap and a large rectangular shade and protective glass plate. The shade appears greenish when not in use, but the moment you start welding and the lens detects UV light, it darkens like sunglasses to nearly black. Don't worry, though, you will still be able to see your weld and the area immediately around it, glowing like a candle flame. Welding helmets have adjustable settings to change the darkness and also the delay (so when you are done welding, the lens stays dark for a few extra moments). Most are battery-powered and have replaceable lens plates. A good practice is to also wear polycarbonate safety lenses for another layer of UV protection.

Welding wire and rods: There are different thicknesses and types of metal welding wire, so buy the appropriate one for your welding machine. For the welder we have at Girls Garage, we use a flux-core .030-inch-diameter wire. Welding wire comes in spools that can be easily loaded onto the spindle in your machine. While I much prefer wire-feed welding, there are also types of welding that require a stick or rod instead of a spool of wire.

Welding pliers (aka welper, for "welding helper," get it?): Welding pliers are a four-in-one tool for all your welding needs! Use the end of the jaws to grab and manipulate welding wire, the cutting jaws to clip wire to the correct length before you start a weld, and the two round grab points to hold and twist both the contact tip and the outer nozzle.

Soapstone: Soapstone is like chalk for metal! It comes in long rectangular or round sticks that can be loaded into a metal pen. Use it to make marks or lines on your metal before welding. It won't conduct electricity and can withstand high temperatures, so your marks won't burn away as you weld.

Wire brush: One of the most satisfying moments for me is the post-weld scrub. A wire brush acts like a toothbrush for your metal pieces, brushing away all your slag and other gunk so you can see your clean welded bead and prepare your finished metal for any paint or sealant. Wire brushes have wood or plastic handles and short, stiff metal wire bristles. They come in various sizes, ranging from a brush head of about 1 inch up to giant ones with 8- to 10-inch brush heads.

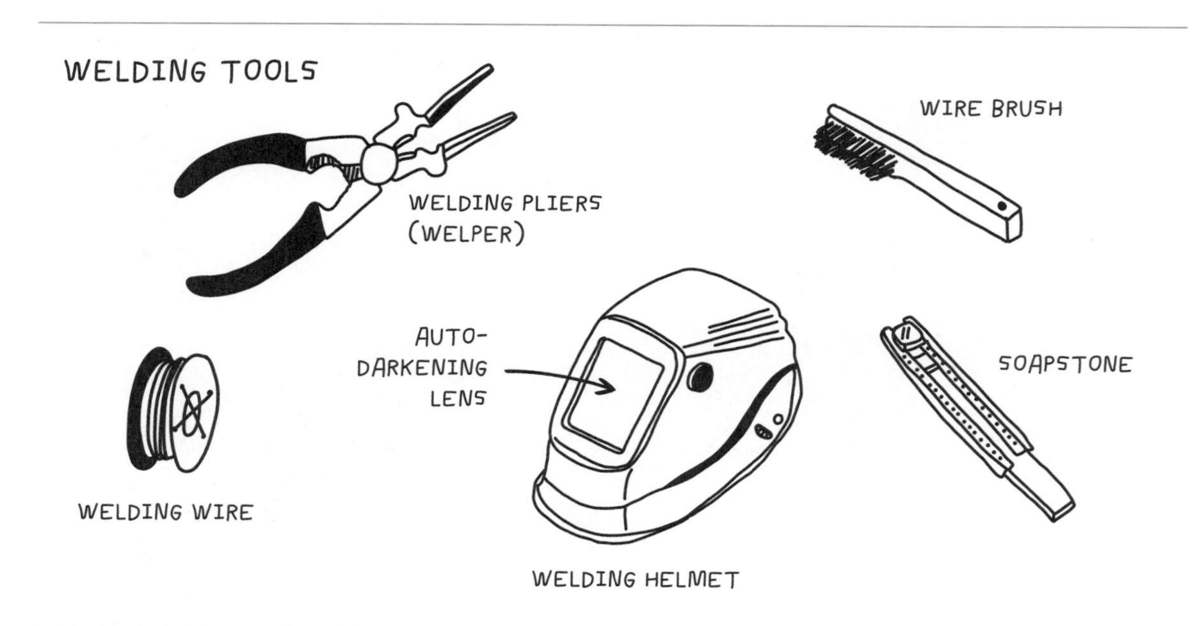

LIISA PINE

Welder, Welding Instructor, Certified Welding Inspector

Oakland, California

In the very first summer session of Girls Garage, Liisa Pine came to visit to talk to us about her work as a professional welder. She walked in covered in dirt and wearing a welding hat and unloaded boxes full of metal creations she had made. We were all awestruck. Liisa is also a welding instructor at a community college in Oakland and an inspiring female leader in the vocational education community. She is also the first person I call anytime we have an issue with our welding equipment that I can't figure out myself!

"I don't recall a time when I wasn't building or making things. Early on, I made houses and clothes for my dolls, and since then I've worked on robots, buildings, a giant waterslide, and even a record-breaking jet car. I like to do work that requires a high level of skill and that is so big and complicated, you can't do it all yourself. I like anything that can take materials through radical phase change, like how welding changes hard steel to molten metal and back again. I love having that power; it feels magical.

"On the one hand, the materials and equipment don't care if I'm male or female, and I love that about them. I tend to forget about my gender when building. On the other hand, since there are still so few women welders, I know I represent more than just myself whenever I show up. It's a big responsibility. Knowing how much I had to stretch to get where I am is a really powerful feeling.

"My typical day is never typical. I head to the community college, where I teach, check inventory on the welding supplies that we use regularly, get a few quotes to purchase materials, work with the tool room assistant to ensure that all the supplies and equipment are ready for classes, and visit the dean's office to discuss the progress for the program. I teach a few classes, from welding science and theory to facilitating lots of hands-on practice in the weld lab. Teaching is great, and welding is so much fun; doing them both together is a lot of work!

"I'm most interested in solving problems for a community, and I also love when people learn by doing. Building requires collaboration. People have to get together and work face-to-face. They need to share ideas and solve problems. We learn so much when we work together to complete something bigger than ourselves.

"My best advice is to make friends! Meet people who do what you want to do, introduce yourself, and ask questions. Go to a class and build with other people. Creating relationships is a huge part of building and can lead to opportunities beyond your wildest dreams."

Other metal tools

Even though welding will always hold a special place in my heart, it's often only one step in any metal project. Other tools are required to finish, grind, and smooth your metal pieces. There are also other ways to manipulate metal besides welding it. We'll learn about some other common metal tools here.

Oxyacetylene torch: While some people use an oxyacetylene torch for welding or brazing, I think it is most useful for roughly cutting through thick metal. I used one with my high school students to cut a door opening in the side of a shipping container—if that gives you any idea of its cutting power!

The oxyacetylene torch uses two tanks of gas—oxygen and acetylene. Each gas flows from its own tank, but they work together, differently, toward a common purpose. First, the acetylene and oxygen flow together and are ignited at the torch's tip, creating a very hot flame you hold over your metal to heat it.

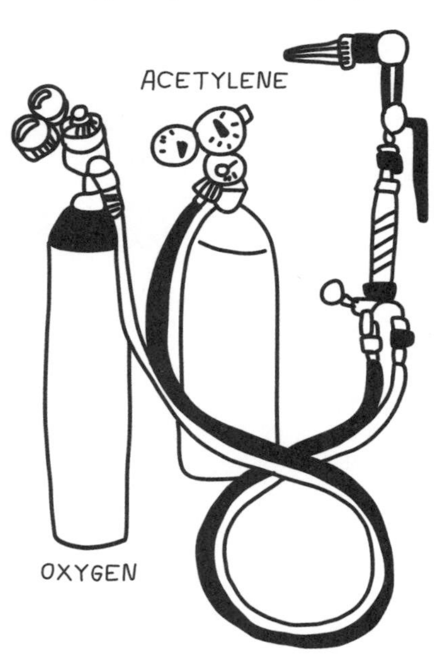

Second, the torch has a trigger that releases a lot more oxygen. This blast of oxygen does the cutting through the metal. The chemical process works sort of like rusting in fast motion, cutting through your metal by quickly corroding it and burning it away.

The oxyacetylene torch is for experienced adult users only. Do not use this tool if you are under the age of 18. Always read the manual for your specific tool (the following tips are general guidelines). Shade-5 safety glasses and personal protective equipment should be worn at all times by both users and observers!

The key to an oxyacetylene torch is using the regulator dials on your gas tanks and the torch handle to make sure gas flows at the proper rate. This is most important when you squeeze the trigger for your blast of oxygen. You need enough oxygen to actually cut through the material, but not so much that it burns away more metal than you want. Read any manufacturer manuals or instructions to identify the recommended valve settings, which are measured in psig (pounds per square inch gauge).

When lighting your torch, turn on the flow of acetylene first and then light the flame. The flame will be orange with black smoke. Then slowly open your oxygen valve on your torch, which will change the flame to a shorter, pointed blue flame with a small white core. Heat your metal first, and then use the trigger on the torch to release the blast of oxygen that will cut through the metal. **When using an oxyacetylene torch, make sure you wear Shade-5 protective eyewear and head-to-toe nonflammable protective clothing!**

Abrasive saw/metal chop saw: An abrasive saw works just like a wood miter saw but has an abrasive disk instead of a metal blade, and a larger metal guard to protect you from sparks when cutting metal. While most people use an abrasive saw for cutting down steel, it can also cut through materials like tile and stone.

The disk of an abrasive saw is actually less like a saw blade and more like the grinding disk you see on an angle grinder. The disk is usually 14 inches in diameter, made from fiberglass with an abrasive coating like aluminum oxide or silicon carbide. You can buy disks for various kinds of material you might want to cut: steel, aluminum, stone, and more.

Do not operate an abrasive saw without the supervision and hands-on support of a skilled and experienced adult. Always read and follow the manual for your specific tool. Safety glasses should be worn at all times by both users and observers!

When cutting metal on an abrasive saw in particular, get ready for a spark show! Set your metal in place and use the fence and built-in clamp to secure it. When you make your cut—**wearing a face shield and gloves, and making sure your whole body is covered**—cut steadily and slowly through the material. When finished, your metal will be very hot where the cut was just made, so handle with care!

Abrasive saws are great for making quick rough cuts in metal stock. If you're looking for more precision, however, a metal band saw (either handheld or mounted on a base) is probably a better choice.

Angle grinder: Angle grinders are amazing multiuse tools for cutting, polishing, and grinding metal. They're like a metal saw blade and sander in a compact handheld package! Use an angle grinder for grinding down the surface of metal (like a weld bead) or, with a cut-off disk, for cutting steel into pieces.

Most angle grinders are relatively compact, with a disk size of 4 inches to 7 inches. They have a wide body with a grinding wheel at one end and a metal guard to protect you from sparks. They also have a side handle so you can hold the tool steady, and the side handle can be moved from one side to the other (for lefties!). The most common types of disks are grinding disks, which are thicker and have an abrasive surface, and

METAL CHOP SAW/ABRASIVE SAW

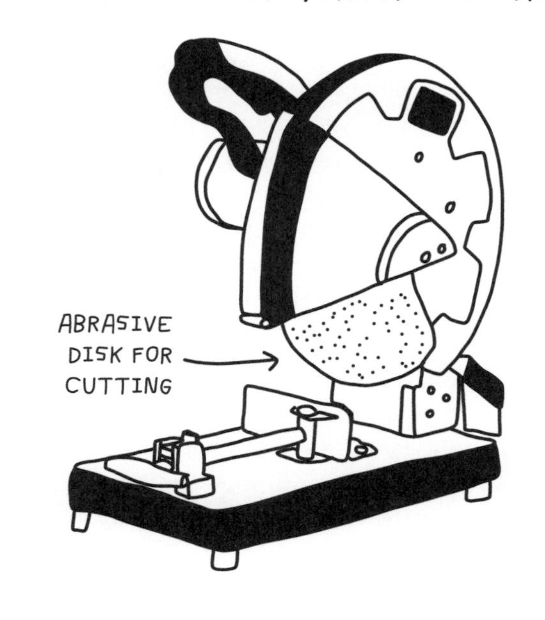

ANGLE GRINDER

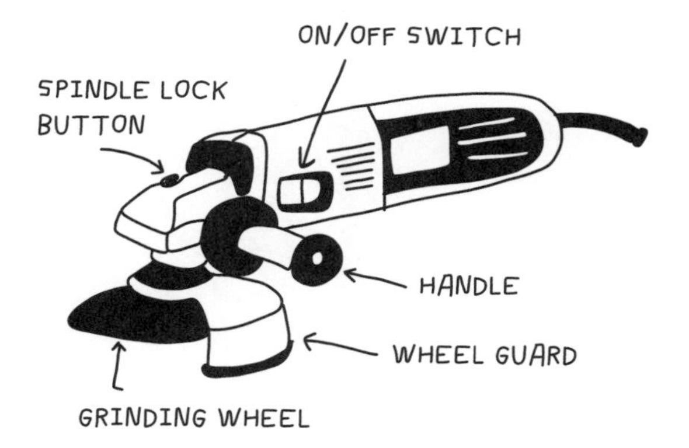

cut-off disks, which are thin like a DVD so you can use their thin edge like a saw blade. Disks can be easily removed and swapped out for another one, using a spindle mechanism that holds the disk in place at its center point.

Do not operate an angle grinder without the supervision of a skilled and experienced adult. Always read the manual for your specific tool (the following tips are general guidelines). Safety glasses and personal protective equipment should be worn at all times by both users and observers!

The disk spins in one direction (usually indicated with an arrow on the tool). This direction is important because sparks fly in the direction of the rotation, and you always want to **orient yourself and the tool so these sparks fly down and away from you** (never upward or toward your face). You should always wear a face shield when using an angle grinder, and hold the tool with two hands at all times. Clamp your material securely in place before grinding or cutting. The tool will do most of the work for you, so no need to push too hard!

SOLDERING IRON

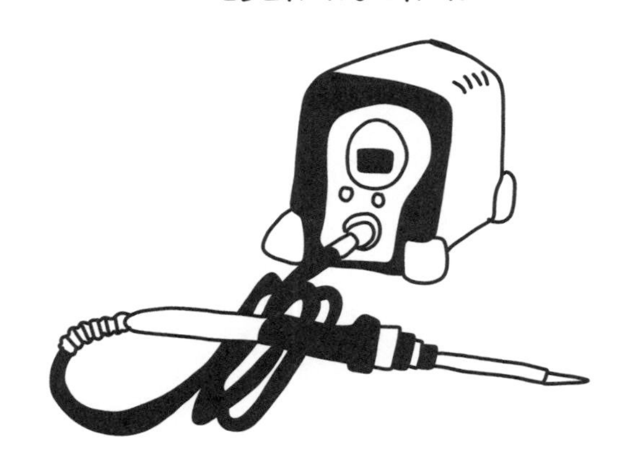

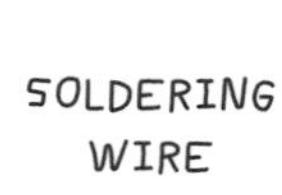

SOLDERING WIRE

Soldering iron: A soldering iron looks like a technical piece of equipment, but it is really just a very hot metal stick. You might use a soldering iron to repair or make an electrical circuit, or to join two pieces of electrical wire together.

Soldering irons come in many different varieties: electric, cordless, and temperature regulating. Regardless of which type you use, the principle is the same: The tip of the soldering iron reaches a very high temperature, melting the wire onto a surface. Soldering is often used to complete circuits on circuit boards. However, you can also use soldering as a technique for jewelry making, using the solder to join different parts. For electronics work, I definitely prefer to use a lead-free solder wire and a digital temperature gauge that also comes with a stand to rest the iron on when you aren't using it. To use a soldering iron, heat the object you want to solder (like two electrical wires you wish to join), NOT the solder wire itself. Hold your iron on the object to heat it up, then touch the solder wire to the object, which will melt onto your object. Keep the tip of your soldering iron clean and free of debris and black oxidation (but wait until it has cooled off before you clean it!).

Do not operate a soldering iron without the supervision of a skilled and experienced adult.

A good rule is to **hold the soldering iron on the plastic handle only.** Because you hold the iron like a pencil, it can be tempting to inch your hand toward the tip, but it's best to keep it back as far away from the hot tip as possible. One last warning: **Don't breathe the fumes, and wash your hands after!**

CLEAN UP

A lot of people who walk into our Girls Garage space say, "Whoa, how do you keep this place so clean?" In my mind, a clean space is a safe space, free of rogue nails you could step on, dust you could slip on, or tools left out all over the place you could trip on. It's also important to dispose of materials safely and into the right refuse channels, so check with your local waste-management company to see if they have specific guidelines for wood, metal, or other construction-related debris. Here are my favorite cleaning tools.

Big metal dustpan

Get one that is durable and wide enough to sweep up piles collected by both your counter brush and your push broom. We have one that is about 18 inches wide and it's great.

Push broom and dust mop

A 36-inch-wide stiff-bristle push broom with a long wooden handle will be your favorite cleanup tool to keep a floor clean. I find it meditative to walk back and forth across the shop floor with my push broom at the end of

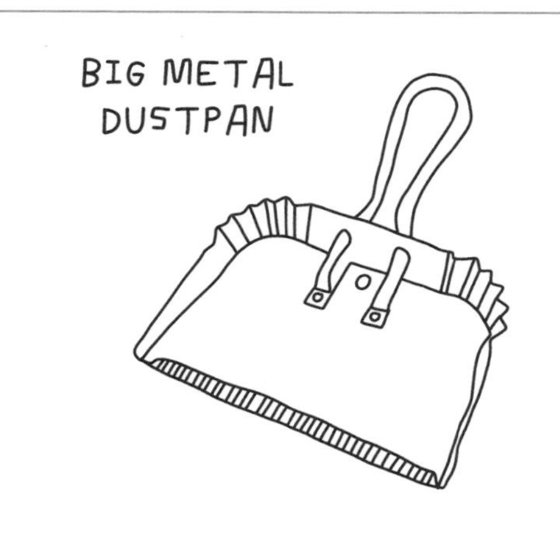

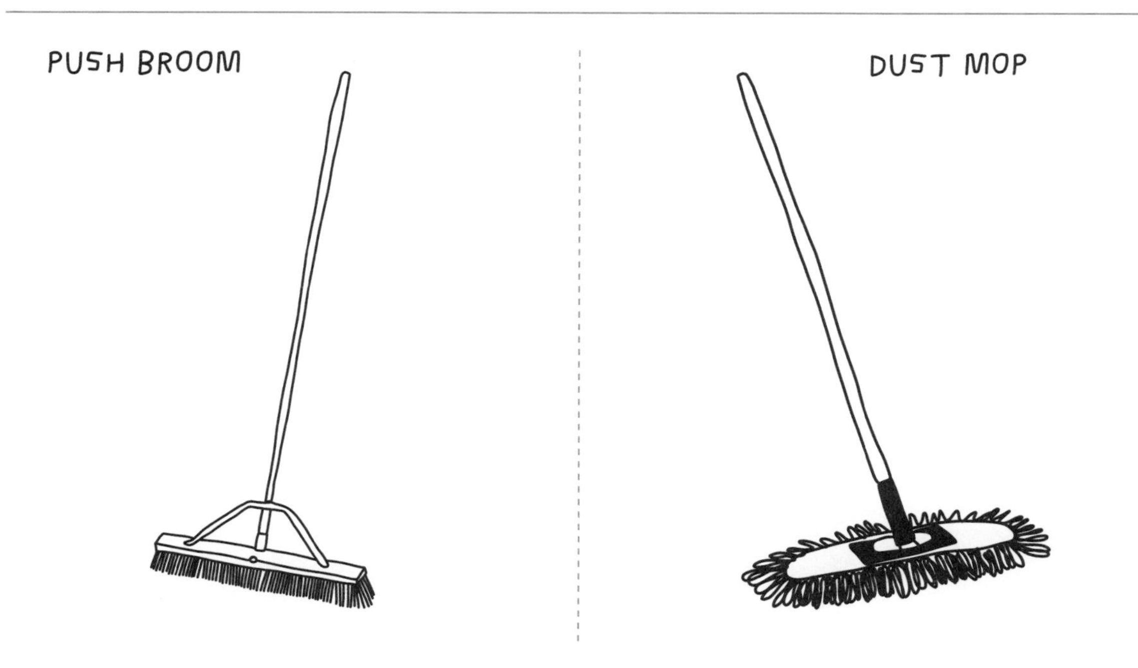

COUNTER BRUSH

USE WITH DUSTPAN TO SWEEP TABLE SURFACES

BLOWER

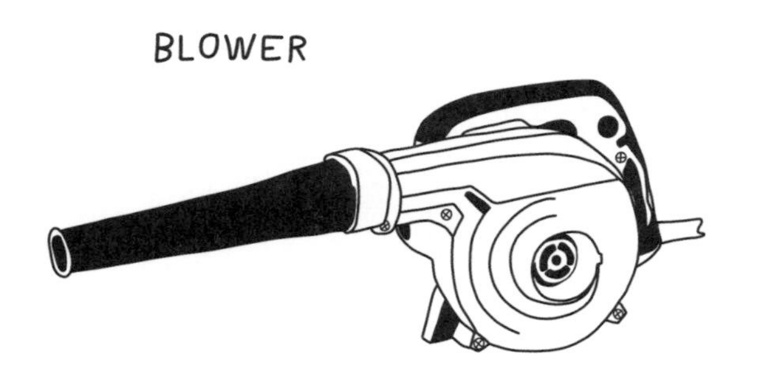

SHOP VACUUM

a workday. We also have a large, flat dust mop, which is helpful for getting those pesty piles of dust from under tables and in corners.

Counter brush

A small counter brush does the trick for all benchtops and tables that a broom can't reach. An 18-inch counter brush is a great size to go with your big metal dustpan.

Blower

My colleague Allison Oropallo (see page 156) is the queen of cleanup, thanks to her trusty blower. At the end of an extra-dusty day, the blower is a great complement to the shop vacuum and push broom. It might seem counterintuitive, but clearing all the dust out of the funny crevices makes it that much easier to sweep up. A handheld battery-operated blower is a huge time-saver!

Shop vacuum

You can collect piles of sawdust in your dustpan, but sometimes you need the big guns. A Shop-Vac is a staple in most work spaces and comes with a few attachments that fit the long hose, each of which is quite useful. You can also connect most shop vacuums directly to power saws to act as a dust collection system!

Magnetic broom

The first time I saw this in action, I was in love. Think of a magnetic broom as a metal detector that actually grabs all your metal for you. On a job site or in a shop where you've been dropping nails and screws, sweep your magnetic broom across the floor to collect any rogue hardware.

Pumice scrub

You'll find orange scrubs in the soap and cleanup aisles of most hardware stores. They're commonly used in auto-mechanic shops because they're great for removing grease and other nasty stuff from your hands. And it smells great!

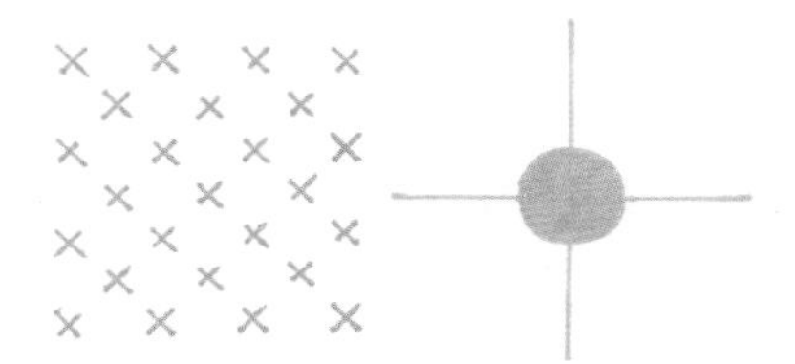

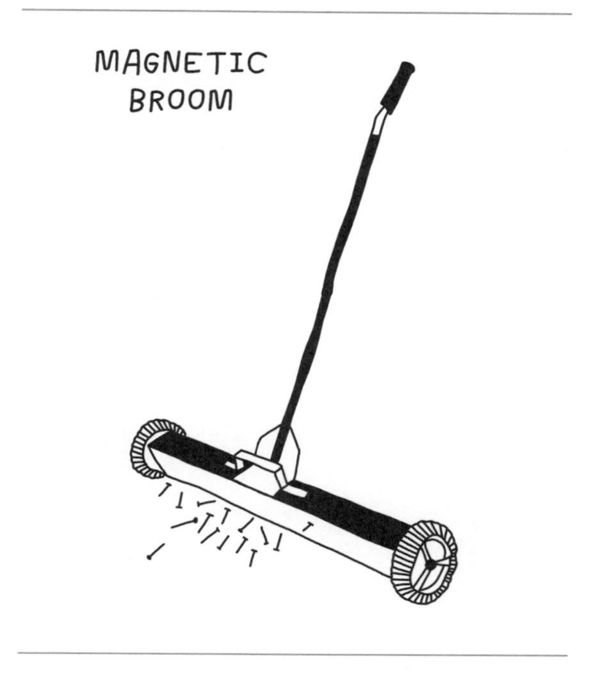

PUMICE SCRUB

ESSENTIAL SKILLS

In the not-so-distant past, my kitchen sink broke. More specifically, the handle on my kitchen faucet broke. For a split second, I thought, "Ugh, now I have to call a plumber," and then almost immediately snapped back to reality and remembered, "I can fix it myself!" And I did, with a little help from some instructional videos and the installation guide for my new shiny faucet.

The skills in this section are meant to cultivate the "I can fix it myself!" mindset for you, too. Some of them are household fixes (like what to do when your toilet won't stop running), and others are just simple ways of solving small problems in the world (like knot-tying and measuring). Some are specifically related to building tasks, and others are just great things to know how to do as women. There are twenty-one of them in total, and if you can master them all, you'll be the hero of any sticky situation!

HOW TO WRITE OUT FEET AND INCHES

Builders have a common numerical system that helps us communicate precise dimensions for cutting and assembly. Some folks might choose to write "two feet" as 2', while others might write 2'-0", and yet others might write 24". Even with these personal variations, there is a common language and format you can follow.

- The measurement in **feet** is always followed by a single mark called a prime, which looks like a straight apostrophe.
- The measurement in **inches** always gets a double mark called a double prime, which looks like straight quotation marks.

If you're showing a number that requires both feet and inches, show the feet first, then a hyphen, then the inches, including any fraction of an inch.

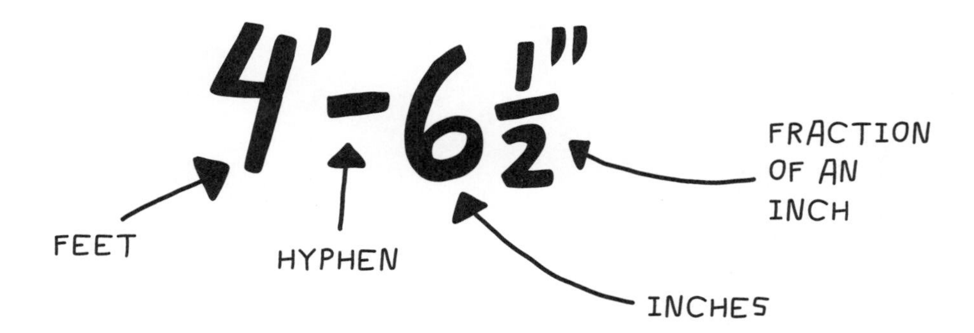

HOW TO WRITE AND SHOW DIMENSIONS

Once you know how to write measurements in "builder language," you can use these measurements to show the dimensions of a two- or three-dimensional object.

Dimensions are shown in **length × width × height**, using the inch and foot notations you just learned.

In some instances, you might only need to show length and width (for instance, if you were indicating the size of a tabletop). On a drawing, you can also show these dimensions using a dimension line that extends from one end of the object to the other, with the measurement in the middle. These dimension lines are common on most construction documents you'll use.

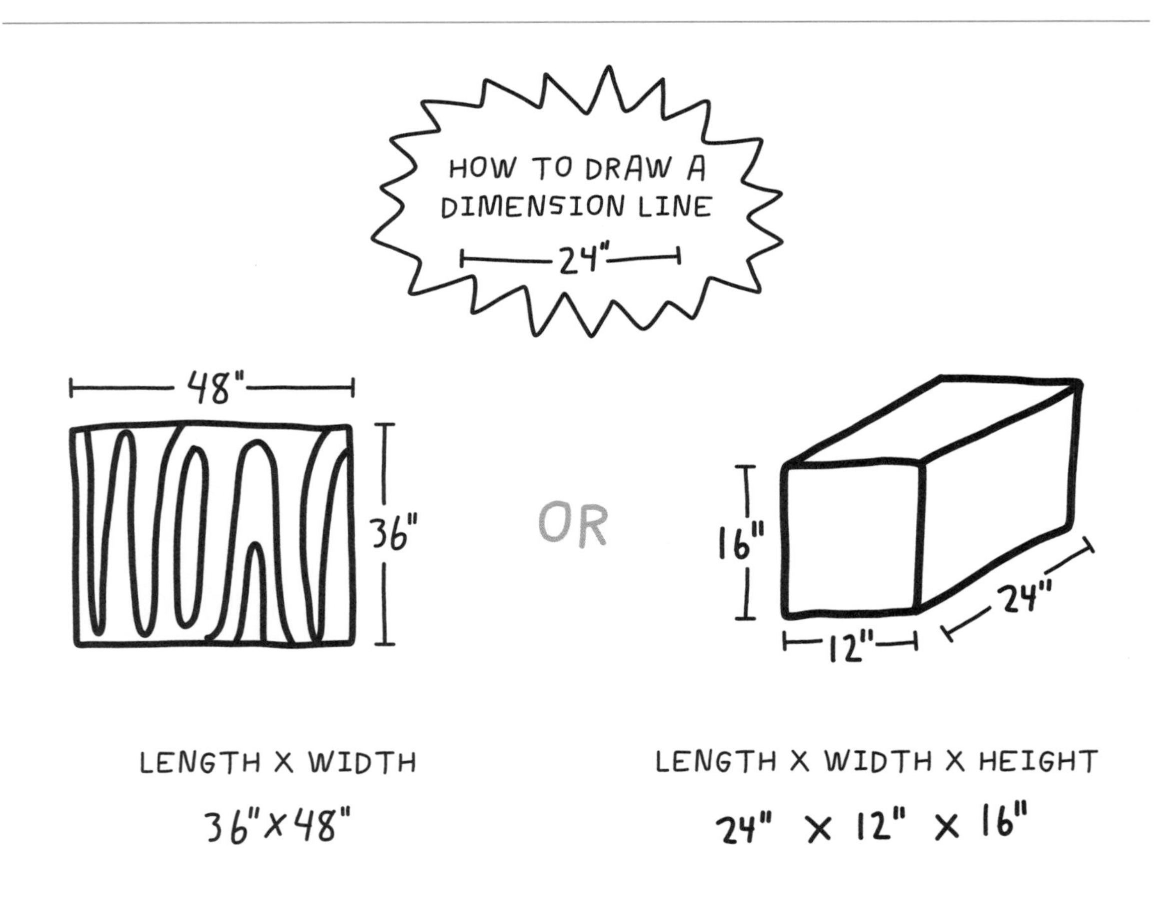

HOW TO WRITE A MATERIALS "CUT LIST"

Let's say you've just designed and drawn sketches of an awesome piece of furniture you want to build. Just like when you're cooking a delicious meal, you need a recipe. For building, this recipe is called a cut list, and it includes all the pieces of material you need and at what size.

My favorite format for writing a cut list is a list with three columns:

- The first column shows the kind of material (e.g., ½-inch plywood or 2×4 or ¾-inch threaded rod).
- The second shows the quantity of cuts you'll need from that material.
- And the third shows the dimensions of cut material you'll need (specifically, the cuts you will need to make).

Once you have your list, you can use it as a shopping list at the hardware store, and then bring it home and check items off as you cut them!

EXAMPLE CUT LIST

MATERIAL	QUANTITY @	DIMENSIONS TO CUT
2X4 LUMBER	8@	36"
1/2" PLYWOOD	4@	12" X 24"
3/4" THREADED ROD	16@	18"
4" X 1/4" STEEL BAR	12@	10 1/2"

HOW TO FIND THE CENTER OF A RECTANGLE WITHOUT MEASURING OR DOING MATH

This sounds like a riddle of some sort, yes? How can you find the exact center point of a rectangle without measuring anything? You'll be surprised at how often you need to do this when building. For example, let's say you're building a birdhouse, and the front of it is a rectangular piece of plywood. You'd like for the little birdie perch and hole to be perfectly in the middle of that rectangle. You COULD measure the length of all four sides, divide by two on each side, make a mark at that midpoint, and then connect the lines in a cross shape to find the center (whew!). That method requires both a measuring device and a mathematical calculation, and even though we're all math nerds at heart, let's save the brain cells for building (plus, what if you don't have a measuring device handy?). Here is a shortcut that is just as precise and requires no math or rulers or tape measures:

Step 1. Using anything with a straight edge (like a piece of wood, a long ruler, or even a table edge), draw a straight line to connect the opposite diagonal corners of the rectangle.

Step 2. Draw a second line to connect the OTHER two diagonal corners of the rectangle.

BOOM! That's it. Where the two diagonal lines meet is the exact center of your rectangle, and you now have a center point to work from. Count your extra brain cells and get back to work!

HELPFUL TIP! This technique works with any four-sided shape with parallel sides, like a diamond or parallelogram. Thanks, geometry!

---

1. DRAW A STRAIGHT LINE TO CONNECT OPPOSITE DIAGONAL CORNERS.

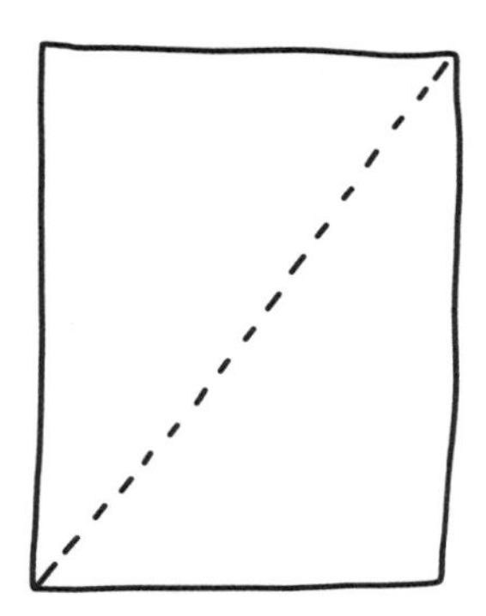

2. DO IT AGAIN WITH THE OTHER CORNERS.

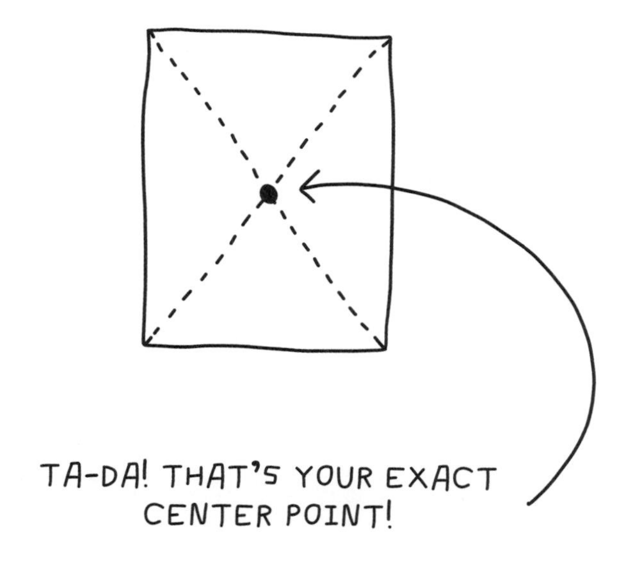

HOW TO READ BASIC CONSTRUCTION DRAWINGS

Just like we learn to read books as kids, we have to learn to read construction drawings as builders. Architects, engineers, and builders use a (mostly) common visual language to help show three-dimensional objects on two-dimensional paper. Whether you're looking at a blueprint that someone has already made or you want to draw your own, here are four basic principles that will help you understand them.

Scale: When you have to draw something as big as a house on a small piece of paper, you can't possibly draw the house at full size, or "full scale." Architects, builders, and engineers use a specific scale to shrink the object they're drawing into a smaller version of itself, and indicate the scale on the drawing. For example, you might be building a shed that measures 12 feet wide by 12 feet long in real life. You might use the scale "half inch equals one foot" (shown as ½" = 1'-0"), and then draw the shed on your paper at 6 inches by 6 inches. You'll have to determine an appropriate scale for your drawings based on the size of the object or structure and the size of your paper. You usually find the scale on a drawing in the lower right or left corner.

Plan: A plan is one of three common types of construction drawings. If you've ever drawn a floor plan of your bedroom, that is a plan drawing. A plan shows an object or a space looking down on it, like a bird's-eye view. Some architectural drawings, however, are actually downward-looking views of a space as you would see it from only 3 feet above the ground. Architects draw this "3-foot-high, looking down" view so they can show the interior of the walls, staircases, etc.

Section: A section is another common drawing, which is a cross section, or slice, through an object or space. Sometimes a set of construction drawings show multiple sections, one cutting through the object in one direction, and another section showing the object sliced the other way. You'll often see a dotted line on the plan drawing that shows where the section was "cut."

Elevation: An elevation is a side view—looking at an object from the side. Elevations are a little funky, though, because they don't show perspective or depth, so it is a flattened-out version of the object's side. Like a section, you might see multiple elevation drawings showing all sides of a house, for example.

PLAN

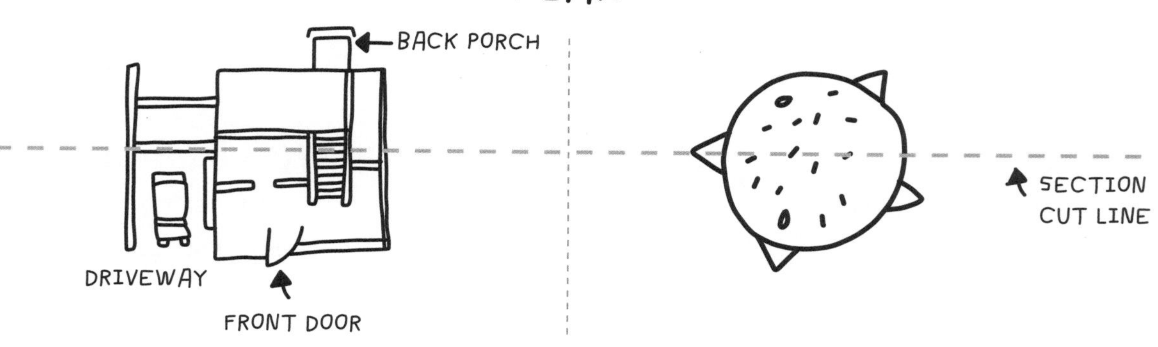

SECTION

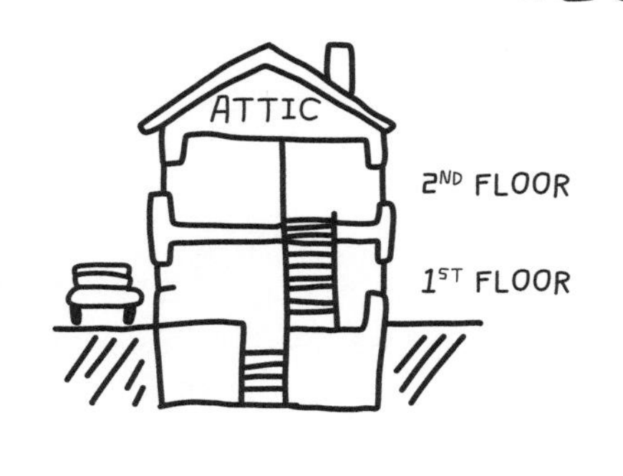

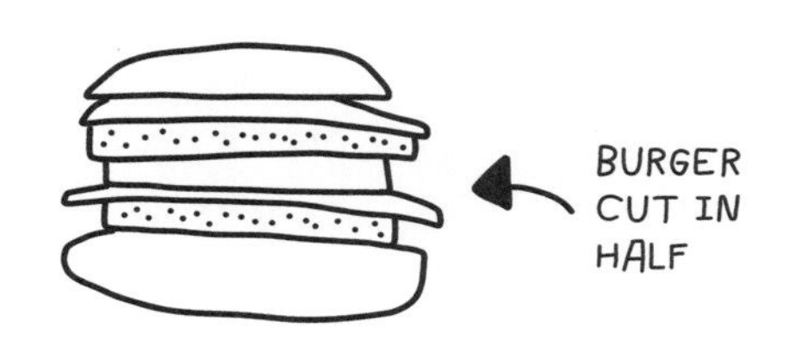

ELEVATION

SCALE

DIMENSION ON PAPER → $\frac{1}{2}"=1'-0"$ ← DIMENSION IN REAL LIFE

HOW TO USE A 3-4-5 TRIANGLE TO MAKE PERFECT SQUARE CORNERS

Math alert! A triangle with sides that measure 3, 4, and 5 will always have a right angle where the 3- and 4-length sides meet. We can look to good old Pythagoras for an explanation of why. The Pythagorean theorem is a mathematical equation that helps us find the lengths of the sides of any right triangle (a triangle with one 90-degree angle in it). It goes something like this:

The 3-4-5 triangle is one of the best examples of the Pythagorean theorem because all the numbers are whole and easy to remember. Since we know a 3-4-5 triangle always has a right angle, we can use this particular triangle in real life to help us lay out perfect corners.

Here's an example: Let's say you're lucky enough to have a big backyard and want to mark out a rectangle as a footprint for a raised garden bed. You'd want your rectangle to have four 90-degree corners. Using a 3-4-5 triangle, two friends, and three tape measures, you can easily do this without much effort, and definitely without mathematical calculations.

Step 1. Choose a spot for the first corner of your rectangle and stand there.

Step 2. Hold the ends of two tape measures in the same spot at your feet. Have one friend walk 3 feet to your left (or right) while extending one tape measure, and one friend walk 4 feet straight in front of you extending the other tape. You should now have one friend

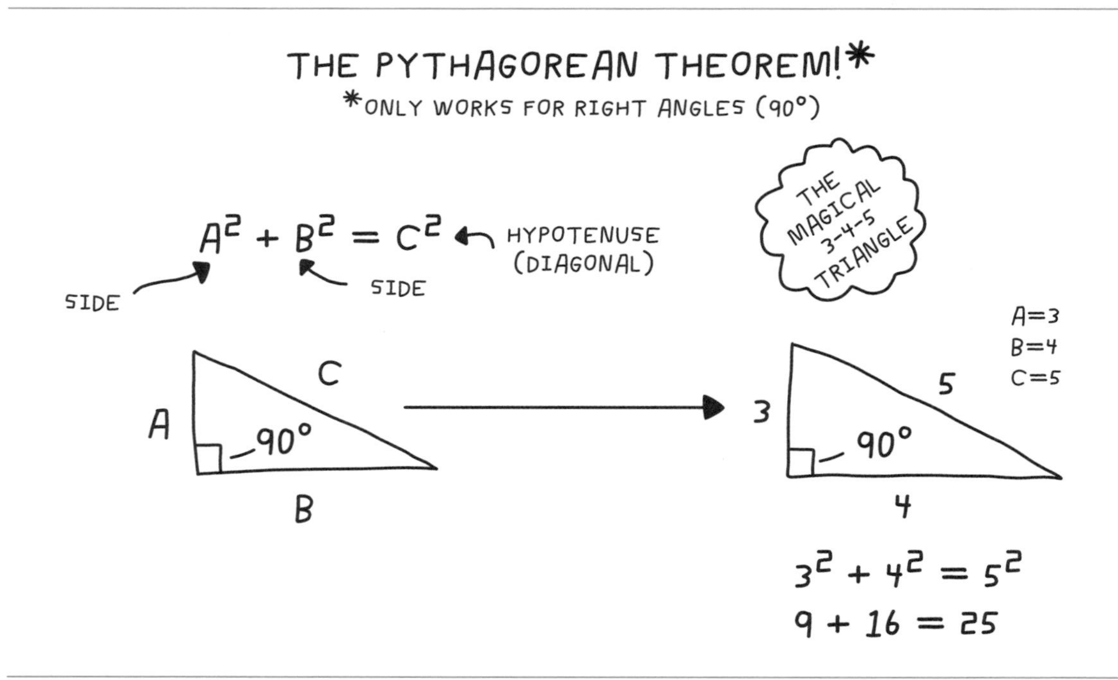

3 feet away from you, and one friend 4 feet away from you.

Step 3. Now have your friends extend a tape measure between them, with the end of the third tape measure touching the 3-foot mark on one friend's tape measure, and the 4-foot mark on the other friend's tape measure. The measurement (hopefully) of this third tape measure should be roughly 5 feet, but probably not exactly.

Step 4. Have your two friends walk toward or away from each other until the reading on that third tape measure is exactly 5 feet. Once you've measured 5 feet, congratulations! The three of you are now standing at the points of a perfect right triangle (and you are standing at the 90-degree corner). You can now mark your friends' locations and connect the lines between you and each of them. Those two lines are the first two sides of your garden bed (and you can obviously extend them to make your planter beds bigger than 3×4 feet).

Step 5. Repeat this process to locate and mark your other three corners, using the lines you've just established.

If you don't have three tape measures, you can do this with one flexible tape measure, in one 12-foot extension that makes a triangular loop back to you (3 feet plus 5 feet plus 4 feet, with your friends holding the two points).

This technique is also helpful for checking the "squareness" of a large object, like a corner where two walls meet. You can measure 3 feet from the corner going in one direction, 4 feet going in the perpendicular direction, and then measure the diagonal connecting those two lines (hypotenuse). If the measurement is NOT 5 feet, your corner is not a perfect right angle.

The 3-4-5 triangle math works in larger scales, too! A 6-8-10 triangle (twice as big) can also be used as a perfect right triangle, as can a 9-12-15 (three times as big), or a 30-40-50 (ten times!).

HOW TO USE 3-4-5 TRIANGLES FOR PROJECT LAYOUTS

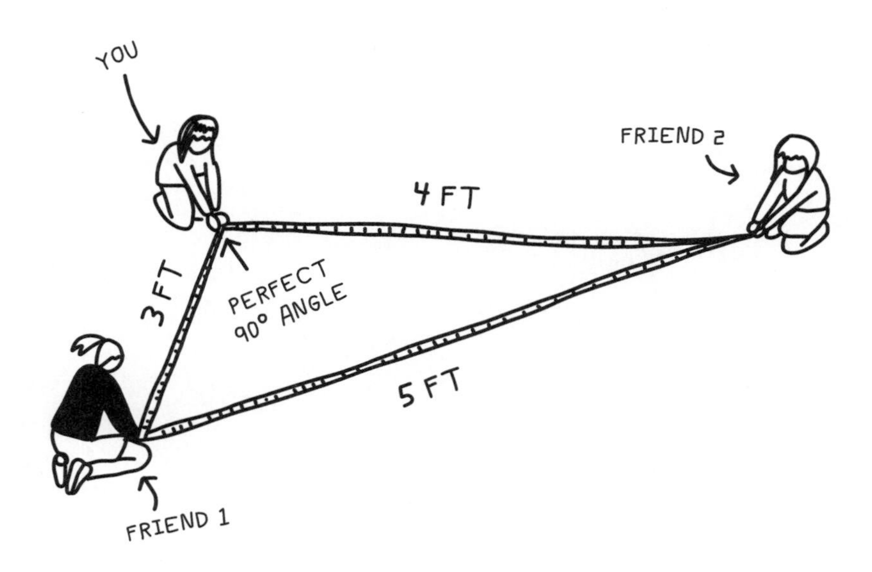

JEANNE GANG

Architect, Founding Principal and President of Studio Gang

Chicago, Illinois

When I was living in Chicago, during and after graduate school at the School of the Art Institute of Chicago, I lived two blocks from Jeanne Gang's architecture office. I passed it every morning walking to the Blue Line train. I knew of her work and secretly bowed down every time I passed her building. Her work has always inspired me because of its balance of beauty and ferocity, detail and boldness; she is also deeply committed to making and to the human hand. Jeanne is a force to be reckoned with and is arguably the most accomplished female architect in the world. Actually, scratch that—one of the most accomplished architects in the world. She is also incredibly generous; she could have easily said no to contributing to this book, but gladly took time away from her many projects to tell her story.

"As a kid, I could often be found outside, building tree houses, or inside, building elaborate structures with blocks. I think being a bit shy as a young girl was something I had to overcome because I wanted to be heard and acknowledged. Sports were a great vehicle for me to break out of my shyness. Through training and playing sports, I built confidence in myself and learned how to work with others. Sports brought out the fighter in me and taught me it was okay to be competitive; and equally important, they nurtured my sense of fun.

"As I got older and started to consider majors in college, I found myself attracted to architecture and design because they combined my passions for mathematics (especially geometry) and the visual arts. When I began to travel to faraway places and discover connections between each new city's built environment and its culture, I was hooked on this lifelong journey of studying and designing architecture.

"I love architecture because it is relevant and creative. Every day, my team and I get to use our collective imagination to design what we want to see become reality in the world, in the form of physical places. We also get to try to understand the world through the eyes of others, who will use the spaces and buildings we create. As a career, architecture is a wide-open field that can be as creative, intellectual, socially engaged, and technical as you want it to be.

"I design buildings at various scales, from community centers to museums to high-rises, as well as large sections of cities. I try to think about how physical spaces can connect people to each other and their environment. Design involves everything from sketching an initial concept and developing how a building is going to look to exploring how people will use it and anticipating how it will function over its life span. It also includes figuring out the details of

how all of a building's materials come together to make it resist gravity, use the least amount of energy, keep water out, and endure over time. Because of the complexity of design, it takes more than one person; it takes a whole team, and each person plays a part.

"I've always been, and continue to be, very curious. I like experimenting with materials—including breaking them—to test their properties and discover new ways they can be used. I also like exploring ideas across many different subjects, from biology to fine arts. At a very young age, my parents showed me how to research and find information; they also taught me to question the sources of that information. This kind of critical thinking is crucial for any profession. Several of my professors in graduate school pointed me toward great works of architecture and encouraged me to study them, and it was through examining and understanding these precedents and then working through my own projects that I developed my design skills. I also benefited from discussions and debates with my fellow classmates. Surrounding yourself with talented, intelligent people is a great way to advance your own thinking.

"I am proud of all the projects I've designed or have contributed to designing, even the very early ones when I was just starting my career. I think I'm most proud of those projects where I feel I have discovered something new, and those that have challenged the status quo, led to improvements in our environment, and benefited underserved communities.

"I love to draw and sketch with a simple soft-lead pencil. This basic low-tech tool is all I need to convey an idea or record an impression. Sketching is a different way of thinking. I like the smudginess of pencil sketches, too. Smudges give the sketch a certain ambiguity that invites multiple readings. I draw every single day, and pencils fit easily into any size bag, or even a pocket. With a pencil, you can draw anytime, anywhere!

"So my advice to others interested in building is this: learn to draw! Drawing is absolutely a learnable skill, and it helps you see the world around you. Draw something to scale and then build it yourself. Take items apart to see how they work. Reflect on the way the spaces you spend time in affect you, and ask yourself how you would remake them to change what happens within them. Read about architecture. Read about everything!"

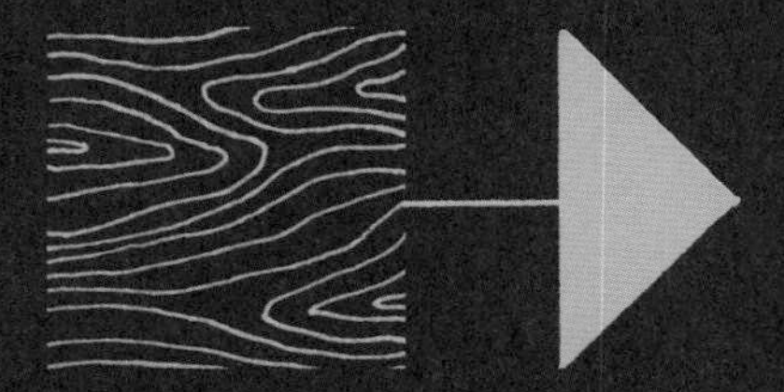

HOW TO CARRY LUMBER

If you're using dimensional lumber, chances are you've got pieces that are 8, 10, or even 12 or more feet long. You might be thinking this is a job for multiple people to carry one piece of wood, but think again: you've got this on your own!

Step 1. Squat down and pick up one end of the long piece of lumber (remember to squat and lift with your legs; don't bend over and lift with your back!).

Step 2. Raise that end up to your shoulder height.

Step 3. Walk forward so the center point is now resting on your shoulder.

Step 4. Use your shoulder as a balance point and let the front end of your piece of lumber come off the ground. Use both hands to hold and stabilize, and off you go! This technique works great because the center point (or "center of mass") falls right on top of your shoulder. Make sure to keep an eye on the ends of your lumber, though, and make any turns very slowly.

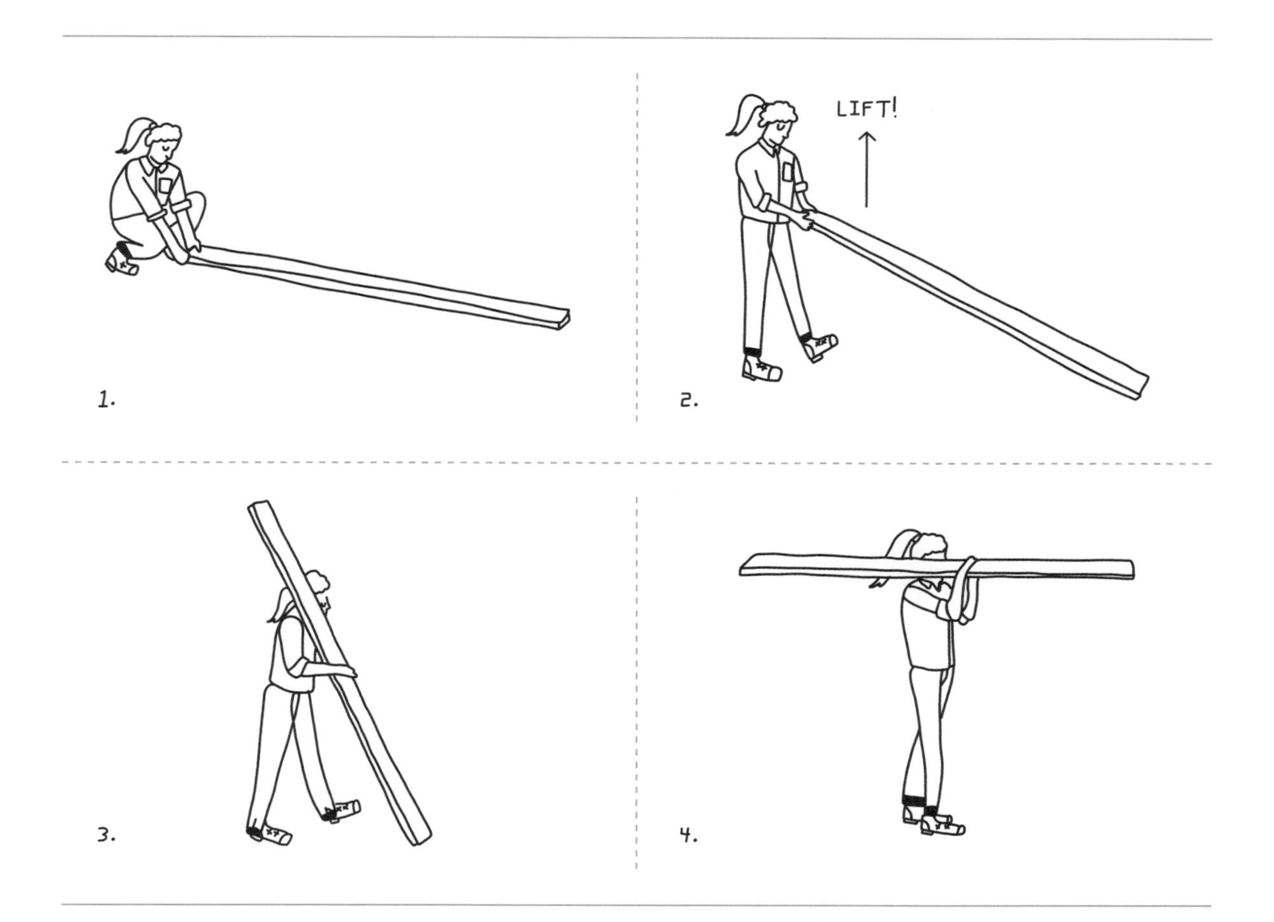

HOW TO "TOENAIL" A NAIL

There might be some instances when you need to attach two pieces (especially lumber, like 2×4s) when you don't have access to one side of your project and have to hammer a nail at an angle, through one piece into the other. This is most common when one piece is already installed against a floor or ceiling, like where the bottom of a vertical 2×4 stud hits a horizontal 2×4 baseboard on the ground. This is called "toenailing," and can take some practice, but is a great skill to have as a builder.

The trick with toenailing is that once the nail has gone through the vertical stud and into the horizontal baseboard, it will pull the stud in the direction you're hammering. So you'll have to compensate by starting with your stud slightly off where you want it to be, so it slides into place as you nail.

Step 1. Mark the location on your bottom base piece where you want your vertical stud to attach. Hold a nail at an angle along the side of it, to make sure the nail is long enough to sink at least 1 inch into your baseboard. Usually, holding the nail at about a 45-degree angle will work, but you can gauge for yourself and adjust accordingly to ensure that the nail will go through one piece of wood and deep enough into the other.

Step 2. Start your nail straight on, not at an angle, and tap it into the stud just enough so it bites and holds the wood.

Step 3. Use your hammer claw to pull it upward to your original angle (about 45 degrees), so when you hammer it, it goes through your stud at an angle and into your bottom board.

Step 4. Hammer the nail and use your toe as a backboard, so the vertical stud slides into place, but not past it.

1.

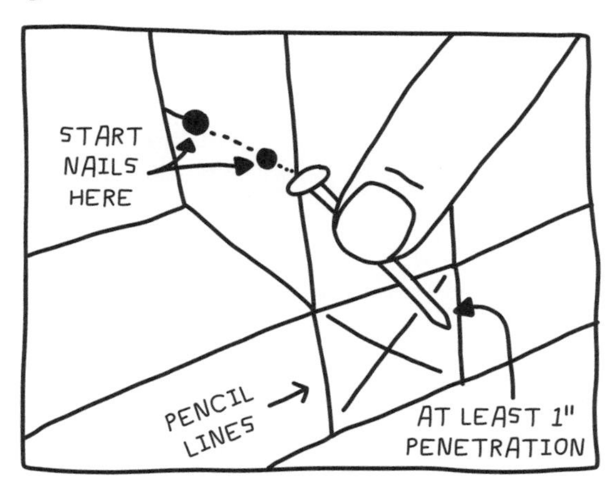

2.

START THE NAIL

START 2X4 HERE

3.

4.

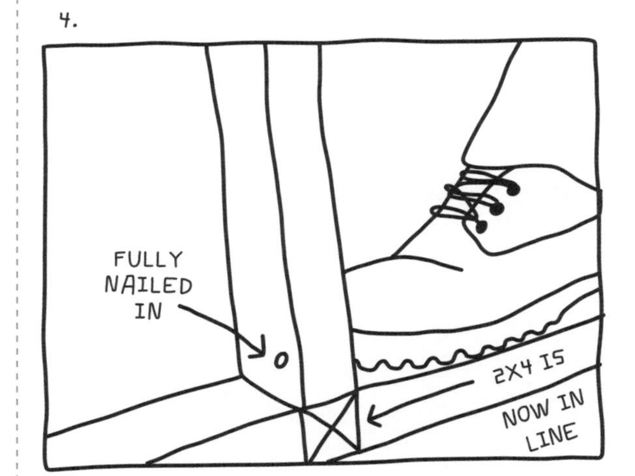

FOUR HELPFUL KNOTS AND WHEN TO USE THEM

Square knot: One of the most common and useful knots, the square knot (sometimes called a reef knot) joins two ends of a rope securely. A square knot will not loosen in any direction as it is pulled on. Square knots are ancient, and are used to secure sashes around Japanese kimonos, to tie down sails on ships, and for medical stitches. The trick to a good square knot is to remember: left over right, then right over left.

SQUARE KNOT

1.

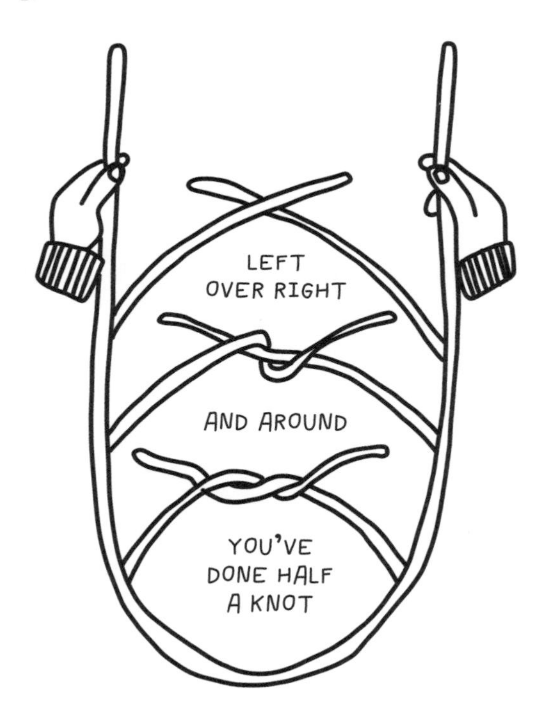

2.

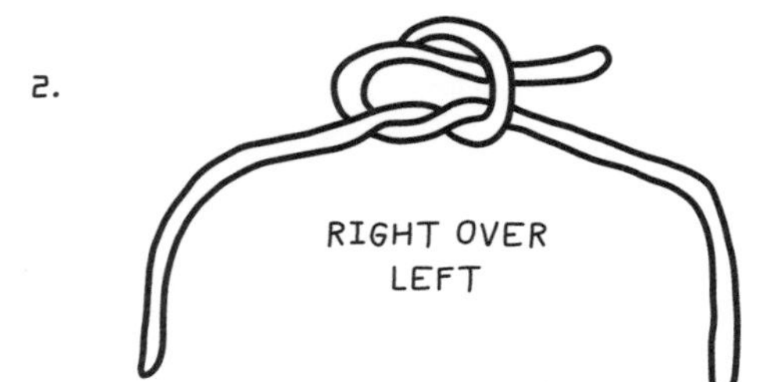

3.

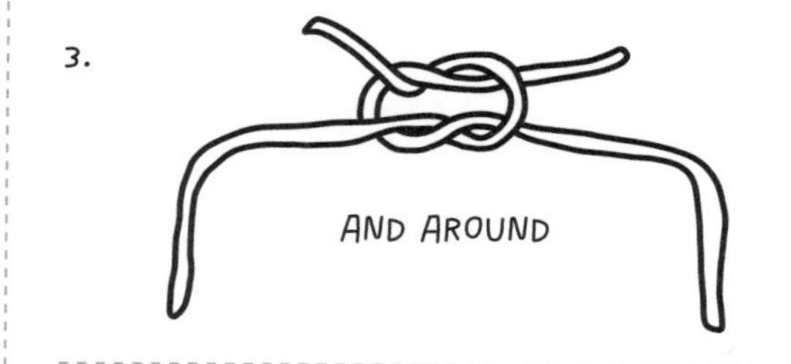

4.

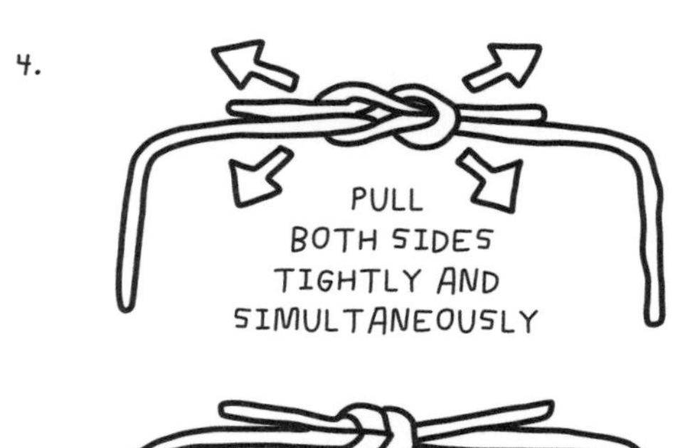

YOU HAVE TIED A SQUARE KNOT!

Double overhand: The double overhand is a great knot to use as a "stopper," kind of like the knots at the end of your hoodie strings (though those are probably single overhand knots). A double overhand is useful anytime you want to tie off a rope to keep it from fraying, or to keep an object from sliding off it, like knots at the end of a rope on the underside of a tree swing.

DOUBLE OVERHAND KNOT

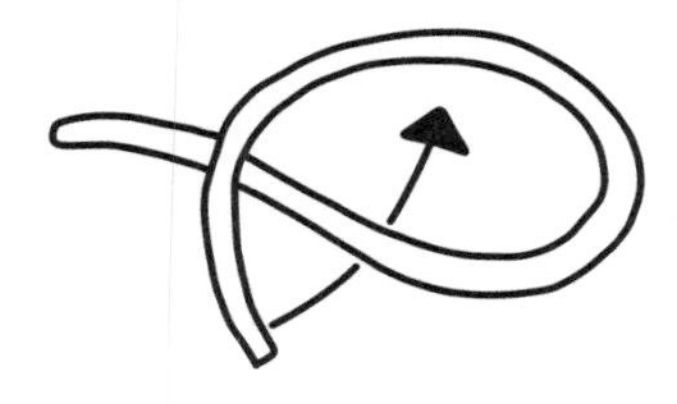

1. MAKE A LOOP AND FEED THE END INTO IT.

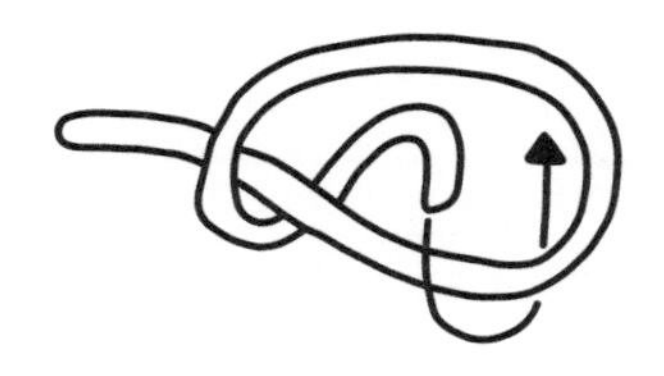

2. FEED THE END THROUGH THE LOOP ONE MORE TIME.

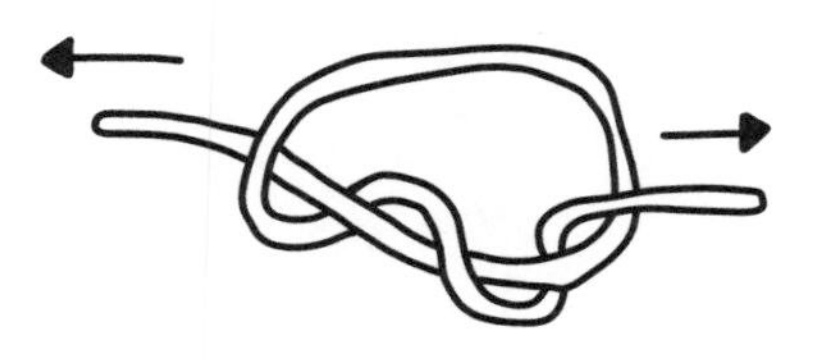

3. PULL BOTH ENDS TO TIGHTEN.

4. TA-DA!

Bowline: Pronounced "bow" (like an arrow's bow) and "Lynn" (like the name), the bowline is often called the king of knots because of its longtime use and versatility. The bowline is a fixed loop, meaning when you tug on either the loop or the other end, the loop will never loosen or tighten. In fact, the bowline can also hold a heavy load in its loop and still be easily undone. It is popular for rock climbing, sailing, and on farms to loop around animals' necks without tightening. You can also use a bowline as a loop on one end of a clothesline, with a taut-line hitch at the other end to tighten the line. I remember how to tie a bowline with the adorable and helpful rabbit story!

BOWLINE KNOT

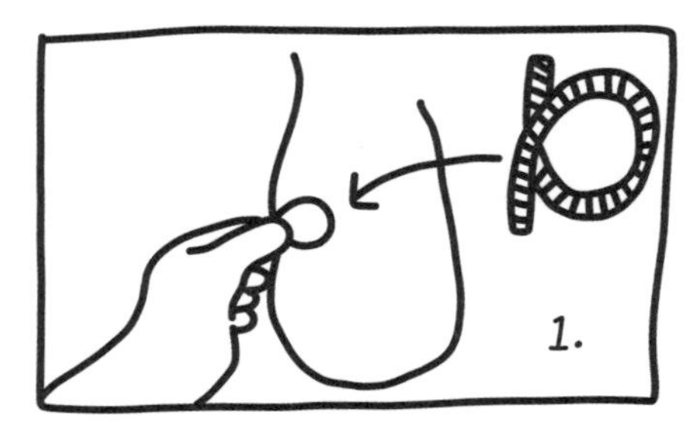

MAKE A RABBIT HOLE.

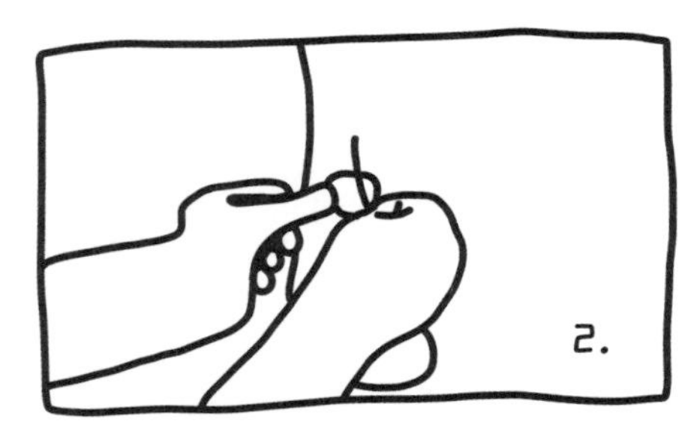

THE RABBIT COMES OUT OF ITS HOLE.

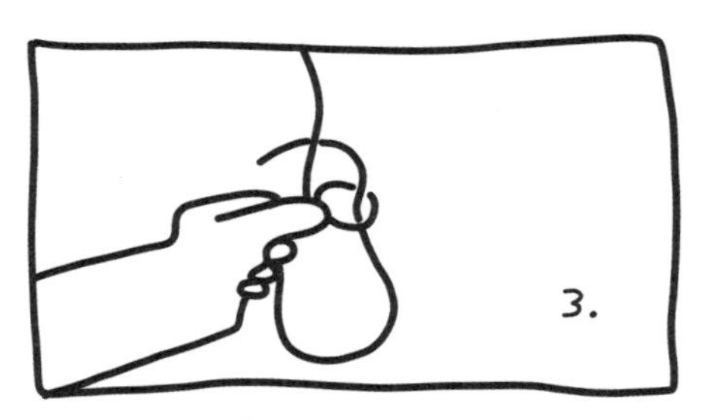

THE RABBIT RUNS BEHIND THE TREE.

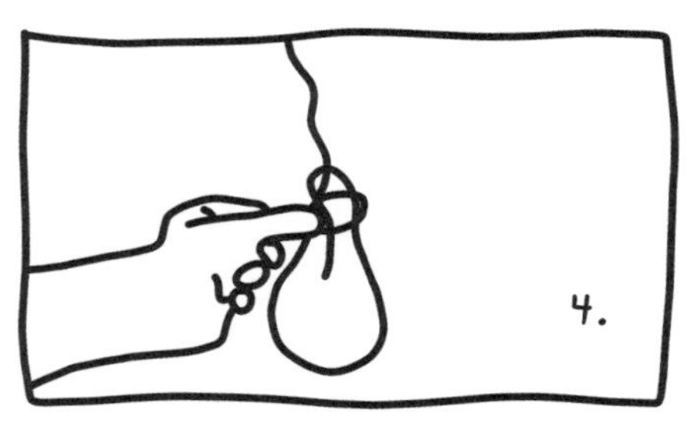

THE RABBIT RUNS BACK DOWN ITS HOLE.

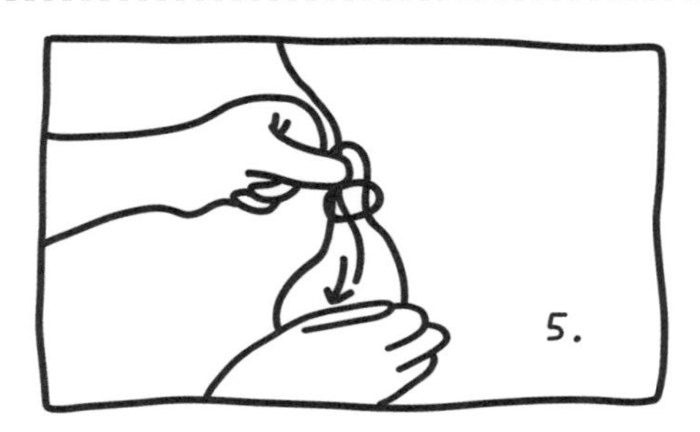

PULL ENDS TO TIGHTEN.

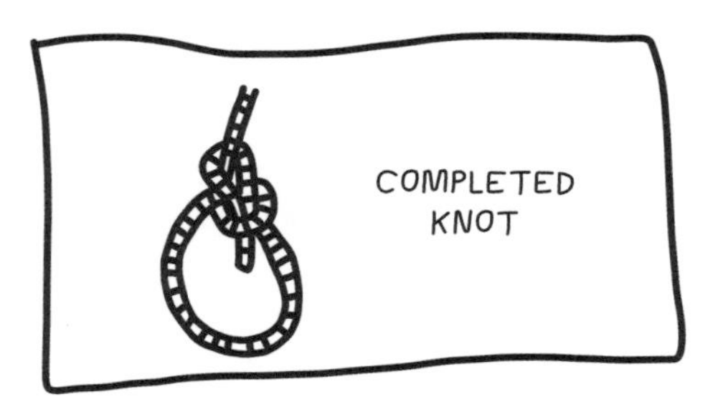

Taut-line hitch: Use a taut-line hitch when you have to pull a line from one point to another and then tighten it. You might use this knot for one end of a clothesline or to tie down a tent or tarp. When you tie a taut-line, tighten it by tugging on the loose end, which will slide the knot up and down the line to tighten it.

TAUT-LINE HITCH KNOT

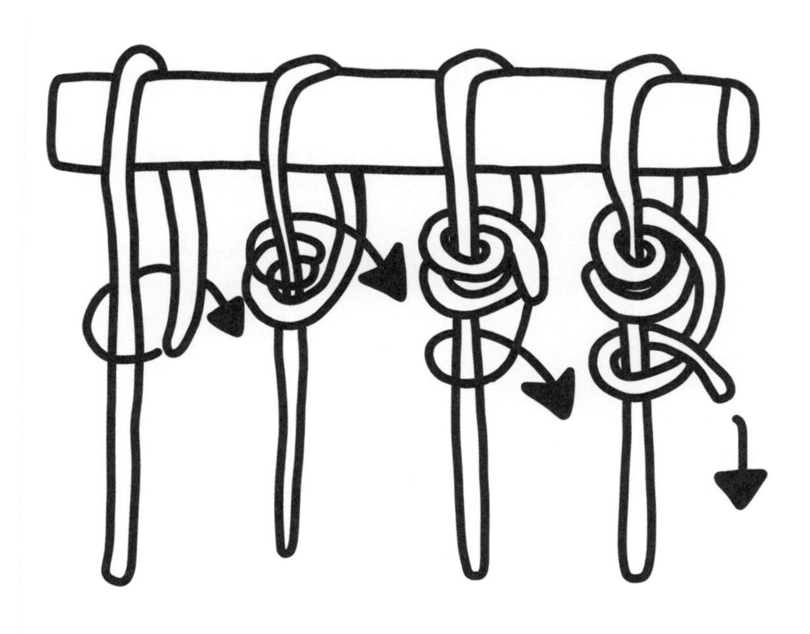

HOW TO WIND UP AN EXTENSION CORD

A rat's nest of cords and wires is a mood killer, especially when you have work to do! If you're using an extension cord, it can be super helpful to wind it up afterward and store it in an orderly fashion.

While you could just make a bunch of circular loops, that technique can result in kinks in the cord when you unwind it. Here's a foolproof way to wind up and store an extension cord so you aren't stuck in a giant tangle when you need it next. This is particularly helpful with extra-long extension cords, like 25- to 100-footers.

Step 1. Hold the end of the cord in your nondominant hand.

Step 2. With your dominant hand, extend an arm's length of cord, then bring it back to meet your nondominant hand as a loop.

Step 3. Repeat, but in the other direction, making a loop on the opposite side (like you're making two bunny ears!).

Step 4. Repeat your loops back and forth until you have about 5 feet of cord left. You should now have a bunch of bunny ears hanging out both sides of your hand.

Step 5. Combine the two bundles of loops together, and wind the remaining cord around the middle to gather and secure it.

Step 6. With about a foot left, bring the end of the cord up through the top loop. You can use that top loop to hang the cord on a wall, or simply store the whole bundle.

To unwind, loop the tail back through the top loop, unwind the middle "belt," and pull the cord as you need it. It should unwind and release smoothly, without knots or kinks!

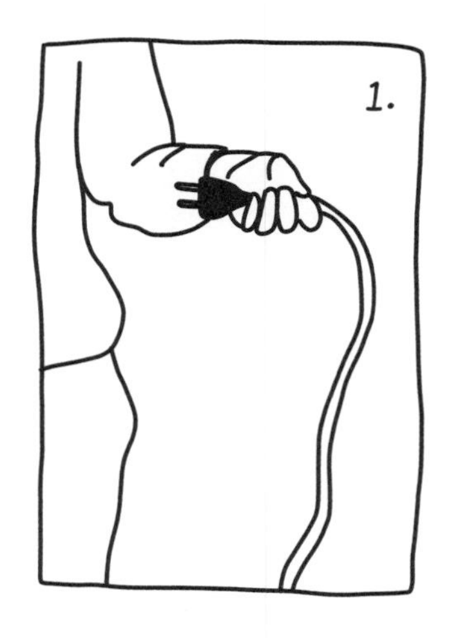
1.

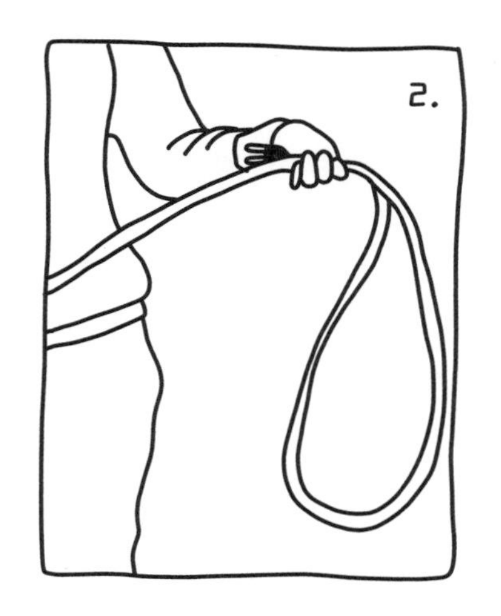
2.

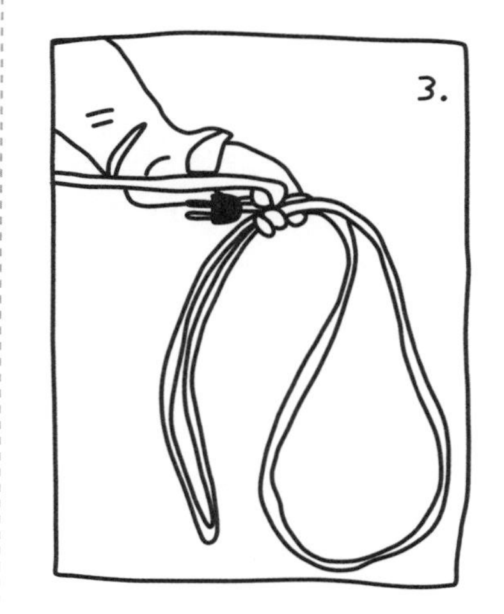
3.

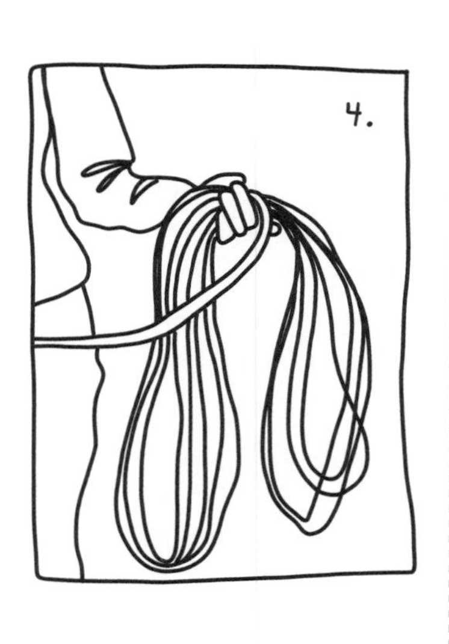
4.

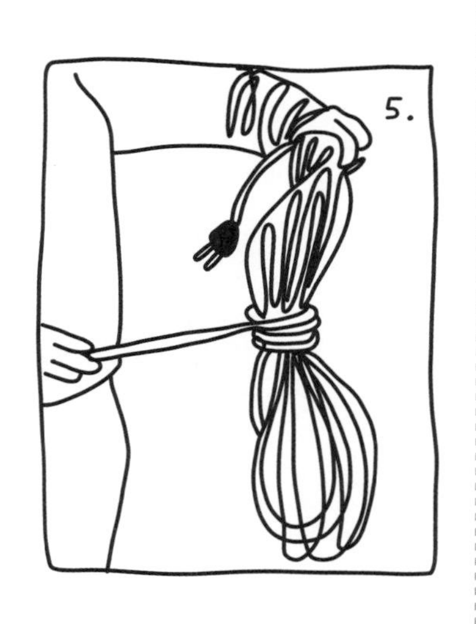
5.

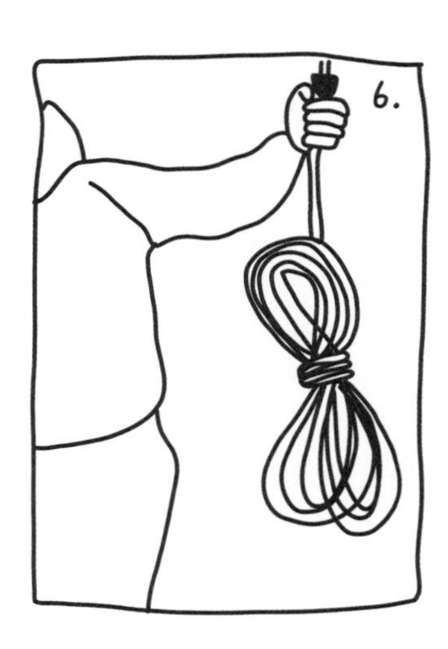
6.

HOW TO FIND STUDS WITHOUT A STUD FINDER

Most homes are built with wood-framed walls, which are then filled with insulation and covered with plywood, drywall, and, finally, paint. But the structure that keeps our walls from falling down are vertical 2×4s called studs (in most commercial buildings, they use metal studs instead). These wooden studs in your home are almost always placed 16 inches apart "on center," meaning you can measure 16 inches from the center "spine" of one to the center of the next. For some projects, especially when you need to hang something on a wall or secure a heavy piece of furniture to a wall, you will want to find a stud to nail or screw into, as it is the strongest part of the wall. If you only nail or screw through the thin layer of drywall, you don't have much to grab on to, and your painting or bookshelf is more likely to come crashing down.

Here's the trick: You can't see the studs through the wall unless you have X-ray vision, so you need a way to locate them using your other skills and senses. Okay, sure, you can go buy an electronic stud finder, which detects changes in the wall's density to tell you where the studs are. Or you can do it the old-fashioned way, which is much more fun and will ensure that you can do it anytime, anywhere, without a gadget to help you. And all you need is a hammer.

Start from the corner, or the end of the wall. The first stud is probably about 16 inches from the end (you can just estimate this). Tap on the wall with the hammer, lightly, moving horizontally as you tap. When you tap a spot without a stud behind it, you'll hear a hollow *tap-tap* sound. When you hit a stud, it will change to a lower, solid *thud* sound. Boom! You've found a stud! Mark this spot with a piece of tape. You can also use a level or plumb bob to extend this spot vertically so you have a whole line showing the stud. Now that you've found one, you should be able to find them all, as they should each be 16 inches apart. You can measure 16 inches and then find the next one, or just keep tapping.

If you're having trouble distinguishing the different *tap-tap* and *thud* sounds, here's another trick. The drywall laid over wooden studs is attached using screws, which are magnetic! Use a refrigerator magnet and slowly run it across the surface of the wall. You'll feel a tug when you run over a screw, meaning you're also over a stud!

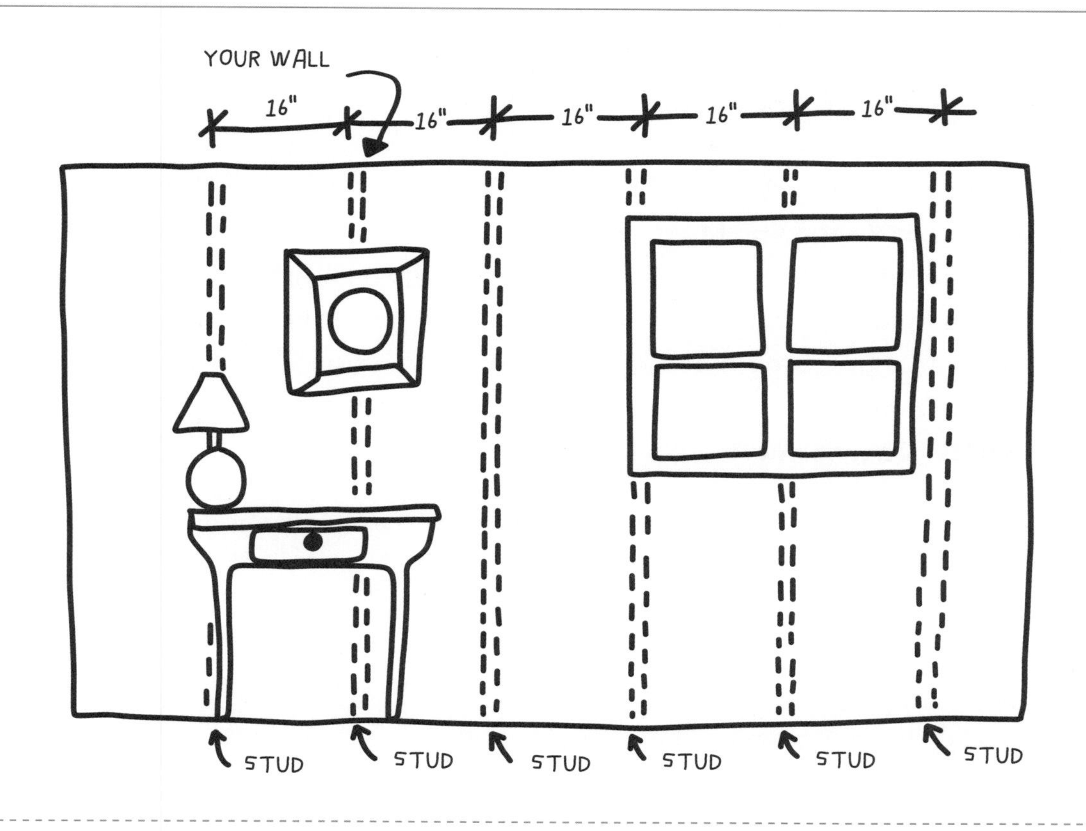
YOUR WALL
16"
16"
16"
16"
16"
STUD
STUD
STUD
STUD
STUD
STUD

"TAP! TAP!"
NOT ON A STUD.
VS.
"THUD."
ON A STUD! YAY!

HOW TO HANG A FRAME ON THE WALL

When I was a young girl, I used to spend weekends rearranging my room (and drawing the accompanying blueprints), framing and hanging artwork, and curating my "art collection" on the walls of my room. I got really good at hanging stuff. There are so many kinds of hardware for hanging items on walls, including some heavy-duty anchors, toggle bolts, and molly bolts (page 65). However, for light-duty jobs such as hanging framed artwork or paintings, these types of hangers should do the job just fine. Here's how to use them:

Step 1. Attach the hanging hardware to your frame. Some frames come with the hardware already attached; if not, add it yourself. You can use a sawtooth hanger in the center of the top edge of your frame, or two D-rings (one on each side of the frame attached with a wire—use your taut-line hitch knot from page 211!).

Step 2. Locate your hang spot on the wall. Find a stud if you can, since you just learned how on page 214! Mark your hang spot with an X.

Step 3. Using a hammer, nail your hook hanger into the wall at your hang spot. There are hook hangers rated for different weights. They have one, two, or three nail holes, depending on the weight they will have to hold.

Step 4. Hang it! If you're using a sawtooth hanger on the back of your frame, just hook the teeth onto the hook hanger (for lightweight frames, you can also forgo the hook hanger altogether and just hang the sawtooth hanger on a nail). If you're using a wire and D-rings, locate the midpoint of the wire and tug it upward, toward the top of the frame, noting the distance from the top of the frame to the top of the wire. This will help you locate where to place your screw or hanger on the wall relative to the top of the frame. After you've installed the screw or hanger in the wall, set the center of the wire on the hanger and adjust until level.

Step 5. Check your level. Use a torpedo level to make sure the top of your frame is level and adjust as necessary. Now stand back and admire your handiwork (and your artwork!).

HEAVY-DUTY TIP! If you're hanging something extra heavy, or oversize, behold the miracle that is the French cleat! The French cleat is a type of hanger that uses two interlocking steel plates—one is attached to the wall, and one is attached to your painting or frame. This is one of the strongest hanging devices. We used one to hang a 3-foot-wide framed etching of a girl holding an axe over the reception doorway at Girls Garage.

HANGING HARDWARE

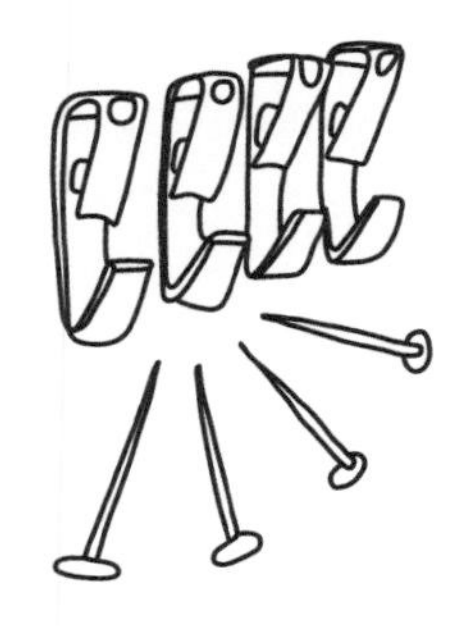

D-RINGS AND WIRE

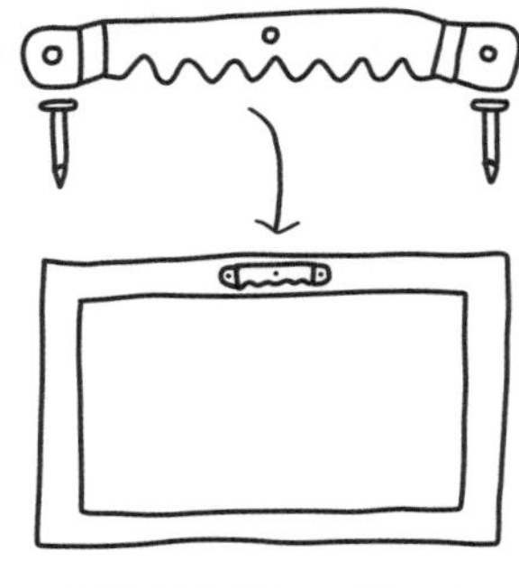

SAWTOOTH HANGER

HOOK HANGERS

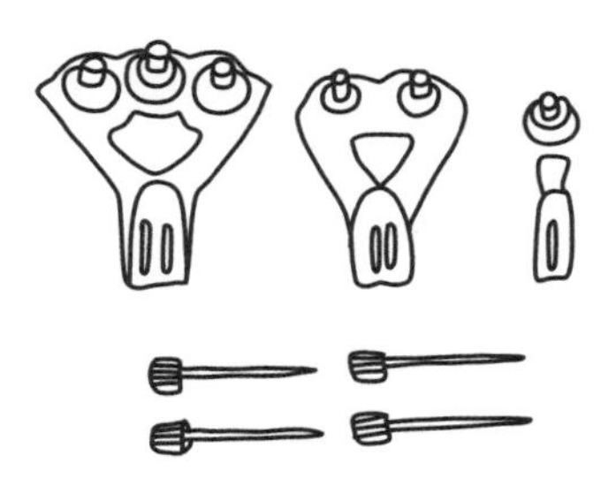

CHECK LEVEL

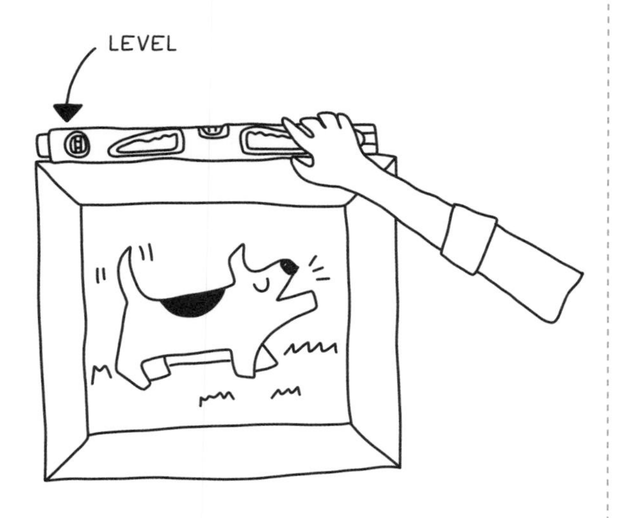

FRENCH CLEAT

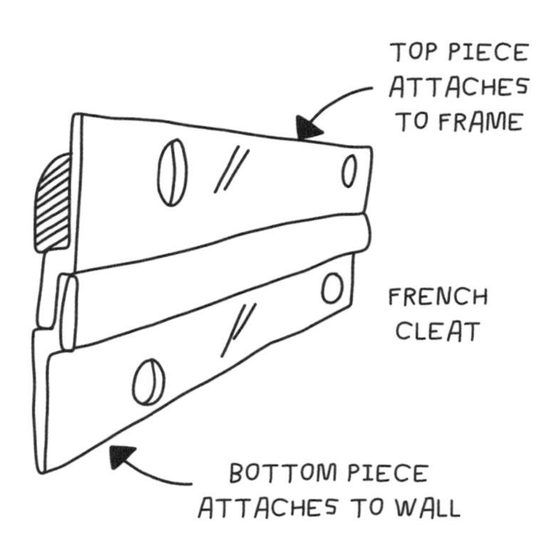

HOW TO PAINT A WALL

When I had my heart broken for the first time, I went to the paint store, bought a gallon of lovely buttercup yellow paint, and painted an accent wall in my bedroom. There's nothing like a fresh coat of paint to brighten up any room (and mood)! Here's how to do it:

Step 1. Pick your paint. I like an eggshell finish and an interior low- or no-VOC grade ("volatile organic compounds," which are toxic). In general, shinier finishes (satin, semigloss, or gloss) are easier to clean and reflect more light. Matte and eggshell finishes reflect less light but can be harder to clean. You can get paint and primer (which a paint store can color-optimize for you), or a paint and primer in one.

Step 2. Use painter's tape to tape off edges, ceiling, and baseboards. I prefer 2-inch-wide blue painter's tape (or the green FrogTape brand) to cover all the edges you don't want to paint. Tape the corners and edges where the ceiling meets the wall, where one wall meets another (if you're doing only one accent wall), any baseboards or molding you need to work around, and over light switches and outlets (if you can, though, it's best to remove the light switch and outlet covers altogether, usually by removing two small screws that hold them in place). Use a fingernail or spatula tool to press the tape down firmly all the way along the line. This ensures that you don't end up with spots where the paint can sneak under the tape.

Step 3. "Cut in" with a paintbrush and paint about 6 to 8 inches from the edge of the wall. A 2- to 3-inch-wide angled trim paintbrush works best for this step. Paint using short strokes that are perpendicular to your tape, and then a few long strokes parallel to your tape. When you get really good at this, you can go tapeless and cut in freehand with a super-steady hand.

Step 4. Paint the rest of the wall with a roller in an up-and-down zigzag pattern. Using a paint roller and paint tray (and a long handle extension so you can reach all the way to the ceiling), paint the entire wall in giant zigzag motions, all the way to the edge of the perimeter you just cut in.

Step 5. Repeat steps 3 and 4 with another coat. Depending on your paint color and type, you may even need a third coat.

Step 6. Remove the tape while your last coat is still wet. If you remove the tape after the paint has dried, the paint might peel off with the tape.

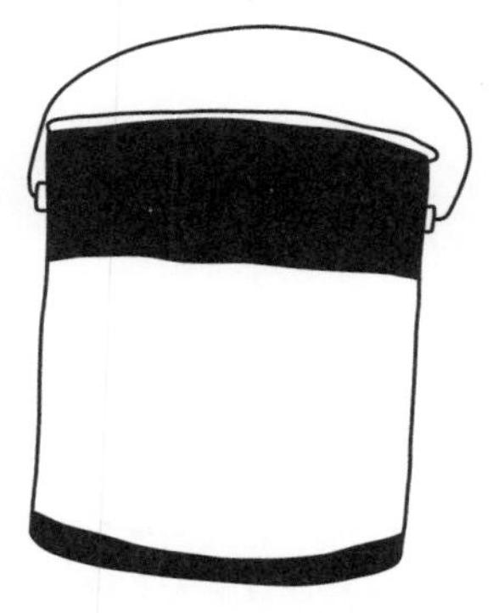

1. PICK YOUR PAINT.

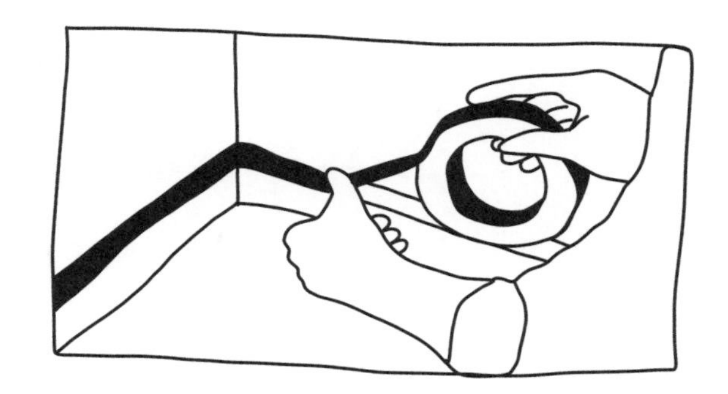

2. USE PAINTER'S TAPE TO TAPE OFF EDGES, CEILING, AND BASEBOARDS.

3. "CUT IN" WITH A PAINTBRUSH AND PAINT ABOUT 6 TO 8 INCHES IN FROM THE EDGE OF THE WALL.

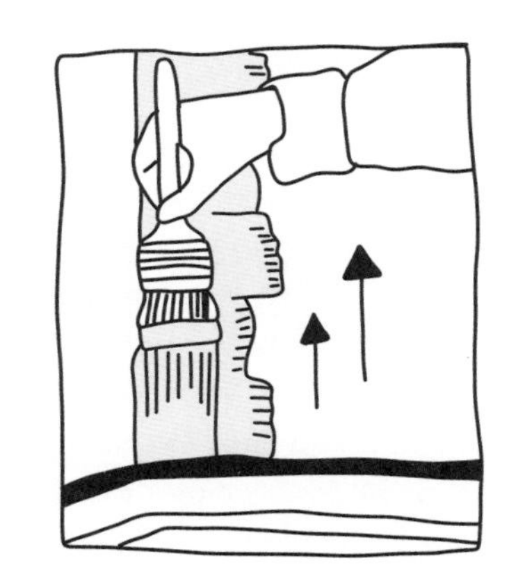

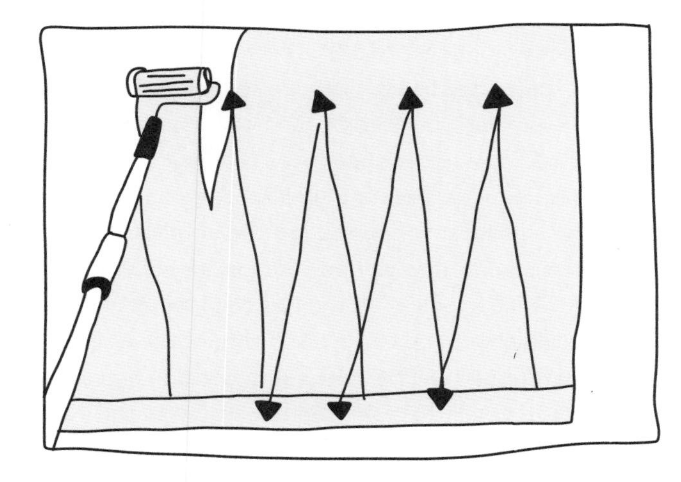

4. PAINT THE REST OF THE WALL WITH A ROLLER IN AN UP-AND-DOWN ZIGZAG PATTERN.

5. REPEAT STEPS 3 AND 4 WITH ANOTHER COAT.

6. REMOVE TAPE WHILE YOUR LAST COAT IS STILL WET.

HOW TO FLIP A CIRCUIT BREAKER BACK ON (AND WHY IT TRIPS IN THE FIRST PLACE)

Remember that time you were drying your hair and you had your phone charging in the same outlet, and all the lights on, and music playing on your computer, which was also plugged in? And then all the power shut off and you were left with wet hair in the dark? Mystery solved: You probably tripped your electrical circuit. Here's why that happens, and how to fix it.

SAFETY ALERT!: Touching or manipulating any part of your electrical system puts you at risk of electric shock or electrocution. Do not touch your electrical panel or outlets without the supervision of a skilled and experienced adult. Don't bring any metal tools near it, and if anything looks burnt-out, smoky, or covered in soot, call an electrician.

All the electricity that powers your house runs through one central box called the electrical panel (or breaker box). If you look around your house, in a closet or garage, or somewhere inconspicuous, you should be able to spot a metal box mounted on the wall, with a door that opens. This is your electrical panel.

Inside the electrical panel, you'll likely see two vertical columns of switches, with odd numbers going down the left side, and even numbers going down the right (yours may vary slightly, but this is the usual configuration). Each switch corresponds to a particular set of outlets, or a room in your house, or a set of appliances, and hopefully these switches are labeled to show you what's what (example: #16 = bathroom).

Each switch is also designed to carry a certain amount of electrical "load." The limit of this load is called its breaking capacity. If you try to run more electricity through a circuit than its breaking capacity can hold, the circuit "breaks" to protect you from electrical fires and other awful consequences. Breaking capacity is measured in amperes (amps), which is the amount of electrical current that passes through a specific point in one second.

You probably have something called a GFCI (ground fault circuit interrupter) outlet in your kitchen or bathroom, which is the most commonly tripped breaker. A GFCI is installed in places where the outlet might (dangerously!) come in contact with water. The GFCI helps regulate the amount of current flowing, and when there is an imbalance, it trips the circuit.

If you have too many gadgets or appliances plugged into a GFCI (or any) outlet, and all those appliances require more electrical power than your circuit can hold, the circuit will blow and all your appliances will shut off. When this happens, first unplug all the appliances you were using that overloaded the circuit in the first place. Then go find your circuit breaker.

Take a look at all the switches, most of which should be in the ON position. The circuit that tripped will be in a neutral position between ON and OFF. To reset, flip the switch to the OFF position first, just to be sure it gets a fresh new connection, and then flip it back to ON.

Your circuit should be back up and running (but remember not to plug in all those appliances at once again!).

Lastly, a tripped circuit can have some of the same symptoms as a short circuit, which is different and much more dangerous. A short circuit happens when a "hot" wire in your outlet touches another hot wire or a neutral wire, which "shorts out" your circuit. This can happen if you have loose or sloppy wiring. You'll know you have a short circuit if there is smoke or soot or evidence of something burning behind or around your outlet covers. If this is the case, don't turn your circuit back on—call an electrician.

FUN FACT!

The number on the switch itself, in this case 15, corresponds to the number of amps the circuit breaker can hold.

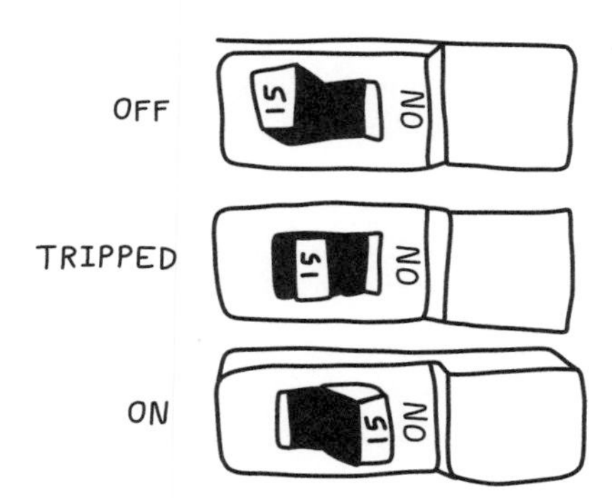

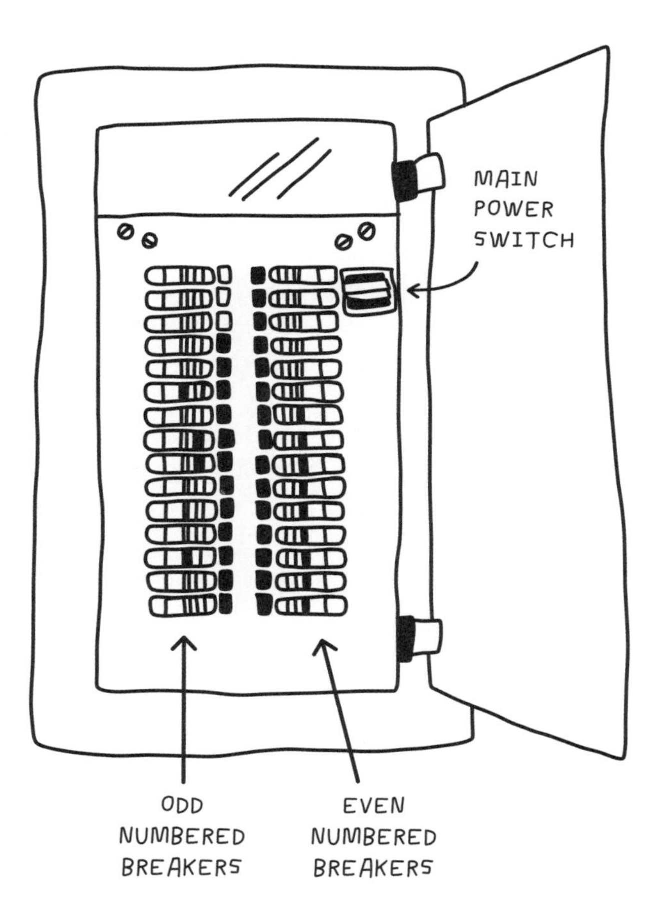

HOW TO RELIGHT THE PILOT LIGHT ON YOUR STOVE OR WATER HEATER

There are some appliances around your house that are not powered by electricity, but by gas. A slow stream of gas flows to these appliances, and that gas burns to create heat. The most common gas appliances are stoves, water heaters, and furnaces, and all three of these kinds of appliances have a pilot light, which is basically an always-burning mini fire so the appliance can start on demand, without having to light it every time. For your stove, this pilot light means that when you turn on a burner, the stove doesn't have to create fire; it just moves your pilot light fire up to your stovetop.

Sometimes this pilot light goes out for any number of reasons: a gap in the flow of gas, a strong gust of air, or little gas goblins (who knows?). In any case, the pilot light is not hard to relight. You just need to find it and also be careful and aware of gas flow and any potential gas leaks (which can cause gas fires and explosions!). Before relighting any pilot light, ventilate the area of any gas in the air by opening windows and doors. **If you can still smell strong gas odors, call your utility company and don't try to relight the pilot light yourself. Do not attempt to repair your appliances without the supervision of a skilled and experienced adult.**

To relight a gas stove: Gas range stoves have multiple pilot lights: one or two just under the burners (usually one pilot light per two burners), and one pilot light for the inside of the oven. You'll need to remove the burners and lift the cooktop, and open your oven door to access the pilot lights. If your surface burner pilot light is out, locate the pilot light gas feed and flame location, which is usually an obvious spot at the end of the gas feed, between two burners. Hold a match or barbecue lighter to this location, and the pilot light should reignite (gas is still flowing, unless you manually turned it off using the shutoff valve). The oven pilot light is usually located underneath or in the back of the oven, and you might have to remove the oven bottom to access it. It should be a similar obvious end to a gas pipe and can be relit with a match or longer lighter. With both the burner and oven pilot lights, blow out your match or lighter as soon as the pilot light is reignited.

To relight a water heater or furnace: Water heaters and furnaces have a different pilot light

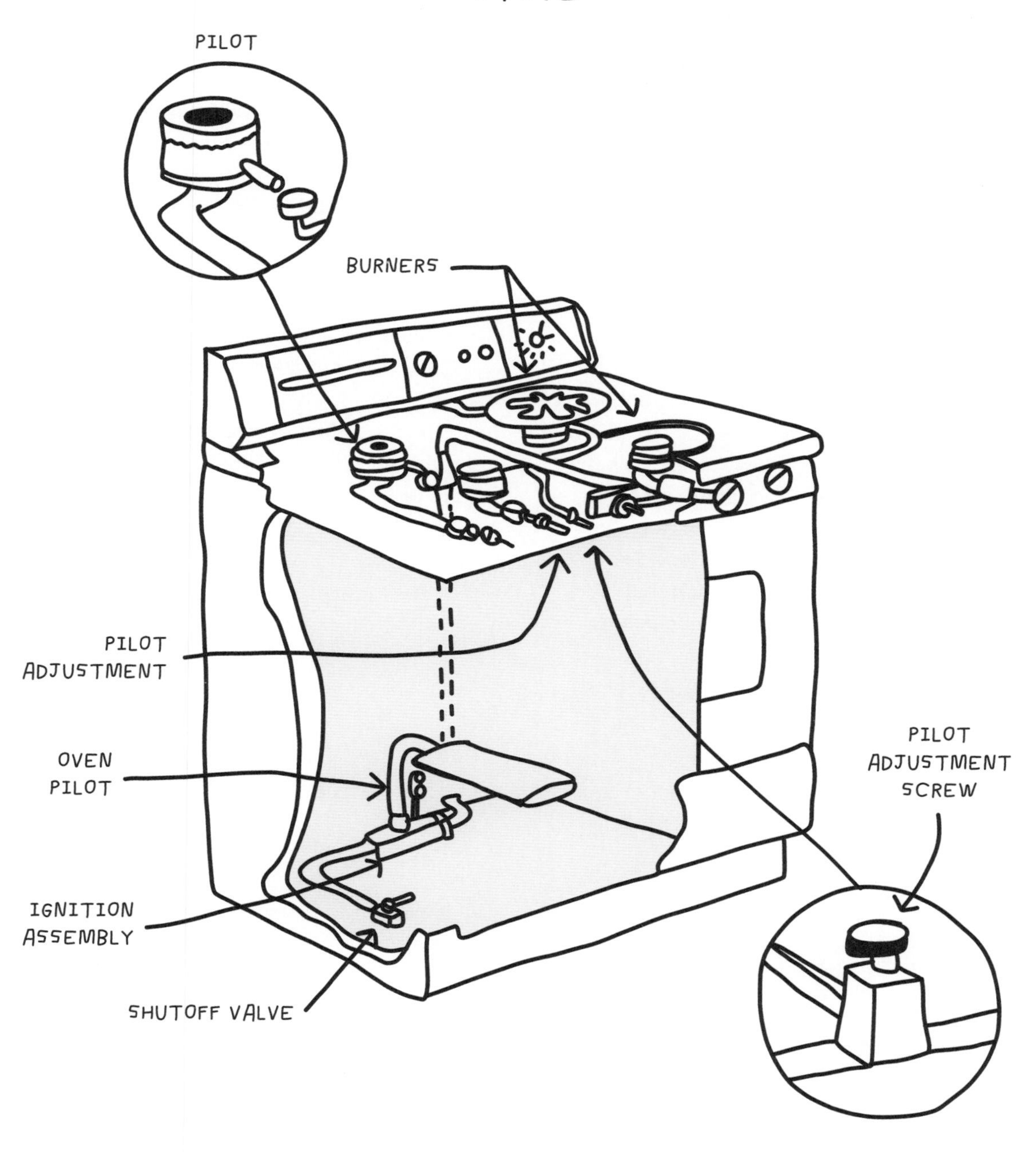
STOVE
PILOT
BURNERS
PILOT
ADJUSTMENT
OVEN
PILOT
IGNITION
ASSEMBLY
SHUTOFF VALVE
PILOT
ADJUSTMENT
SCREW

configuration than a gas stove. The exact buttons and gauges and dials vary from model to model, but the basic arrangement is the same.

First fact to know: The pilot light comes out of a small metal pipe (sometimes with a bend in it) and is next to its friend, the thermocouple. The thermocouple's job is to feel and measure the heat of the pilot light flame, and then signal the gas to flow to the burner. This pilot light and thermocouple unit should be easily accessible by removing the door or access panel toward the bottom of the heater or furnace unit. For some (older) water heater or furnace units, you'll need to light this pilot light manually. On other (newer) units, there is an automatic ignition button that will light it for you.

Go take a look at your unit. You'll probably see a dial or knob (the gas-supply knob), and you might also see a pilot-primer button and/or a pilot-ignition button (probably red). Some units might have a gas supply knob that is ALSO the primer button and can be pushed down. If you don't have an ignition button, remove the access-panel door so you can manually access the pilot light to light it with a long barbecue lighter.

Your gas-supply knob will likely have the settings ON, OFF, and PILOT. Turn the knob to OFF and wait five minutes (this clears out any lingering gas). Then turn it to PILOT.

If you have a pilot-primer button AND a pilot-ignition button, hold down the primer button to start a small flow of gas, then hit the IGNITE button. The IGNITE button will click to ignite the flame.

If you have to light the pilot light manually, make sure your knob is turned to PILOT, hold down the pilot primer button, if you have one, and then light the pilot light with a long barbecue lighter.

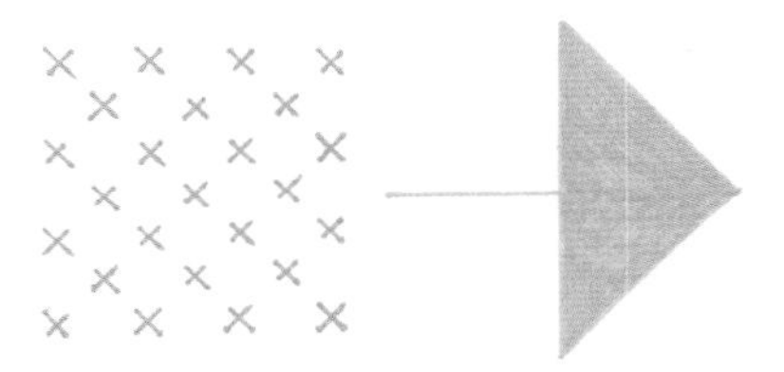

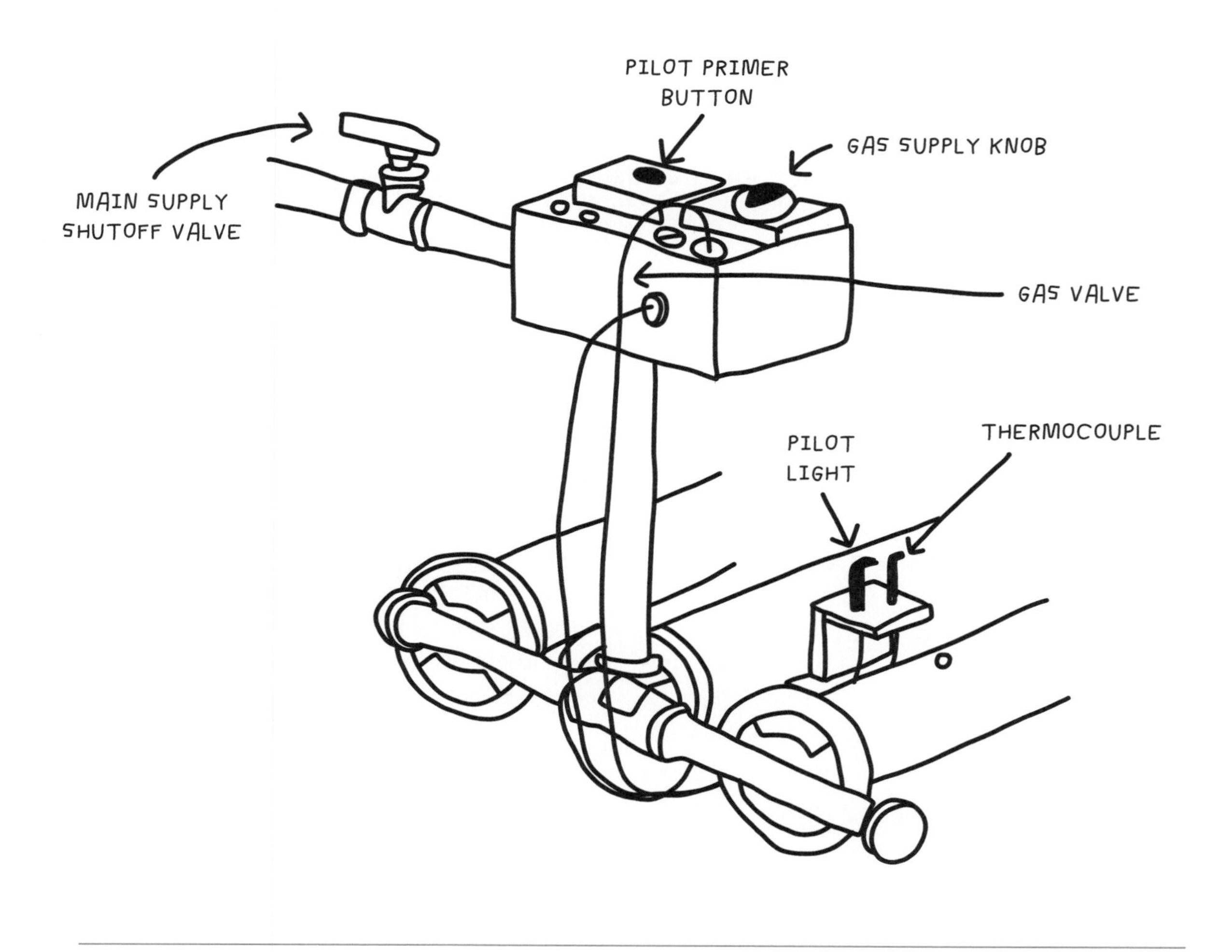
FURNACE OR WATER HEATER
PILOT PRIMER BUTTON
GAS SUPPLY KNOB
MAIN SUPPLY SHUTOFF VALVE
GAS VALVE
PILOT LIGHT
THERMOCOUPLE

HOW TO PATCH A HOLE IN THE WALL

Maybe you pulled a screw out of the wall and now there's a ragged hole. Maybe you opened a door too quickly and the knob punched through the wall. Or maybe, like me, you did a handstand in your living room and misjudged your distance from the wall, came down, and put your knee straight through the drywall. Anyone?

The good news is that holes in drywall (the most common wall material in most houses) are pretty easy to repair and patch. Whether it's a small hole or dent, or a doorknob- or kneecap-shaped hole, here's how to fix it.

Gather your tools and materials. You'll need a drywall spatula (also called a drywall knife), spackle (a tub of white plasterlike stuff) or joint compound (stronger than spackle, and sometimes pink), and a sanding sponge (120- to 150-grit).

For small holes (½ inch or smaller): This is such an easy fix! With your finger or the corner of your drywall spatula, smear a small amount of spackle on top of (and into) your hole. Use the spatula to smooth the surface of the wall so the spackle lies flat. Let it dry, then sand lightly with a sanding sponge. And here's a pro tip: If your hole has frayed edges (from the top paper layer of the drywall), make a dent in the hole with the back of a screwdriver, creating a small depression that your spackle will fill. This will leave you with a flatter-finished surface without evidence of that pesky frayed drywall paper.

SMALL HOLE

For medium holes (½ inch up to about 4 inches): Get a drywall patch kit (available at most hardware stores in various sizes, though most are about 6 inches by 6 inches). The patch kit is like a drywall bandage, made up of a square piece of metal mesh covered in a fabric mesh tape that is slightly larger than the metal part. Before you apply the patch, remove any extra debris from the edges of the hole.

Step 1. Place the drywall patch over the hole, sticky side down. Make sure that the entire hole is covered by the patch.

Step 2. Cover the patch with joint compound. Use a drywall spatula and joint compound (probably pink) to cover the entire patch, using a diagonal crisscross pattern and extending the compound a few inches beyond the patch. Let the joint compound dry. (Apply another coat if it still looks thin and you can see the patch through it.)

Step 3. Sand with a sanding sponge. Give the area a quick sand (not too much!) until smooth, and paint over it to match the rest of the wall.

MEDIUM HOLE

1.

PLACE DRYWALL PATCH OVER HOLE, STICKY SIDE DOWN.

2.

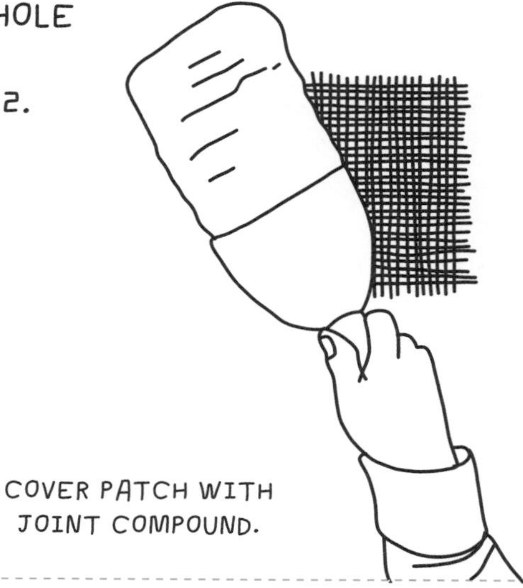

COVER PATCH WITH JOINT COMPOUND.

3.

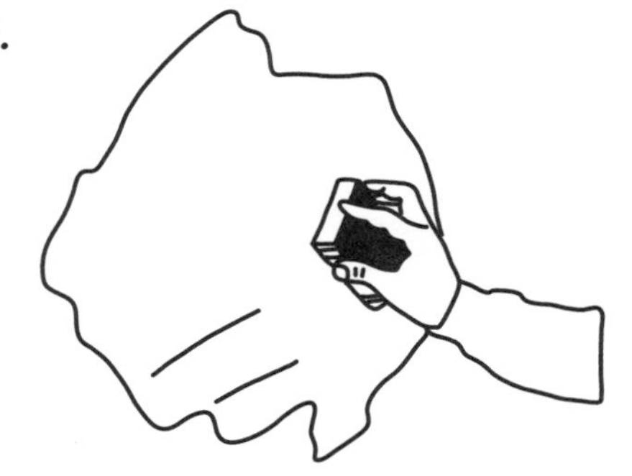

SAND WITH SANDING SPONGE.

For large holes (4 to 8-ish inches): Use a technique called the California patch. You'll need to buy or scavenge a small piece of drywall to make the patch, although some hardware stores may sell a premeasured California patch kit that includes this square of drywall. The drywall has white gypsum in the center, with paper-tape layers on either side, one of which you'll be peeling back later.

Step 1. Use a sharp box cutter and a straightedge to cut a piece of drywall that is about 2 inches larger on all sides than your hole (add 4 inches to both the length and the width of your hole's dimensions). So if you have a 4-inch by 4-inch hole, cut an 8-inch by 8-inch piece of drywall.

Step 2. Take that piece of drywall and measure in 1 inch from the edge, so you have a 1-inch frame around the edge. Score these lines with your box cutter.

Step 3. Snap off this 1-inch frame but leave the bottom layer of paper backing intact (you'll need it shortly!). You should now have a square piece of drywall with a 1-inch paper-backing perimeter. The size of the drywall piece that's left should be slightly larger than your hole.

Step 4. Now place the drywall patch over your hole facedown. Trace the shape of your drywall square onto the wall, around your hole (not including the paper backing). Then use a drywall saw or box cutter to cut that shape out of the wall.

Step 5. It's time to patch, and you should have a piece of drywall ready and a hole in the wall that matches its shape. Smear some joint compound *on the paper backing only*, then stick the drywall patch into the hole in the wall so the paper backing sits flat on the wall. Press the paper edges onto the wall around your hole. Using your spatula, cover the entire patch with your joint compound, smoothing the edges and making the patch (mostly) disappear. You might need two coats of compound (let it dry between coats). Sand and repaint!

LARGE HOLE

1.

2.

3.

4.

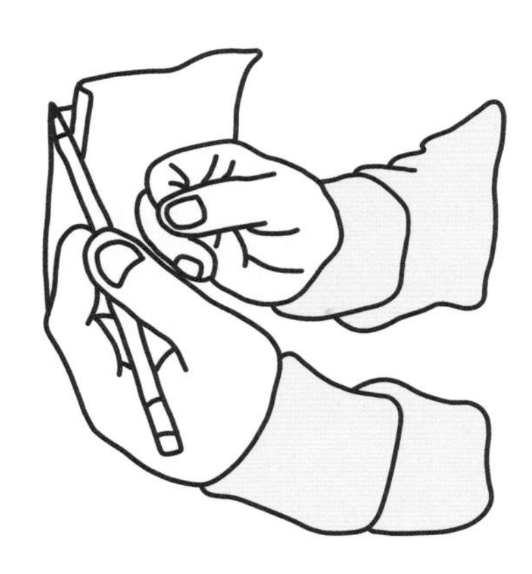

5.

HOW TO FIX YOUR RUNNING TOILET

We've all heard it: that incessant running water sound that won't quit, hours after your last flush. If your toilet is running, it won't flush properly, and it's a huge waste of water (and money). And it's also an easy fix! First let's look at how a toilet works.

Your toilet is made up of a handful of parts: the bowl, the tank, and the mechanisms inside the tank. The toilet tank holds your "per-flush" amount of water—let's say 1 gallon. When you flush the toilet using the handle, water is released from the tank into the bowl, and the bowl's contents are released using gravity into the sewage system. Then new water flows in from your water pipes to refill the tank.

Inside the tank are a few parts that help start and stop the flow of water to refill your tank with the correct amount. Let's look at the mechanics of a single flush.

First, you flush using the toilet handle. The handle is connected to a chain inside the tank, which pulls up on a flapper at the bottom of the tank, lifting it to release the tank water into the bowl. When the tank is empty and the bowl has refilled, the flapper closes again and seals shut.

Now your tank needs a refill. Water flows in from your water line, into the fill valve and into the tank. It will also flow through the fill valve into the overflow tube. The overflow tube keeps your tank from overflowing and moves any extra water into the bowl. As the water level in the tank rises, the float rises. Your float may be a big ball, or a smaller black collar around your fill valve. Once the float rises to the correct water line, the flow of water should stop.

When the flow of water does not stop and your toilet keeps running, it's because there is a leak somewhere in this system. There are a few reasons this might be happening. Here's how to diagnose and fix it:

First, don't mess with your toilet without the supervision of a skilled and experienced adult. Safety glasses are definitely recommended, and make sure to wash your hands thoroughly after you've fixed the problem.

Step 1. Check your flapper. If your flapper isn't creating a tight seal, water can keep running through your tank and into your bowl. Your flapper might be discolored or warped, and if so, it can easily be cleaned or replaced.

Step 2. Check your chain. The chain that attaches your flush handle to the flapper might be disconnected, or it's attached to the flapper but too short or too long. If the chain is too short, the flapper will always be held up, unable to seal itself. If it's too long, the extra chain length might be wedging itself between the flapper and the seal. If the chain length is wrong, unhook it and move it up or down one chain-link length.

Step 3. Check your float ball. If you have a float with a ball on it, it may be that your float ball position is off. When the tank operates properly, the water should fill up to a point just below the top of the overflow tube. Look at where your fill line is. If the fill line is above that point and water is flowing into the overflow tube, your float ball is too high. Lower it by bending the rod it's attached to, or by adjusting the float screw if your toilet has one.

Step 4. Lastly, check your fill tube and fill valve. Your fill tube should be securely connected to your fill valve on one end and your overflow tube on the other. If that's not the problem, check to see if your fill valve is either leaking or dirty or covered with debris. You can remove the fill-valve cap, or the entire valve tube, clean it or replace it, and reinstall it with tight screws.

Hopefully one of these strategies does the trick! There are other less-common problems that might lead to a leaky toilet, like cracks in the actual tank or bowl, or specific hardware that needs to be replaced, but, in my experience, it's usually a quick fix.

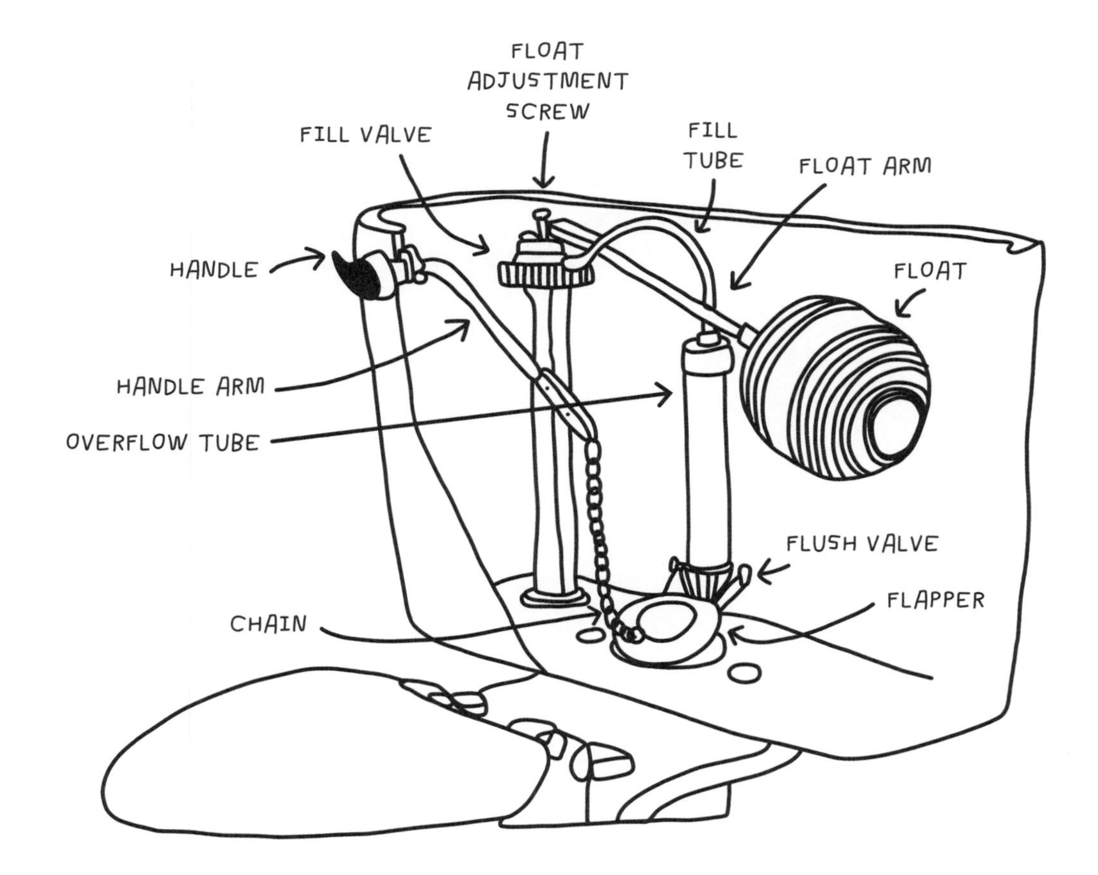

HOW TO START A SMALL ENGINE (LIKE A LAWN MOWER)

Lots of helpful garden tools, like lawn mowers and weed trimmers, are powered by small engines and motors. These engines run on fuel that you can refill yourself. Both the lawn mower and the weed trimmer have a motor that rotates and spins a blade or string that cuts your grass or weeds.

The engine of a lawn mower or weed trimmer is like a mini version of a car engine and works using the same principles. The engine has a combustion chamber, which needs fuel and air to flow into it, a spark to ignite it, and a way for exhaust and fumes to get out. Fuel and air enter the combustion chamber through the carburetor, which sprays a mixture of fuel and air. When the engine is running and burning fuel, it powers a motor that turns the blades.

Do not operate or repair a lawn mower without the supervision of a skilled and experienced adult. Always read the manual for your specific tool (the following tips are general guidelines).

Step 1. To start a lawn mower or weed trimmer, find the helpful starter pull cord. The starter cord usually has a handle that is attached to the long handlebars of the machine. You'll probably first need to push the primer button on your machine, which releases a small amount of fuel into the carburetor to start your engine.

Step 2. Next, release the handle of your pull cord from the handlebars. It will naturally want to recoil or wind up. This "recoil start" action is a common way to start many engines, including chain saws and other small machines like go-karts and small ATVs.

Step 3. Allow the cord to wind itself back up so the handle is close to the engine. In one motion, pull the cord quickly and firmly away from the engine. You may need to do this a few times. As the cord pulls away, it rotates a crankshaft to start the engine turning fast enough for the ignition to generate a spark. Once you've generated that spark, it will ignite the fuel, and you're up and running!

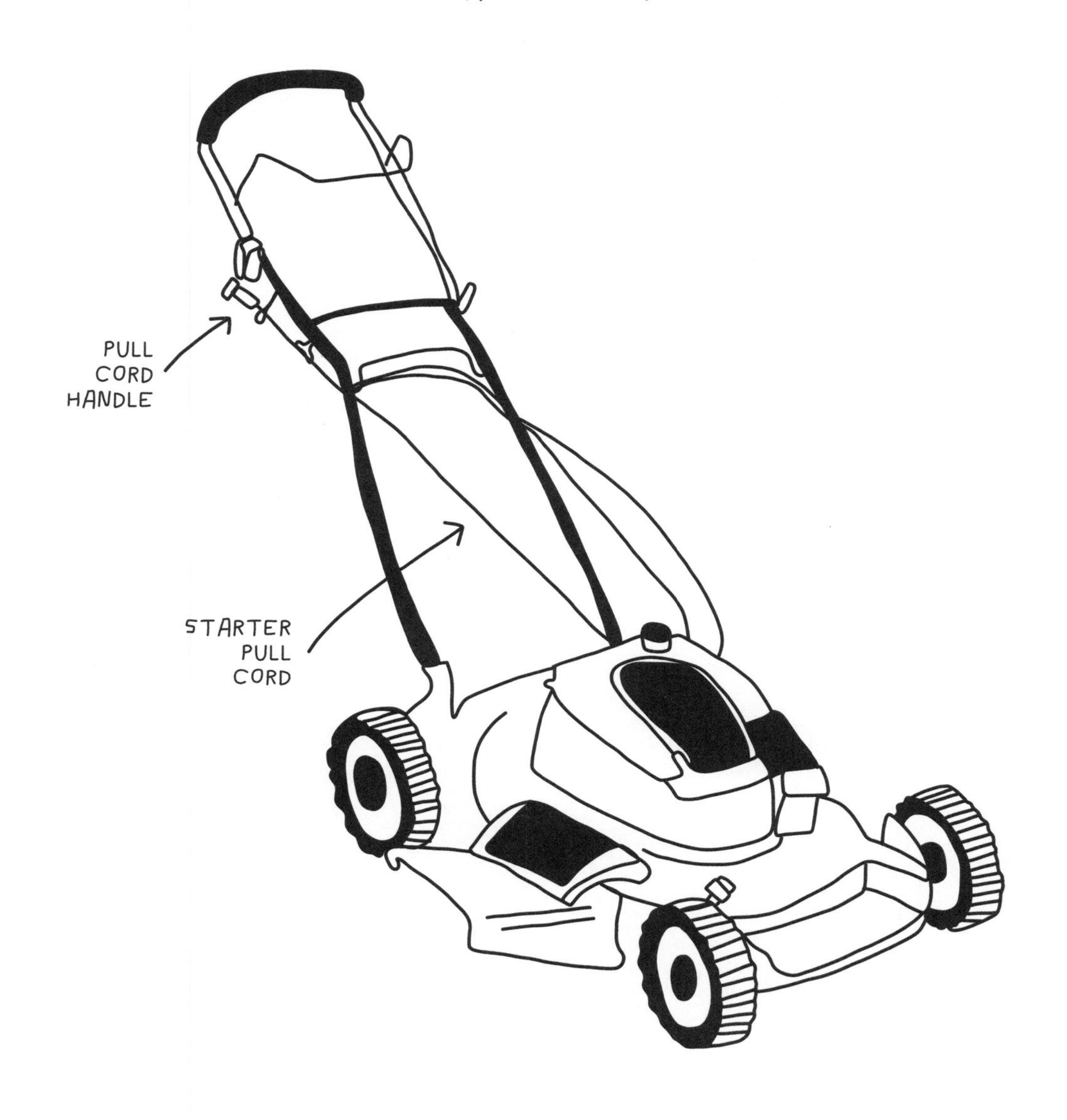
LAWN MOWER
PULL CORD HANDLE
STARTER PULL CORD

HOW TO PICK A LOCK (FOR GOOD, NOT EVIL!)

There are so many kinds of locks out there, and if you're a professional locksmith, you could likely rattle off twenty different types without even thinking—and know how to pick them. But the most common lock you'll find in most doorknobs and padlocks is called a pin tumbler. This particular type has two sets of pins (like top and bottom rows of teeth) that must be lined up in order to be turned by a key.

Here's how it works. A pin tumbler lock is a cylinder, or barrel, with a few important parts inside. The most common pin tumblers have four to six vertical chambers, each one holding a spring, a driver pin, and a key pin. The pins move up and down against the spring in the chamber. The driver pins are like the upper row of teeth; the key pins are like the lower row of teeth.

When there's no key pushed inside the cylinder, the driver pins and key pins are pushed down by the springs, below the "shear line," which is the point where the cylinder rotates to unlock the lock. When the correct key is inserted, the ridges in the key's profile push these sets of pins up so that the line where the "top teeth" (driver pins) meet the "bottom teeth" (key pins) is exactly at the shear line, and you can rotate the cylinder to unlock the lock. With an incorrect key, the pins won't line up with the shear line and you can't turn the lock.

But since we're talking about picking a lock, we can assume you don't have the correct (or any) key and have to find another way to get those pins lined up with the shear line. And that's what lockpicks are for! There are lockpicks in many shapes and sizes, with tips shaped like diamonds, balls, snowmen, squiggly snakes, zigzag rakes, and more. You'll also need a tension wrench, which is the tool you'll use to turn the lock once you've used your picks to line up all the pins. Your job with the picks is to feel the pattern of the pins, and push them up manually to line up with your shear line so you can turn the lock. And you have to do this blindly—you can't actually see the pins! A great way to practice this is with a transparent practice padlock, in which you can stick your picks and see how they manipulate each pin.

Now might also be a perfect time to mention the ethics of lock-picking. As smart and compassionate women, it should go without saying that (a) lock-picking is an awesome skill and a helpful exercise in understanding how the mechanical world works, and (b) we shouldn't use lock-picking for criminal purposes, only for good ones (like breaking your sister out of the closet she accidentally locked herself in).

INSIDE A PIN TUMBLER LOCK

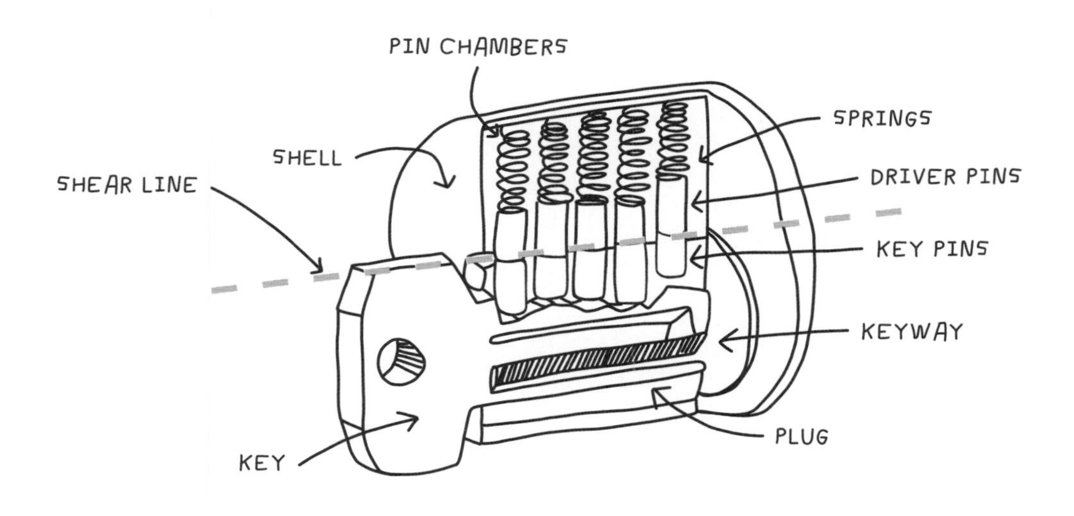

HOW A PIN TUMBLER LOCK WORKS

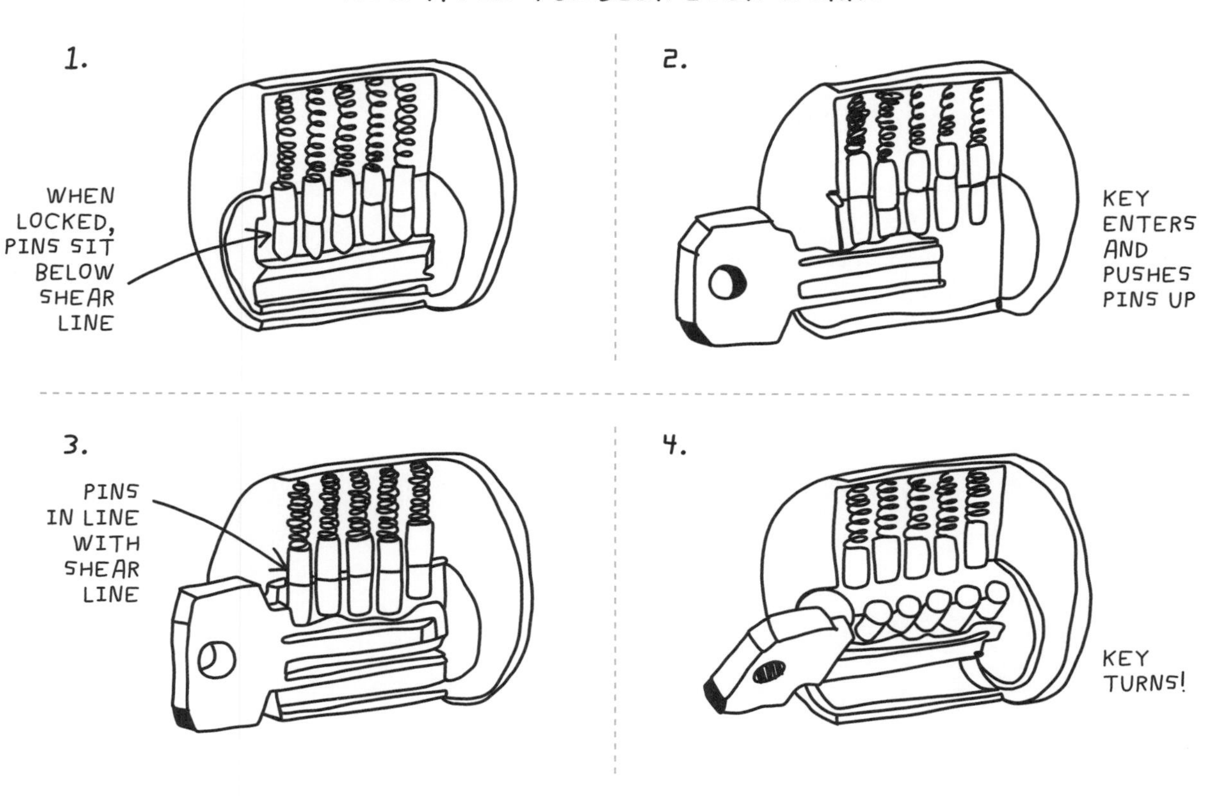

Once you've mastered the transparent practice lock, you're ready for the real world! You can start with small padlocks or doorknobs and work your way up to harder locks. Here is a helpful checklist for how to crack a lock in real life:

Count the pins. Use a diamond or ball pick and simply feel for how many pins you're dealing with. Insert the pick into the lock and run it from the back of the inside of the lock to the front, counting the clicks along the way. Odds are you've got four to six pins to conquer.

Make a mental map. Once you know how many pins you have, go back in, from back to front, and try to figure out which pins are hanging the lowest (and need to be pushed up to the shear line the most). You can feel which are hanging lowest by the tension on each pin: the lower ones will be harder to push up against their spring. Make a mental map—pictured like a mountain range—of the pin pattern, based on what you feel.

Pick your picks! I personally like to use two picks at a time: a diamond-shaped one for the lowest-hanging pins, and a rake or snake to try to even out the others. You'll also need a tension wrench, which is usually a small L-shaped or Z-shaped flat tool that comes in a lock-picking kit, to turn the lock.

Crack it! This is done mostly by trial and error (and with patience). First, get your tension wrench in place, inserting one end of it into the lock. Then grab a few picks and go to town. I like to start with my lowest-hanging pins, and try to raise and hold them using a diamond or ball pick to what feels like the shear line. With a rake or snake pick, I fiddle with the levels of the remaining pins, trying to turn the tension wrench as I go. When the pins are lined up with the shear line, your tension wrench will turn and pop the lock!

One beautiful but frustrating fact about lock-picking is that even if you follow these steps perfectly, there is a certain amount of luck involved. With practice, you will, of course, get better at feeling the subtle differences in pins. You might get frustrated, but keep at it!

FUN FACT!

Actually, a history lesson! The first pin tumbler lock can be traced back to ancient Egypt, around 2000 BCE. Wooden versions of it were used in Egypt to secure valuables and homes. And the pin tumbler hasn't changed much since then; it still works on the same principles! In the mid-1800s, Linus Yale and his son Linus Yale Jr. invented and patented the modern pin tumbler that we still use today. You might notice that many locks still say YALE on them—thanks, Linus!

TRANSPARENT PRACTICE LOCK

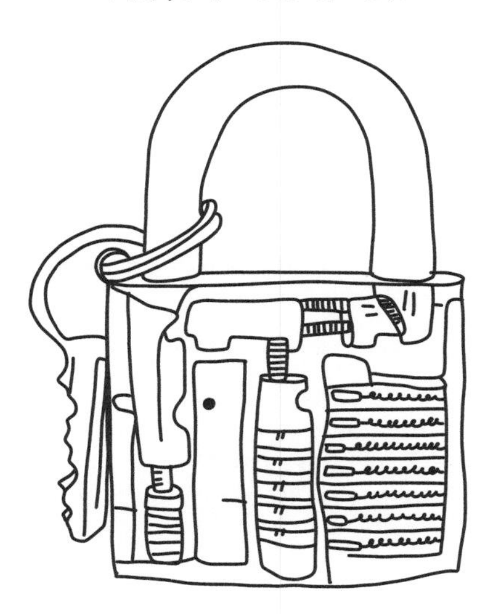

HOW A PICK WORKS

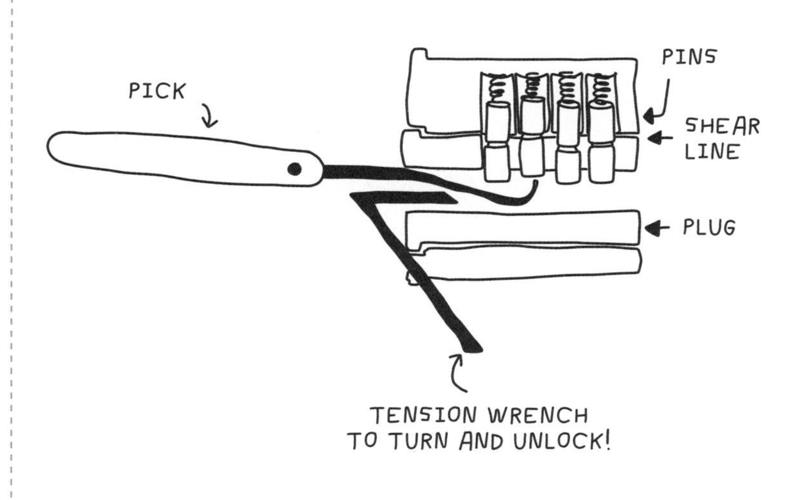

TYPES OF PICKS

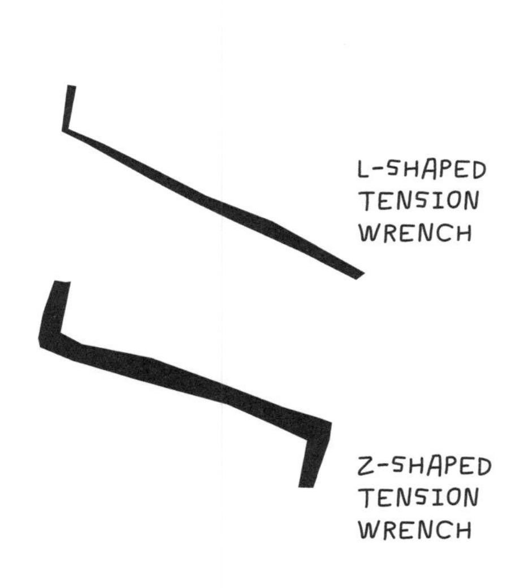

HOW TO JUMP-START A CAR

Some of you readers may not have a driver's license yet, but that doesn't mean you can't be a car mechanic for your friends and family! Anyone who has ever left a car door open overnight (and thus the overhead light on) knows that waking up to a car with a dead battery is a serious bummer. But jump-starting a car is not hard at all. You only need a friend with a working car and a set of jumper cables. And of course, make sure you have a grown-up buddy for supervision and safety.

I like to keep a set of jumper cables in my car at all times so I can get myself out of a dead-battery situation and also help others who need a jump. Jumper cables are a simple gadget: they have a red cord (positive charge) and a black cord (negative charge) and metal clips at each end that make a complete circuit when connected to a power source. The idea here is to use the jumper cables to take the power from a live battery and cycle it into the dead battery using the full circuit formed by the cables.

SAFETY ALERT! Touching or manipulating your car battery and jumper cables puts you at risk of electric shock or electrocution. Do not attempt to jump-start a car without the supervision of a skilled and experienced adult.

Find a helpful friend, or flag down a generous passerby who is willing to lend you their car battery for a few minutes. Park both cars nose to nose, as close as possible so your jumper cables can reach from one car to the other. Turn off both cars. Pop the hoods and locate the battery in each car and the positive and negative points on each one (they should be labeled). The positive point on your battery might be labeled in red, and might also have a cover that needs to be removed or lifted to access the metal point.

The jumper cables need to be attached in a specific order and always to metal.

Step 1. Clip red (+) to the positive (+) on your dead car. A good way to remember this is with the rhyme "Red, dead."

Step 2. Clip red (+) to the positive (+) on your helper's car.

Step 3. Clip black (-) to the negative (-) on your helper's car.

Step 4. Clip black (-) to a bare metal part on the dead car like an unpainted screw or knob, but do not clip to anything near or on the actual battery. This grounds the connection.

Step 5. Once all your clips are attached, start your helper's car, and let it run for a few minutes.

Step 6. Next, start your dead car (which should now start!). Let both cars run for a few minutes. Your dead car should be purring like a kitten!

Step 7. Last, it's safest to remove your cables in the reverse order you attached them (4-3-2-1).

CAR BATTERY

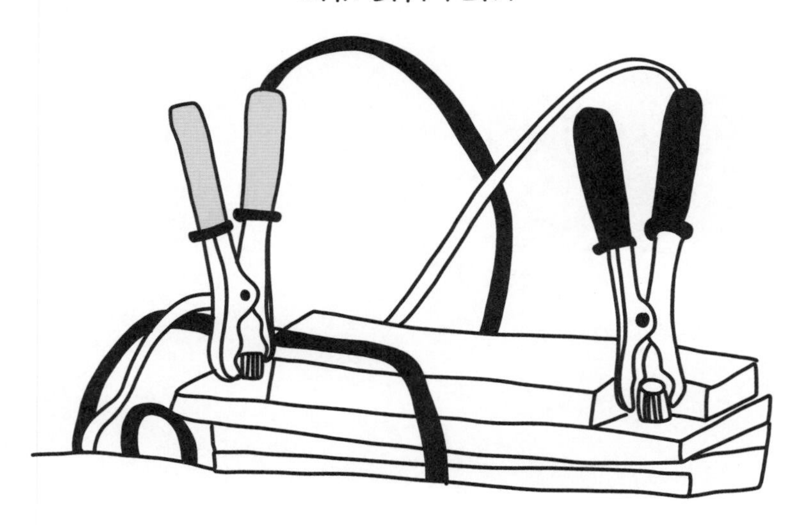

HOW TO CONNECT JUMPER CABLES

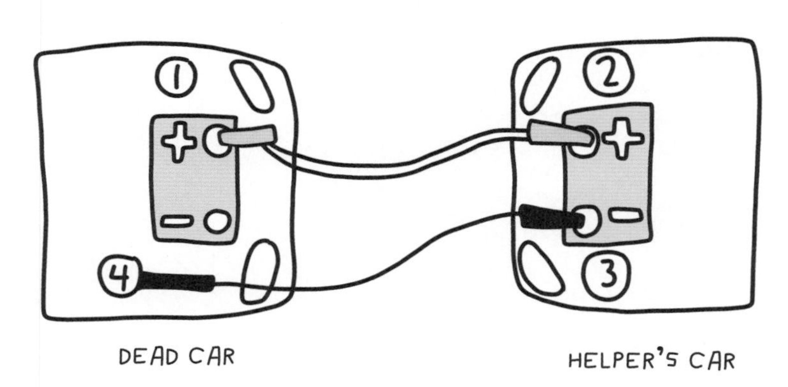

1. CONNECT RED TO DEAD +
2. CONNECT RED TO HELPER +
3. CONNECT BLACK TO HELPER -
4. CONNECT BLACK TO BARE METAL ON DEAD CAR (BUT NOT ON OR NEAR THE BATTERY!)
5. START HELPER CAR. WAIT A FEW MINUTES.
6. START DEAD CAR.
7. DISCONNECT IN REVERSE ORDER (4, 3, 2, 1).

PATRICE BANKS

Founder of Girls Auto Clinic

Upper Darby, Pennsylvania

I came across Patrice's work and thought, *Now, there's a brave woman who dropped everything to pursue the crazy idea she had.* In her own life, she used to describe herself as an "auto airhead," without any knowledge of how a car worked or how to fix it. And she hated the experience of taking her car in for service and being treated like an idiot because she was a woman. So she quit her job as an engineer at DuPont, enrolled in automotive school, and opened her own garage called Girls Auto Clinic, catering specifically to women and their cars. I love her grit and vision, and that she takes so much time to speak about her work and inspire other women and girls. She's the author of a great book called *Girls Auto Clinic Glove Box Guide,* and she will leave you wanting to take on the world (and also change your own motor oil).

"I was once that woman who didn't know who to take my car to for repairs and service or which mechanic I could trust. I felt taken advantage of plenty of times by repair shops, regardless of size and name.

"But I was always interested in understanding how things work and how they are put together. I got started working with cars in 2010 because I didn't know how and I wanted to see if I could do it. I was tired of feeling powerless. I decided to find out why it was so intimidating. I wanted to prove to myself that women can build and fix things.

"So I decided to take control of the situation. I went back to school at night for automotive technology and started working as a mechanic at a garage. I opened my own garage, Girls Auto Clinic, to solve a big problem in women's lives, including my own. I want to change the relationship women have with their cars by giving each and every woman driver a unique and memorable automotive experience!

"Now I take apart cars to find problems, and put them back together to make sure they last as long as possible, as safely as possible. We use so many tools here in the shop: a wide variety of wrenches, hammers, drills, and screwdrivers. We also use different types of oil and fluids, as well as lifts for the cars.

"Along the way, the only person who challenged me or held me back was me. Anytime I felt unable to do something, it was because I wasn't approaching it correctly. I don't believe anyone can stop me. I am always trying to grow and take on new challenges.

"So believe! And know you can do it. Don't let them stop you. Find people who believe in you, who will mentor you. They are out there. Just keep talking to people until you find them. And don't forget what you are here for. If people give you problems, just remember you aren't here for them. Never lose confidence in yourself. Know, like you know the sun's going to rise, that you can do it."

HOW TO CHANGE A FLAT TIRE

Who needs roadside assistance when you can do it yourself? (Even if you aren't of driving age yet, you'll be a huge asset to your friends and family the next time you find yourself on the side of the road with a flat tire.) The process for swapping out a flat tire with a spare is not that complicated, and everyone should know how to do it! Make sure you have an adult for supervision and safety before beginning.

Step 1. Park your car on a flat surface and make sure the emergency brake is on. If you are on the side of a road or highway, pull safely to the shoulder as far away from traffic as possible. Turn off your car, turn on your hazard lights, and put the emergency brake on. Find your spare tire (usually in the trunk) and flat kit (which should include a jack to raise the car and a tire iron to loosen and tighten the lug nuts on your wheel).

Step 2. Loosen the lug nuts using your tire iron, but don't take them off yet. There should be five total lug nuts on each tire. If you have a

TOOLS FOR CHANGING A TIRE

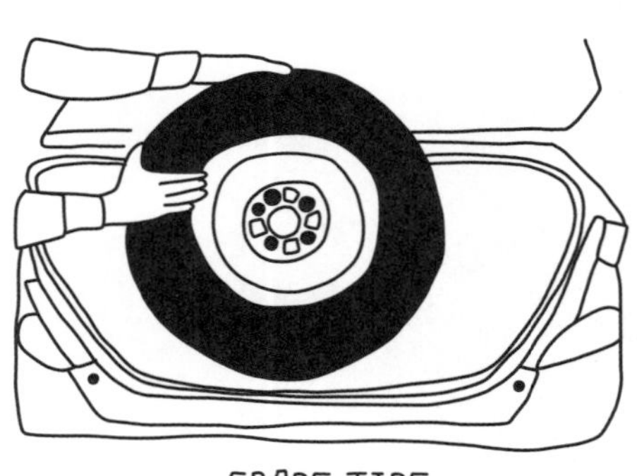

SPARE TIRE

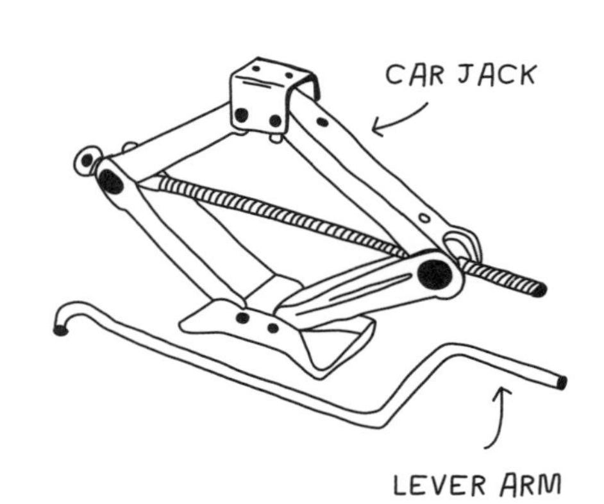

TIRE IRONS

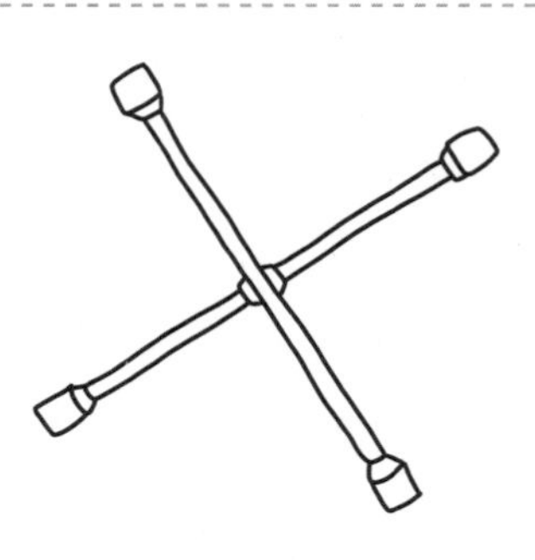

hubcap, you will need to take it off before you can access the lug nuts.

Step 3. Jack it up! To raise your car using the jack, you'll want to place the jack in a secure spot on your car's frame, close to the flat tire. Your best bet is to look at your car's user manual to determine the best spot for the jack. Your jack is most likely a scissor jack, which is a diamond shape that expands to raise the car. It should have a lever arm, which you can attach to it and crank or turn to open the jack and raise the car. Raise the car so the bottom of the flat tire is about 6 inches off the ground. **Don't stick any part of your body under the car!**

Step 4. Remove the lug nuts and the flat tire. Now you can undo the nuts entirely, and keep them somewhere safe! You'll need them in a moment to put on the spare. Remove the flat tire by grabbing the sides of the tire and pulling it directly toward you.

[continued]

HOW TO CHANGE A FLAT TIRE

1. PARK YOUR CAR ON A FLAT SURFACE AND MAKE SURE THE EMERGENCY BRAKE IS ON.

2. LOOSEN THE LUG NUTS USING YOUR TIRE IRON, BUT DON'T TAKE THEM OFF YET.

3. JACK IT UP.

4. REMOVE THE LUG NUTS AND THE FLAT TIRE.

Step 5. Put the spare tire on. Line up the holes with the bolts, push the spare tire all the way on, and then reattach your lug nuts by hand. Don't fully tighten the lug nuts just yet.

Step 6. Lower the car and remove the jack. Use the lever arm to lower the jack until the new spare tire is touching the ground. Then go ahead and remove the jack and put it safely back where you got it.

Step 7. Replace and tighten your lug nuts in a star-shaped order. Use your tire iron and tighten all your lug nuts, working in a star pattern so you don't overtighten two adjacent nuts (this can make the tire crooked and dangerous on the road). Most tire irons have a long arm you can step on and really put your weight into it. You want your lug nuts tight! If you have a hubcap, you can put that back on now.

You've done it! Before you drive off, it's good practice to check your tire pressure in the new spare tire as well. Reference your user manual to determine what pressure it should be and use a tire gauge to check. It may also indicate on the spare tire's wheel rim what the recommended pressure should be. Also remember that spare tires, sometimes called donuts, are not the same as regular tires, so they're not meant to be driven on for too long. Don't go above 50 miles per hour or more than 50 miles in total distance before replacing it with a real tire.

5. PUT THE SPARE TIRE ON.

6. LOWER THE CAR AND REMOVE THE JACK.

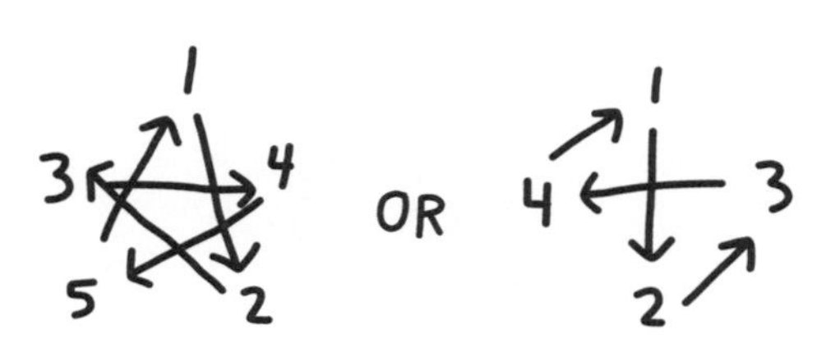

7. REPLACE AND TIGHTEN YOUR LUG NUTS IN A STAR-SHAPED ORDER.

BUILDING PROJECTS

One of the hardest parts of assembling this book was curating a short list of awesome projects that take advantage of your newfound excitement about tools. I could have included a hundred projects!

The eleven projects included here are all doable by YOU, with inexpensive and easy-to-find materials, and you have the option of using power or hand tools. They use many of the essential skills you've learned and combine various materials and hardware, as well as an array of tools. I've done all these projects with my Girls Garage gals already, so they are girl-approved for your enjoyment. Many of them integrate larger building concepts (like stud framing in the doghouse project, or the principles of formwork in the concrete planter project), so I hope they have a ripple effect that inspires you to come up with your own bigger projects.

If I had it my way, I'd be standing there next to you as you embark on these projects. But I'm there in spirit, and don't be afraid to ask for help, or to work with friends or family members to solve problems along the way.

YOUR OWN GO-TO TOOLBOX

Every builder needs a toolbox! Not only is a toolbox a great way to organize and consolidate all your essential tools, but it's a perfect first project to practice basic building skills.

Materials

- x 6-foot-long 1×10 board (you'll only need about 4 feet total, but it's best to buy a 6-foot board in case you need to recut a piece)
- x 6-foot-long 1×6 board (same as above: you'll only need about 3 feet, but over-buy just in case)
- x 1×2 board (you'll need 16½")
- x 1⅝-inch wood screws
- x Paint or paint pens and a paintbrush for decoration, if desired
- x Wipe-on polyurethane and rags (if you want to clear-coat your wood)
- x Disposable nitrile gloves to protect your hands when applying polyurethane

Tools

- x Pencil
- x Tape measure
- x Speed square
- x Miter saw or circular saw
- x Jigsaw or handsaw/backsaw
- x Clamps
- x Sanding block or sponge and sandpaper
- x Drill
- x 3⁄32-inch drill bit for pilot holes
- x Driver and bit to match your screws

SAFETY CHECK!

- ♦ An adult builder buddy
- ♦ Safety glasses (wear at all times!)
- ♦ Ear protection, if using a power saw
- ♦ Dust mask
- ♦ Hair tied back
- ♦ No dangly jewelry or hoodie strings
- ♦ Sleeves pushed up to your elbows
- ♦ Closed-toe shoes

Step 1. **Cut your bottom, end, and side pieces.**

Get your pencil, tape measure, and speed square ready!

Starting with your 1×10 board (which is really only 9¼ inches wide, not 10 inches), cut about 1 inch off the end with the chop saw or circular saw. This first cut doesn't have to be accurate—you're just cleaning up the end of the board so you have a nice square edge to start with.

Mark and cut 3 pieces, 1 measuring 16½ inches (bottom of the toolbox), and 2 pieces at 13 inches (tall ends of the toolbox). Remember: Measure and cut one piece, then measure and cut the next.

[continued]

Cut the long sides of the toolbox out of your 1×6 board (which is really only 5½ inches wide, not 6 inches). Just like you did with the 1×10 board, cut off about 1 inch from the end so you have a clean edge. Then cut 2 pieces, each 16½ inches long. These will be the long sides of your toolbox, which are the same length as the base. **Note:** If you don't have a miter saw, or just want to try something else, you can make these same cuts with a circular saw, jigsaw, or handsaw (just try to make the cuts as STRAIGHT as possible!).

Step 2. **Cut off the corners of your end pieces.**

The tall ends of your toolbox will taper toward the top, making a triangular(ish) shape. To achieve this, you'll need to cut off two of the corners. You just cut these pieces out of your 1×10, and they should measure 9¼ inches wide by 13 inches tall.

Set your piece on a table in front of you, like a piece of paper, with the 9¼-inch edge on top and the 13-inch edge on the sides. From one of your top corners, measure 3 inches from the corner on the top edge and make a mark and 6 inches down the side and make a mark. Connect the marks: this is your corner cutoff line. Do the same on the other top corner.

Use a jigsaw if you have one, or a handsaw or backsaw if you don't, to cut off those corners. With either saw, clamp your material, go slowly, and make your cuts as straight as you can. If you're using a jigsaw, remember to keep the flat "shoe" of your saw flat and in contact with your wood.

After you've made your cuts, use a sanding block or belt sander to smooth any crooked edges.

Now that you have all your toolbox pieces (except the handle, which you'll cut perfectly to size later), it's a good time to sand all your pieces before assembly. Run some sandpaper over the edges (but not the faces) until they are smooth to the touch. It's harder to get to these edges once your wood pieces are attached, so it's better to do it now! If you have access to a random orbital palm sander, or—even better—a disk sander or stationary belt sander, you can use that, too. Just don't take off TOO much material, or your measurements will be short.

Step 3. **Attach your sides to the base, then your ends.**

Set your base piece (9¼ inches by 16½ inches) on the table. Before you attach anything, now is a good opportunity to make sure all your pieces fit together nicely. Take your two side pieces (5½ inches tall by 16½ inches wide) and set them on their long edges on top of your base piece, one on each side, so you now have two sides to your box. Check that everything is in line, and that no piece is too long or too short—you don't want pieces sticking out past the ends of the others.

If something is too long or short, recut any pieces you want, or sand it down if it's just a touch too long.

Now attach the side pieces, which is best done upside down, with the base resting on the two sides, so you can screw through the bottom

[continued]

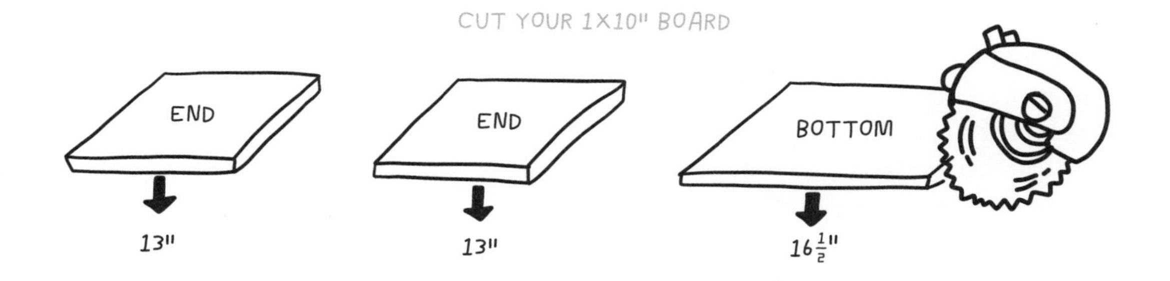

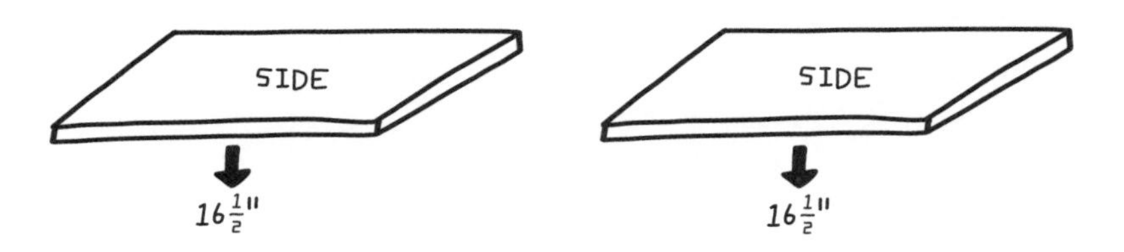

1. CUT YOUR BOTTOM, END, AND SIDE PIECES.

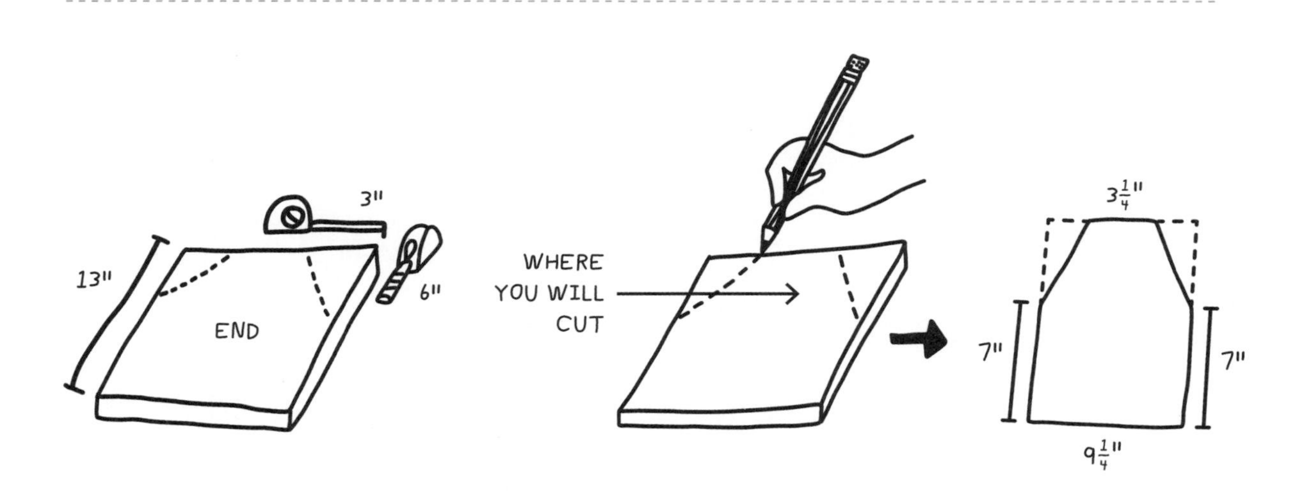

2. CUT OFF THE CORNERS OF YOUR END PIECES.

piece into the side pieces. Use clamps, a buddy, or a table vise to line up and hold everything together for your first couple of screws.

You'll want to place a screw every 4 inches or so, and it's helpful to measure and mark where your screws will go with a pencil first. Place these marks so they go through the bottom piece and into the middle of the width of the side piece—not too close to the edge!

Then use your drill and your $\frac{3}{32}$-inch bit to drill a pilot hole (keep those pilot holes nice and vertical!), and immediately put in the 1$\frac{5}{8}$-inch screw using your driver and the driver bit that matches your screws. Repeat until all your screws are installed along the bottom of your toolbox and both of your side pieces are connected. Turn the whole assembly over and you should now have a toolbox base with two long sides attached.

Now you're ready to attach the ends (the tall triangular pieces you cut with the jigsaw). The ends should cover the entire end of the toolbox (like the bread on a sandwich). Turn your half-done toolbox on its end so you can drill your pilot holes vertically again, and then install your screws.

Put 1 screw at each corner, and 1 in the middle of the bottom edge (for a total of 5 screws). Do the same with your other end.

Be mindful of where you're placing these screws, so they don't collide with the screws you used to attach your side pieces, or split the ends of your wood.

Step 4. Measure, cut, and install your handle.

You waited to cut your wooden handle until now because it has to fit snugly between the tops of the end pieces, and you want the measurement to be SUPER precise!

Take your tape measure and measure the distance between the two end pieces of your toolbox, where the handle will go. This dimension should be close to 16½ inches, but it might be a little longer or shorter. Measure this as precisely as possible.

Measure and mark that same dimension on your wooden 1×2 board.

Cut the 1×2 using a miter saw or coping saw. You could also use a backsaw and miter box (see page 81).

Set your handle in place. It should be snug between the two end pieces.

Drill 2 pilot holes and drive 2 screws to connect the 1×2 to the end of the toolbox. To make sure these screws are tight, hold your handle firmly in place for a long time! Repeat with the other end of the handle.

Step 5. Sand, decorate, and seal.

Even though you already sanded your pieces before assembly, you might want to run a sanding sponge over your edges now that the toolbox is complete. Dust off, and then, if desired, decorate with paint.

Finally, use wipe-on polyurethane and a rag or soft towel to give your toolbox one good coat of sealant to protect it from weather and staining. Now let's get to work!

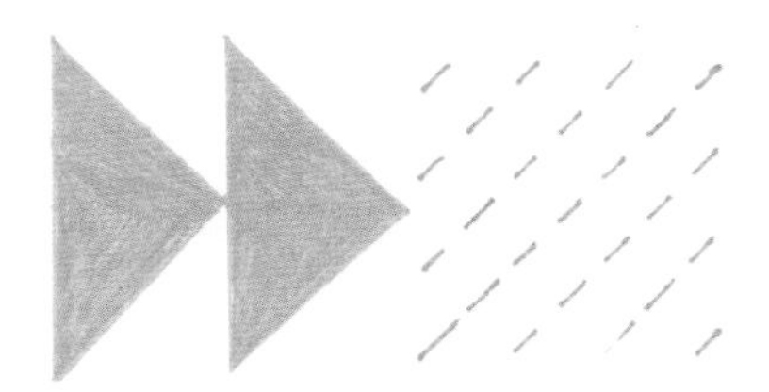

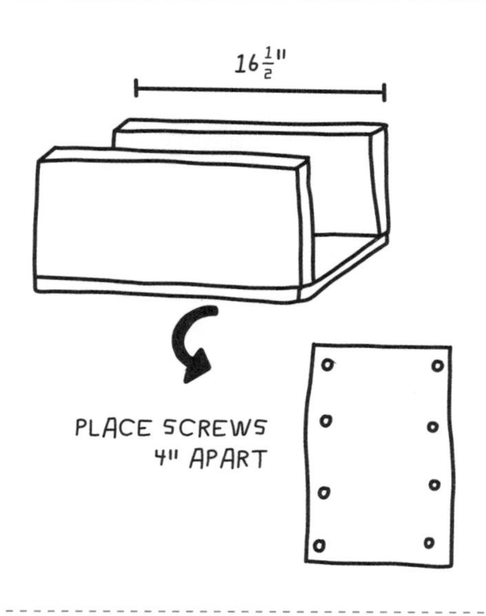

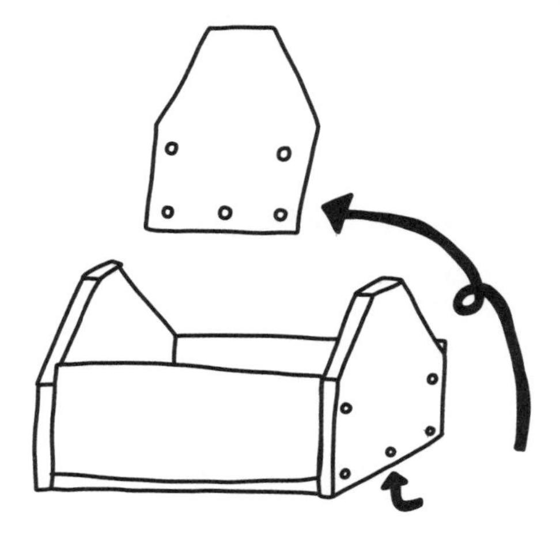

3. ATTACH YOUR SIDES TO THE BASE, THEN YOUR ENDS.

4. MEASURE, CUT, AND INSTALL YOUR HANDLE.

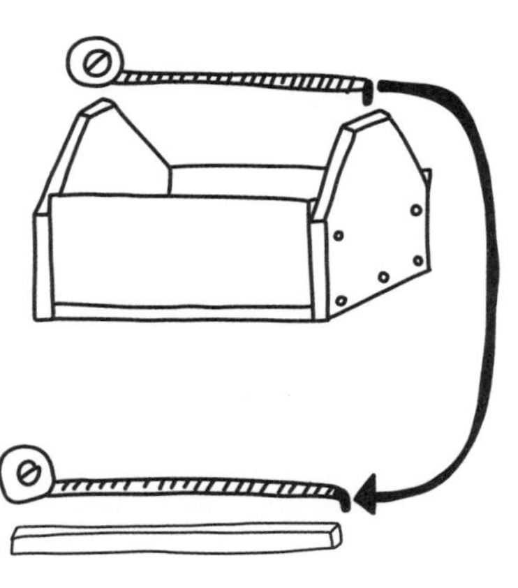

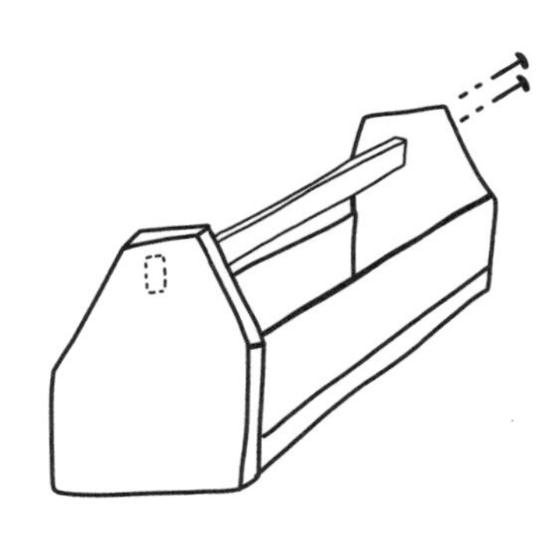

SCREW IN YOUR HANDLE.

5. SAND, DECORATE, AND SEAL.

SAWHORSES

Just like with the toolbox project (see page 249), I love the idea of making your own tools. For any project where you have to cut large or long pieces of material with a circular or handsaw, a good pair of sawhorses will be so useful. Here is a super-easy plan to make your own out of only 2×4s and screws.

Materials

- Twelve 8-foot-long 2×4s (this is enough to make 2 sawhorses)
- 2½-inch and 3½-inch wood screws (deck or construction screws, and I recommend using a T25 star drive if you can get them!)

Tools

- Pencil
- Tape measure
- Miter saw or handsaw
- Speed square
- Driver and bit to match your screws
- Bar clamps

SAFETY CHECK!

- An adult builder buddy
- Safety glasses (wear at all times!)
- Ear protection, if using a power saw
- Dust mask
- Hair tied back
- No dangly jewelry or hoodie strings
- Sleeves pushed up to your elbows
- Closed-toe shoes

Step 1. **Measure and cut all your 2×4 pieces.**

A miter saw is ideal, but you can also use a handsaw. Here is your cut list (for 2 sawhorses):

- 6 pieces at 36 inches
- 8 pieces at 34 inches
- 4 pieces at 30 inches
- 4 pieces at 21 inches

Step 2. **Assemble the I-beam top.**

Three of your 36-inch-long pieces will make up the I-shaped top bars of the sawhorse, which will support the weight of your work material and also attach to the legs.

Arrange 3 of your 36-inch pieces into an I-beam shape, with the top piece lying flat, the middle one vertical, and the bottom one lying flat. The piece in the middle should be sandwiched between the top and bottom pieces, and you should try to center it along the width of the top and bottom 2×4s.

There's no need to drill pilot holes for this project, as it's common (and sturdy) to use 2×4s and 2½-inch screws for wood framing without pilot holes. Use your driver to put about five or six 2½-inch screws through the top horizontal piece into the vertical piece, and then flip over and attach the bottom horizontal piece to the vertical piece, too.

Repeat with the remaining 3 pieces for the second sawhorse.

[continued]

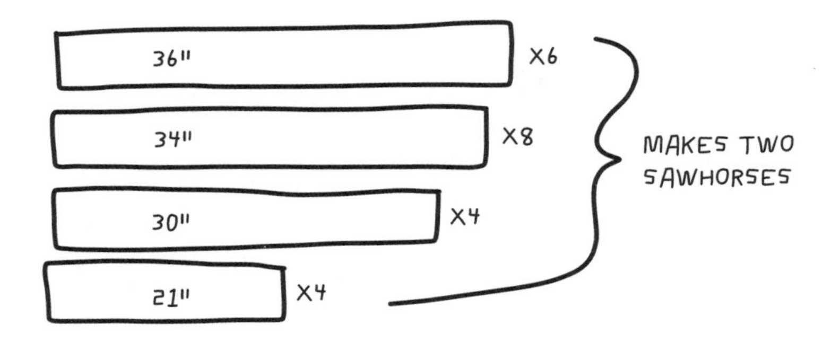

1. MEASURE AND CUT ALL YOUR 2X4 PIECES.

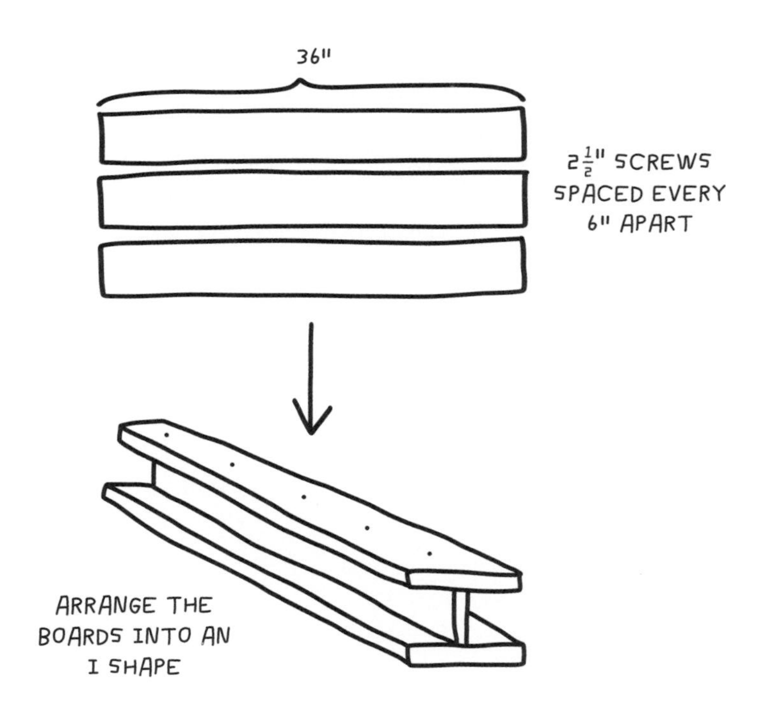

2. ASSEMBLE THE I-BEAM TOP.

Step 3. Assemble the legs.

For each sawhorse, you'll need 2 sets of legs that will attach to the I-beam at an angle. Let's assemble each side separately. For each side of the sawhorse, you need 2 of your 34-inch pieces for the vertical legs and one 30-inch piece for the crossbar.

On the floor, lay the 34-inch pieces parallel to each other, about 30 inches apart (you'll adjust this in a minute).

Lay your 30-inch crossbar on top, perpendicular to the 34-inch pieces, and adjust the exact positioning so that the ends of the 30-inch crossbar align with the long outside edges of the 34-inch pieces.

Position the 30-inch crossbar so it is 22 inches from the top of your 34-inch legs. Use your tape measure and speed square to get everything just so: in the correct vertical position, and at a right angle (square) to each other.

When you're happy with the positioning, use your 2½-inch screws to attach them, using 2 screws at each joint, driving the screws straight down through the 2 pieces.

Repeat this same process 3 more times to make a total of 4 leg configurations (2 for each sawhorse).

Step 4. Attach the legs.

And now for the tricky acrobatic part! You'll probably need a helper to hold everything in place while you screw the parts together.

Place 2 legs standing upright, with the 30-inch crossbars closer to the floor.

With the legs parallel, tilt the tops together so they form an upside-down V-shape.

Where they meet at the top, place 1 of your I-beams between the legs, so the tops of the legs sit underneath the top piece of the I-beam. Your legs should make contact with the I-beam in two places: underneath the top piece, and on the side of the bottom piece.

Use your 3½-inch screws and driver to attach the tops of the legs to the middle vertical piece of the I-shaped bar (3 screws per leg in a triangle pattern). The angle is a little awkward, and you'll have a gap where the screw goes through the leg and into the I-beam, but that's just fine.

Repeat these steps for the second sawhorse.

Step 5. Attach the side crossbars.

You should now have 2 sawhorses with each of their I-beam tops attached to a diagonal set of legs! You just need to attach crossbars on the sides of your sawhorses to keep the 2 legs from spreading apart when there is weight on them.

Grab two 21-inch pieces and place them on the ends of the sawhorse at the same level as your 30-inch pieces. The 21-inch pieces should span the width of the sawhorse to form a square "ring beam" support around the entire base.

Use your 2½-inch screws to attach them to the legs, 2 at each connection point. That's it!

Repeat to finish that second sawhorse so you can use them as a pair! What will you make with them?

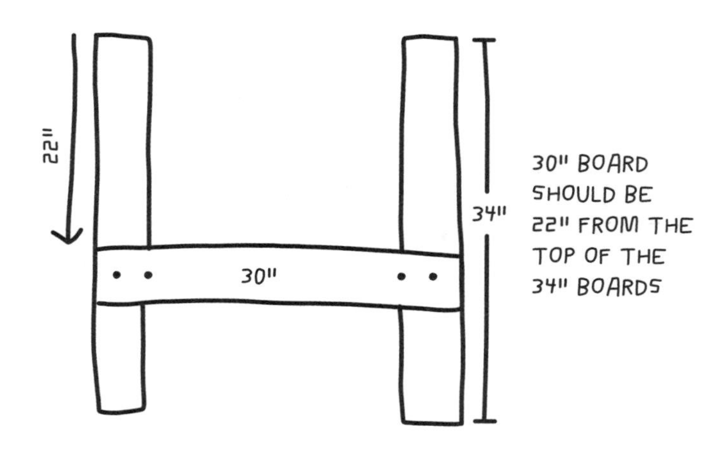

3. ASSEMBLE THE LEGS.

4. ATTACH THE LEGS.

5. ATTACH THE SIDE CROSSBARS.

HAND-ETCHED STEEL RULER

One of the simplest and most gratifying projects I do with my young girls is make our own rulers. Using basic steel stock, you can engrave your inch markings and then use the customized ruler you made yourself for all your future projects.

You can get flat bar steel from your local steelyard, or check your local hardware stores—some of them carry common dimensions. The dimensions can vary, and you can choose your own ruler width (1 inch or 2 inches, for example), and choose a thickness that isn't too floppy.

Materials

- One 12-inch or 18-inch piece of flat bar steel, ideally 1 inch or 2 inches wide and ⅛ inch thick
- Nail polish remover
- Rags or paper towels
- Metal stamping ink, if desired
- Clear coat, such as a spray can of water-based polyurethane
- Disposable nitrile gloves to protect your hands when applying clear coat

Tools

- Metal chop saw, hand hacksaw, or angle grinder with cut-off disk (if you need to cut the length of your metal)
- Bar clamp or C-clamp or vise
- Metal file
- Ruler
- Thin Sharpie or other permanent marker
- Engraving bit to fit your rotary tool (I recommend a diamond-tip ball bit)
- Rotary tool (Dremel)

SAFETY CHECK!

- An adult builder buddy
- Safety glasses (wear at all times!)
- Lightweight rubber-dipped gloves
- Hair tied back
- No dangly jewelry or hoodie strings
- Sleeves pushed up to your elbows
- Closed-toe shoes

Step 1. **Cut your metal to 12 inches or 18 inches.**

Flat bar mild steel (what is commonly sold at hardware stores) is best, but you could also choose a stainless steel, aluminum, or any other metal, for that matter. Mild steel has a black layer on it called mill scale, which is made from the impurities that rise to the top as the steel is "hot rolled" at the time of production. This mill scale makes for a nice finish when you etch through it, making your numbers sparkle and pop. If you want to preserve this look and prevent the etched area from rusting, make sure to use a few layers of clear coat. You might also be able to find a piece of scrap steel to use.

[continued]

1. CUT YOUR METAL TO 12" OR 18".

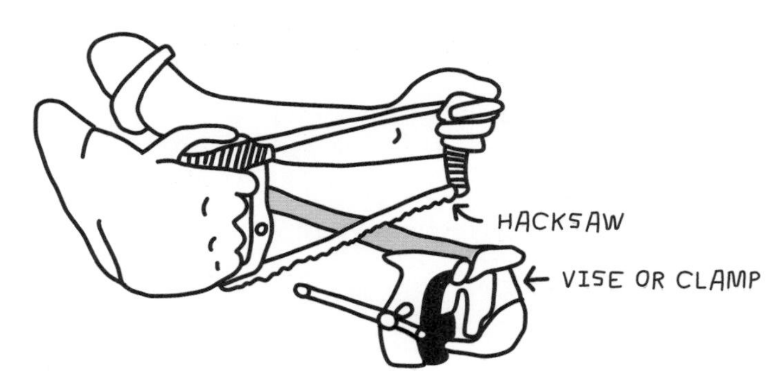

2. FILE YOUR EDGES.

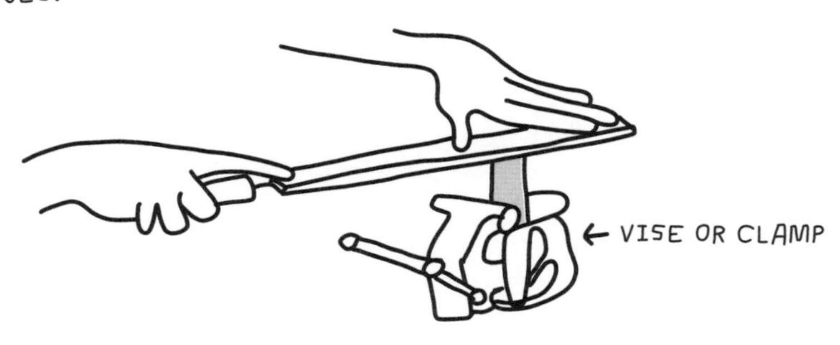

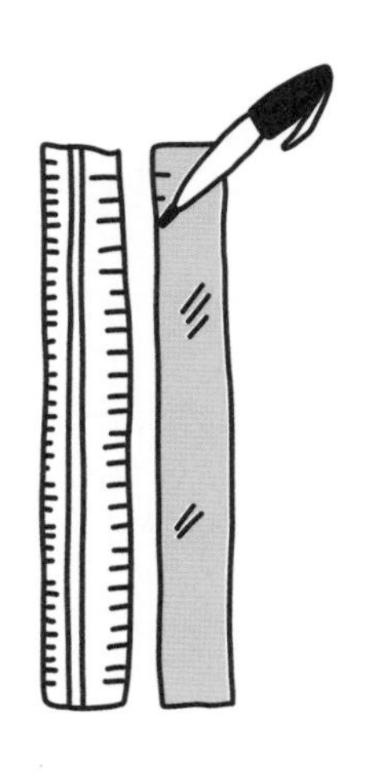

3. MAKE YOUR RULER MARKS.

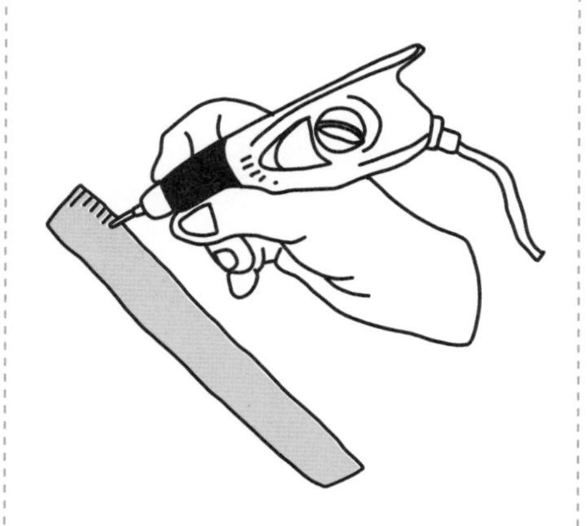

4. ENGRAVE YOUR RULER MARKS.

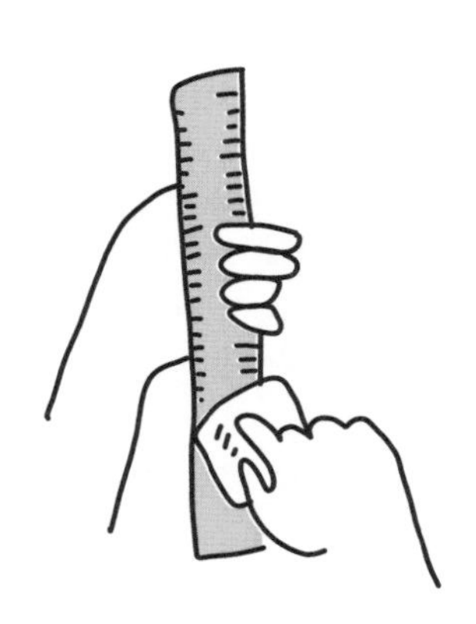

5. CLEAN IT AND INK YOUR RULER MARKS, IF DESIRED.

In either case, you'll likely need to cut it down to size, either 12 inches long or 18 inches if you're going big.

If you have access to a metal chop saw, use it! If not, cut the steel bar by hand with a hacksaw (clamp your steel firmly to a work surface or use a vise!), or use a cut-off disk on an angle grinder (a less precise cut, but you can file it later).

Step 2. File your edges.

Use your file of choice (page 166) to smooth and round out any sharp edges on your metal, especially the cut ends where there might be "burrs." It's easiest to clamp your material securely to a table using a clamp, or in a tabletop vise, so you can file freely. If you have a polishing or grinding bit for your rotary tool, you can use one to smooth your edges as well.

Step 3. Make your ruler marks.

Line up an actual ruler with your piece of steel so the zero point on your ruler matches up with the end of your piece of steel.

With a thin permanent pen, make your inch marks all the way down the length of your steel.

Now do your ½-, ¼-, and ⅛-inch marks, too. Notice that the smaller the increment of the inch, the shorter the line on the ruler.

Label your inches 1 to 12 (or 18 if you went for a longer ruler).

Step 4. Engrave your ruler marks.

Put your engraving bit in your rotary tool (the rotary tool usually has a wrench included so you can loosen and tighten the chuck to install and take out different bits). You might want to practice engraving precisely before you work on your actual ruler. Get a feel for the tool, holding it like a pencil so you can accurately make your marking (see page 159 for a refresher).

Go over all your pen marks to make your engraved marks.

Step 5. Clean it and ink your ruler marks, if desired.

If your pen marks are still showing, use an acetone nail polish remover and a rag or paper towel to wipe them off. Even if you don't still have ink on the surface, give your ruler a quick wipe to take off any dirt or debris. Depending on which kind of metal you chose, you might find that the engravings don't stand out much from your non-engraved surfaces. Mild steel will have much shinier engraved parts than stainless or aluminum, for example, because of that layer of mill scale.

If you want to make your engraved markings pop, use a metal stamping ink. Just wipe the ink over the entire surface, and then wipe it off, leaving ink only in the recessed engraved areas.

Step 6. Clear-coat it and measure away!

Once your ink dries, if you used it, you're ready to add a clear coat! For stainless steel and aluminum, this step isn't all that necessary because those metals are more resistant to rust. If you used steel, though, clear-coat your ruler so that your freshly etched shiny markings don't become rusty. Use a spray-on water-based polyurethane in a can, and give it a light coat on both sides. Let it dry, and do it again.

Ta-da! A handmade ruler. Now you can go forth and measure proudly!

6. CLEAR-COAT IT AND MEASURE AWAY!

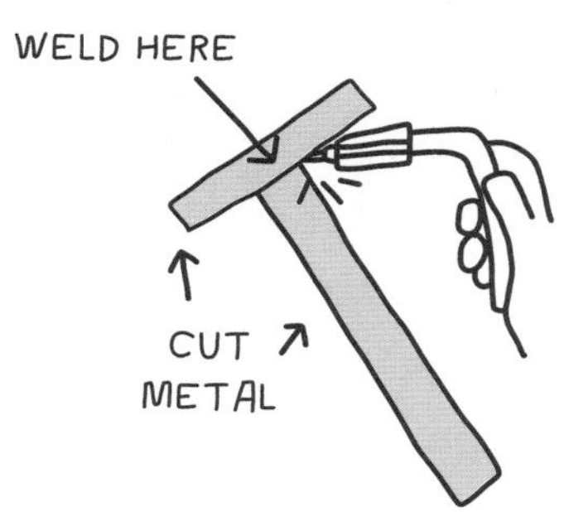

BEYOND. WELD 2 PIECES TO MAKE A T-SQUARE

+ + BEYOND + +

For an additional challenge, make a T-square or carpenter's square (L shaped) by welding two pieces of metal together. Etch the two pieces with your inch measurements, and then lay them out in either a T shape or an L shape to make a square tool. Weld the two pieces where they meet at a right angle as a butt-joint bead (a weld that sits on top of the surface and bridges the seam where the pieces meet), then grind down the weld using an angle grinder if you want to. For a refresher on welding, turn to page 171.

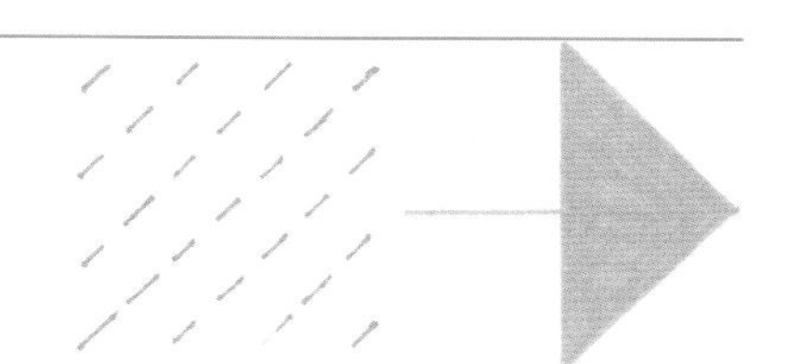

WHATEVER-SHAPE-YOU-WANT WALL CLOCK

This would be a great project to do with a group of friends, for a birthday party or girls' night, because it's always fun to see what shapes people come up with. It's also really easy and can be done with either hand or power saws, using pretty basic and inexpensive materials.

Materials

- ¼-inch-thick plywood or MDF (medium-density fiberboard), about 12 inches square, depending on the size of your desired clockface
- Battery-powered clock movement kit (includes the ticking-clock mechanism in a battery case, and hour, minute, and second hands). For ¼-inch plywood, you'll need a clock movement that can accommodate a *maximum dial thickness* of at least ¼ inch. A common shaft length is 11⁄25 inch, which will work just great. This dimension corresponds to the length of the threaded shaft that goes through your clockface and is secured with a nut on the other side.
- Spray paint, paint pens, or art/craft paint and paintbrush
- Spray-on water-based polyurethane
- Disposable nitrile gloves to protect your hands when applying polyurethane
- Batteries as called for in the clock-movement kit (usually 1 or 2 AA)
- Hanging hardware (nail and hook), if needed

Tools

- Pencil
- Band saw, scroll saw, or coping saw
- Bar clamps
- Sanding block and sandpaper, or belt/disk sander
- Tape measure or ruler
- Drill and drill bit set
- Wood burner, if desired
- Adjustable wrench or needle-nose pliers
- Scissors, if desired
- Hammer

SAFETY CHECK!

- An adult builder buddy
- Safety glasses (wear at all times!)
- Ear protection, if using a power saw
- Dust mask
- Hair tied back
- No dangly jewelry or hoodie strings
- Sleeves pushed up to your elbows
- Closed-toe shoes

[continued]

1. DRAW YOUR SHAPE.

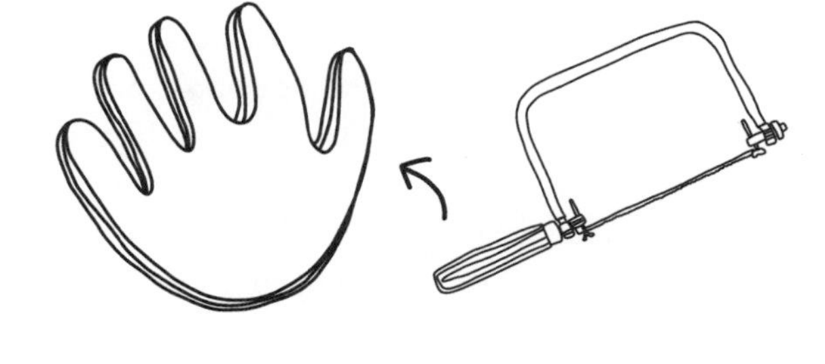

2. CUT OUT YOUR CLOCKFACE.

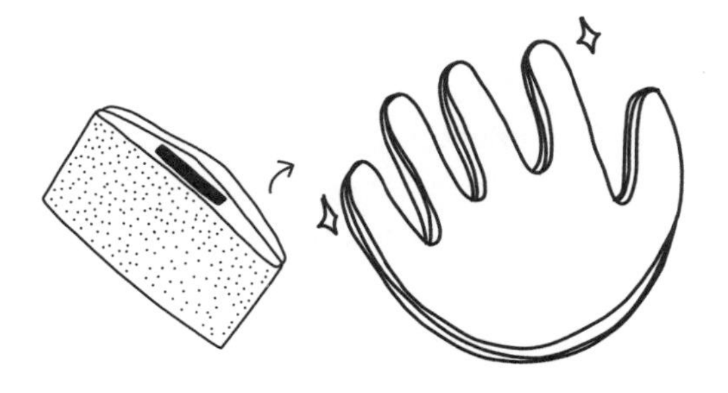

3. SAND YOUR EDGES.

4. LOCATE AND DRILL YOUR HOLE FOR THE CLOCK MOVEMENT.

5. PAINT YOUR CLOCKFACE.

6. ATTACH THE CLOCK MOVEMENT AND HANDS.

7. ADD A BATTERY AND HANG!

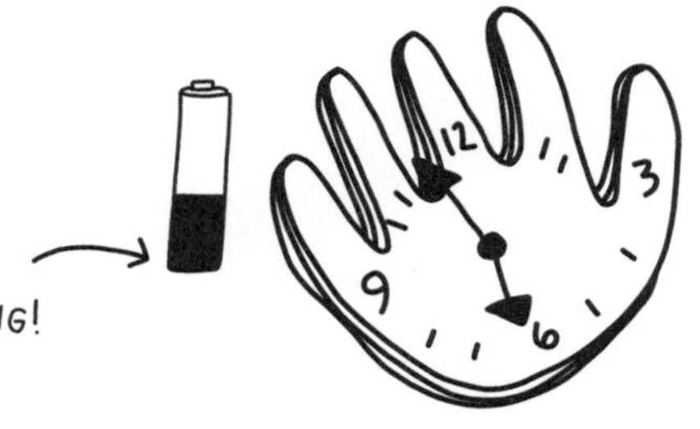

Step 1. **Draw your shape.**

Decide on an awesome shape for your clockface! I've seen some great ones: stars and letters and dog faces and fist bumps and more. Choose a shape that's large enough to accommodate the clock hands (you can trim your clock hands a bit if you need to, but you don't want to lose too much length). Once you've sketched and figured out your shape, draw it in pencil on the face of your plywood.

Step 2. **Cut out your clockface.**

You have many saw options to accomplish this. If you have access to a band saw or scroll saw, that's definitely your best bet. It is also absolutely workable to use a hand coping saw, and it might even be more gratifying to muscle your shape by hand. (If you go with the coping-saw option, make sure to use your bar clamps to secure your wood to a tabletop before cutting.) Whichever saw you choose, take your time and cut your curves and profile as precisely as you can. You can sand it to perfection later, but try to get it close to its final shape. With all these saw options, you might have to get creative with your cutting paths, taking notches or pie pieces out a little at a time, to get to hard-to-reach spots.

Step 3. **Sand your edges.**

Use whichever sanding device you prefer; either a belt/disk power sander for curved areas or just a sanding block with sandpaper will work just fine. Get into all those nooks and crannies until you are happy with your final shape and don't have any rough edges that might give you splinters.

Step 4. **Locate and drill your hole for the clock movement.**

The clock-movement kit should tell you the diameter of the shaft (the post that goes through the face of your clock).

If not, measure it (it's probably between ¼ inch and ⅜ inch in diameter).

Find the same-size drill bit, or one that's 1⁄16 inch larger (this will make it easier to insert the clock-movement shaft).

Decide where you want the center point of your clock to be (where the hands will be attached), and drill a hole through your plywood.

Sand the edges of the hole so you don't have any jagged splinters.

Insert the clock-movement shaft to make sure the hole is big enough, but then remove it and set it aside (you're going to paint first before installing it!).

Step 5. **Paint your clockface.**

Before you install your clock movement, now is your chance to paint to your heart's content. You can paint the background solid or with a pattern, and you might want to consider painting on or attaching some small numbers around the edge of the clock to help tell the time. Make sure to give your clockface a good dusting off before applying any paint. If you have a wood burner, you could also burn your numbers or other patterns into the wood. Or use spray paint or any kind of art/craft paint you'd find at your local art store. Sequins, sparkles, or any other ornamentation are always welcome! Once the paint is dry, give it a spritz of a clear coat, like a water-based polyurethane in a can.

Step 6. Attach the clock movement and hands.

Your specific clock movement will likely show you the exact order you should install each of the components, but most likely, you will follow these steps:

- Place a rubber washer on the bottom of the shaft of the movement's battery box.
- Insert the shaft from the back of your clockface through to the front, so that the movement's battery box is on the back side of the clockface. If your clock movement has a built-in hanger, install the mechanism with the hanger facing up, toward the ceiling and the 12 o'clock mark.
- Secure the shaft in place with the metal washer and nut on the clockface side. Use an adjustable wrench or needle-nose pliers to tighten this nut, and hold the battery box stationary while tightening, so that the shaft does not just spin in place.
- Place and push down your hour, then minute, then second hand on the top of the shaft in that order (you'll see that they each fit onto a certain spot on the shaft, with the second hand serving as a sort of "cap" on top). You can cut down the length of your hands with scissors, if they are longer than the width of your clockface.

Step 7. Add a battery and hang!

Your clock mechanism likely takes a small battery, like AA, which you should pop into the back at this time. If you don't have a hanger built in (see step 6), you'll need to attach some hanging hardware, like a D-ring and wire. Choose your wall location, tap a small nail into the wall, and hang your clock!

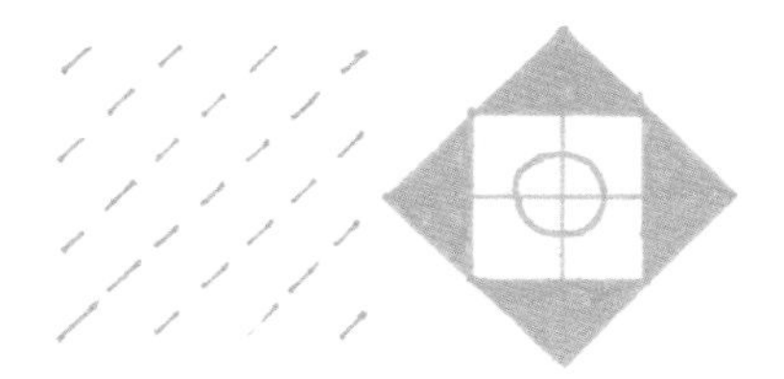

WOODEN SPOON

This is a project you can do over and over and end up with a gorgeous collection of kitchen spoons. Depending on the wood you choose, the spoon's purpose, your tool selection, and more, the wooden spoon is an always-different and always-evolving project. I definitely recommend doing your first spoon entirely by hand (no power tools), and choosing an easy-to-carve wood, such as poplar, soft maple, or—my personal favorite—if you can find it, European beech.

Materials

- 1×4 piece of carving wood, 6 to 12 inches long (depending on what kind of spoon you're making)
- Vegetable or mineral oil and rag

Tools

- Pencil
- Clamps or table vise and bench dogs
- Gouge chisels (a few sizes)
- Rubber or wooden mallet
- Coping saw or band saw
- Wood files and rasps
- Spokeshave
- Other carving tools, if desired (hook knife and straight knife)
- Sandpaper (medium- and fine-grit, 120 and 220)
- Chop saw, handsaw, or backsaw

SAFETY CHECK!

- An adult builder buddy
- Safety glasses (wear at all times!)
- Ear protection, if using a power saw
- Dust mask, if using a power saw
- Hair tied back
- No dangly jewelry or hoodie strings
- Sleeves pushed up to your elbows
- Closed-toe shoes

Step 1. **Cut a length of your chosen wood.**

For your first spoon, try something like poplar or European beech, which are both easy to carve, especially if they still have some moisture in them and are not super dried out. Decide on the rough size and depth of your spoon based on the type of spoon you want to make (serving spoon? ladle? spatula-ish device?), and find a piece that will accommodate it. A 1×4 board should work for most spoons, unless you are going for an unusually deep spoon bowl, in which case you might need a 2×4 piece. Cut a length of about 12 inches, using a chop saw, handsaw, or backsaw. Give your piece of wood a good long stare. Your spoon is in there—you just have to bring it out!

[continued]

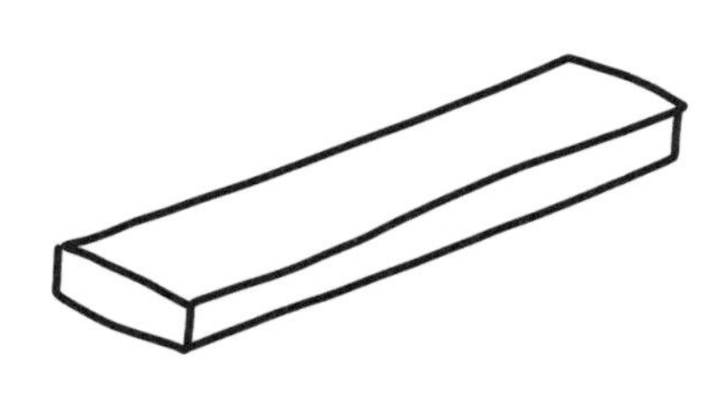

1. CUT A LENGTH OF YOUR CHOSEN WOOD.

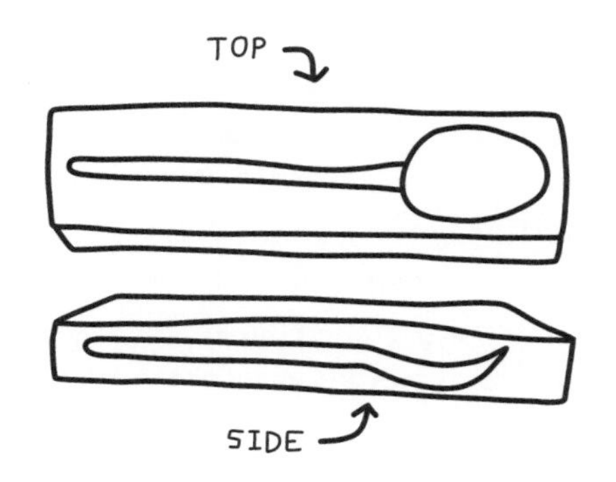

2. DRAW YOUR SPOON PROFILES (TOP AND SIDE).

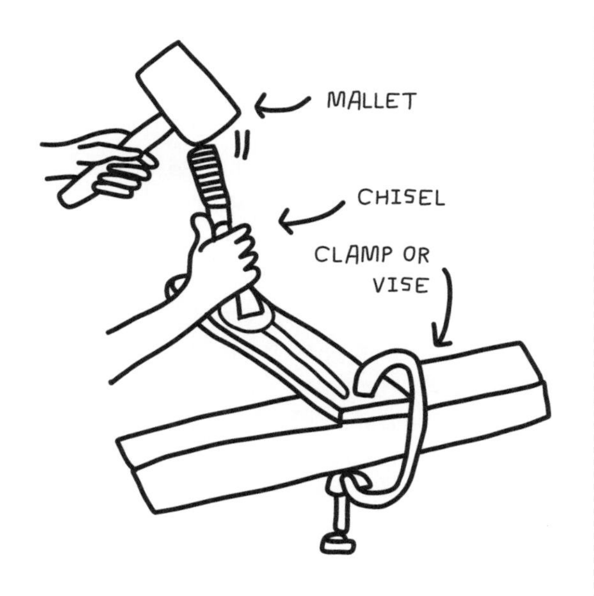

3. CARVE OUT THE BOWL OF YOUR SPOON.

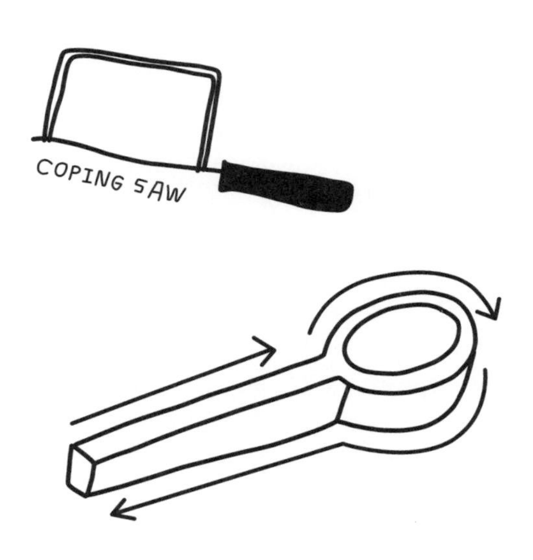

4. CUT OUT THE TOP PROFILE OF YOUR SPOON.

Step 2. Draw your spoon profiles (top and side).

On the top of your piece of wood, draw the top profile of your spoon (handle and round bowl shape). If you're worried about symmetry, you can draw just one half (the right or left side) of the spoon's shape on a piece of paper, cut that shape out of the paper, lay it down and trace it onto your wood, and then flip it over as a mirror image to draw the other side for a perfectly symmetrical spoon.

On the thin edge of your wood, draw the side profile, including the depth of your bowl and the shape of your handle. Take note of the wood's grain; your handle should run in the direction of the grain (parallel to the long grain lines).

Step 3. Carve out the bowl of your spoon.

Secure your piece of wood to a work surface. If you have a workbench with a table vise and bench dogs, use it! If not, you can clamp both ends of your wood firmly to a table.

Grab your gouge chisels and mallet. Work from the outside of the bowl in, toward the deepest part, starting with a wider gouge chisel and working toward smaller and more detailed tools (see page 118 for a refresher on chisel technique). **Don't point the sharp end of the chisel at yourself!** Carving will take some practice, and you'll need to regulate the depth of your carvings so they are deeper toward the middle of the bowl, getting shallower toward the lip. Be careful not to carve so deep that you go through the entire thickness of your wood to the other side (unless you want a hole in your spoon!). You don't have to perfect the bowl now, but get it close to the shape you want it.

Step 4. Cut out the top profile of your spoon.

Use a coping saw or a band saw to cut the outside shape of the spoon (along the sides of the handle and around the bowl). This cut doesn't have to be perfect, but stick as close to your pencil lines as possible. Go all the way around so you now have a rough spoon shape. Your side profile will still be rectangular, but we'll get to that next!

Step 5. Carve the side profile and underside of your spoon.

Next you will round out the underside of your bowl and start shaping your handle. This is the most difficult step and requires the most patience. If you want, you can first use a file or sanding tool to round out the corner edges of the spoon's profile, which will help you start your work with the spokeshave. Let's work on the underside of the bowl first, and then tackle the handle.

Clamp your spoon to the table, upside down, so that you have access to all sides of the bottom of the bowl. If you're using a table with a vise, tighten the vise around the spoon's handle with the bowl sticking upright.

Use a spokeshave on the underside of the spoon's bowl to begin to form it into a round shape. First take some material off the sides, then the bottom, slowly, a little bit at a time, until your rough bowl shape emerges.

For the handle, clamp the bowl shape in a table vise or to the side of a table, so the handle is accessible.

Use your spokeshave in long strokes down the handle to shave off a bit at a time, until it is rounded out (or it becomes whichever shape you've chosen).

[continued]

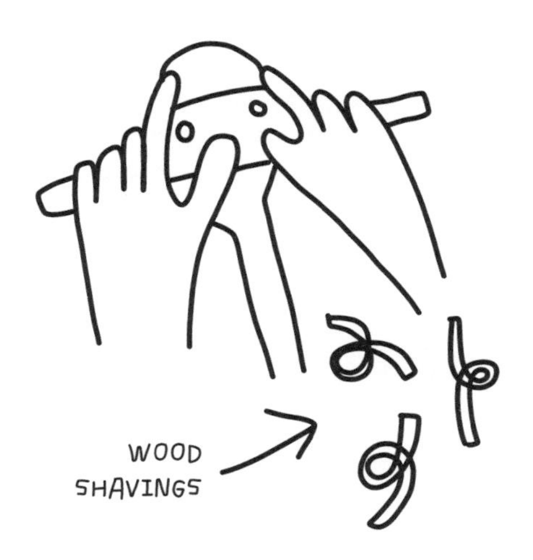

5. CARVE THE SIDE PROFILE AND UNDERSIDE OF YOUR SPOON.

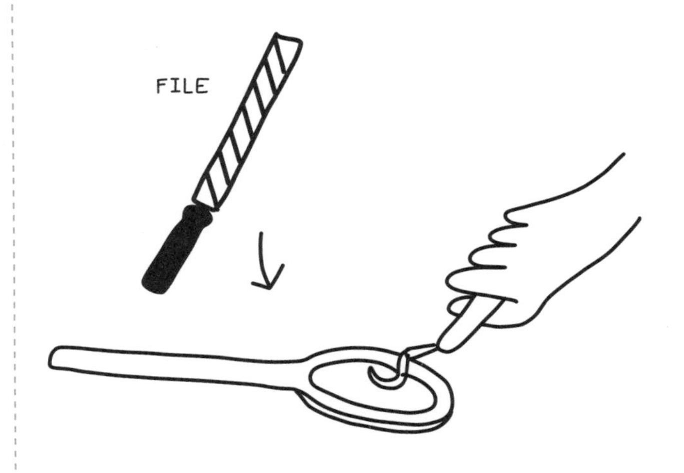

6. ROUND OUT THE BOWL, FILE, FINISH, AND SAND.

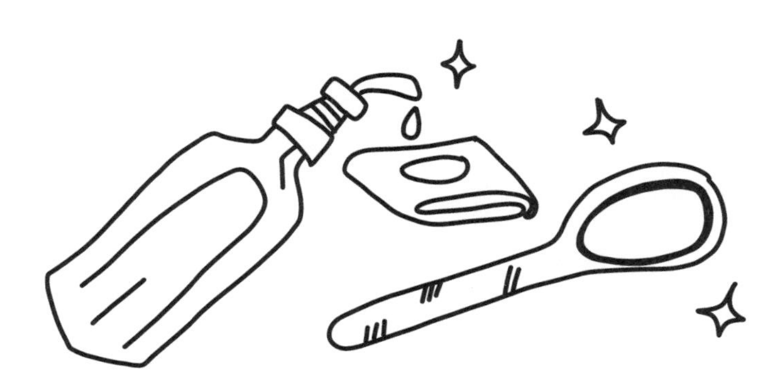

7. OIL AND SERVE!

Step 6. Round out the bowl, file, finish, and sand.

At this point, hopefully your spoon looks spoonish—but it probably still has some rough edges. Now is your chance to finish all those edges and get your spoon as close to perfection as possible.

A hook knife will help you smooth out the inside bowl of the spoon (kind of like a melon baller).

Use flat and round wood files to smooth out the sharp lines and edges on the handle and outside of the bowl.

Finally, use a medium-grit sandpaper, like 120, to start sanding, then move up to a finer grit, like 220, to finish your spoon.

Step 7. Oil and serve!

Your spoon is nearly food-ready! Mineral oil is your best option to coat and seal your spoon: it's food-safe, cheap, and available at most grocery or drugstores. Dust off your spoon completely, then give it two good coats of mineral oil, letting it dry in between each coat. Your spoon is ready for food service and can be washed and used for years to come! You can reapply some mineral oil every few months to keep it in great shape, too.

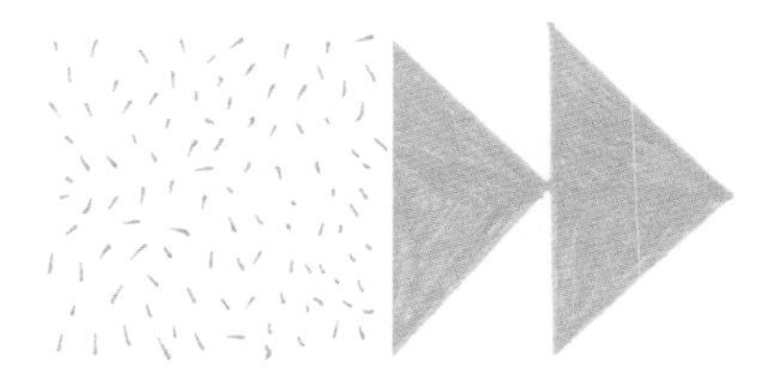

TIARRA BELL

Furniture Designer

Germantown, Pennsylvania

When I began making my list of potential women and girls to feature in this book, I reached out to some friends who work in similar community design and building capacities. One of the first responses I got was from my friend Alex Gilliam from Public Workshop, who said, "You HAVE to talk to Tiarra Bell." A few days later, I found myself on the phone with Tiarra, whose sweet demeanor and fierce passion undeniably came through the airwaves. Her story is one of total determination and optimism, and she discovered building as the vehicle to take her to incredible places—from urban Philadelphia to the best art and design school in the country.

"I fell in love with building in high school! One day, Alex Gilliam, founder of a local community building project called Public Workshop, came to my school and told us he could help us build any project we wanted to at our school. Some girls and I decided we wanted to build mobile hangout pods where we could sit and do homework, talk, and eat. For the next nine months, we designed and built them ourselves, with Alex's help as a mentor. Along the way, some of the boys from school saw what we girls were doing and joined in the construction of the project. Ever since that first project, I haven't stopped working with my hands to design and build, including community projects around Philadelphia.

"Soon after that first school project, I went to a furniture show in Philadelphia and met a furniture designer named Charles Todd, who owns a furniture-making studio in Mount Airy, Philadelphia. I asked for an apprenticeship during the summer of 2015, and he said yes! That summer, he taught me how to make furniture using traditional woodworking skills.

"The following year, my senior year in high school, Charles helped me build my own line of furniture for my senior project. I made all the furniture using traditional woodworking methods because I want to preserve the tradition of beautiful furniture being made by hand. I designed and built a contemporary-style chair and side table, along with a traditional-style stool. This project was the beginning of 'the new Tiarra' because for the first time, I started to see how everything I'd learned over the past couple of years connected to me as a person and to my future.

"By the time I had to apply to colleges, I had so many items in my portfolio. Three years of nonstop designing, building, asking questions, internships, and trying and failing all led to my acceptance into my dream school—the Rhode Island School of Design! Even though I was confident in my building skills, I was uncertain that my design skills would cut it at RISD. When I got there, I lost a lot of confidence in myself at first. However, something inside me

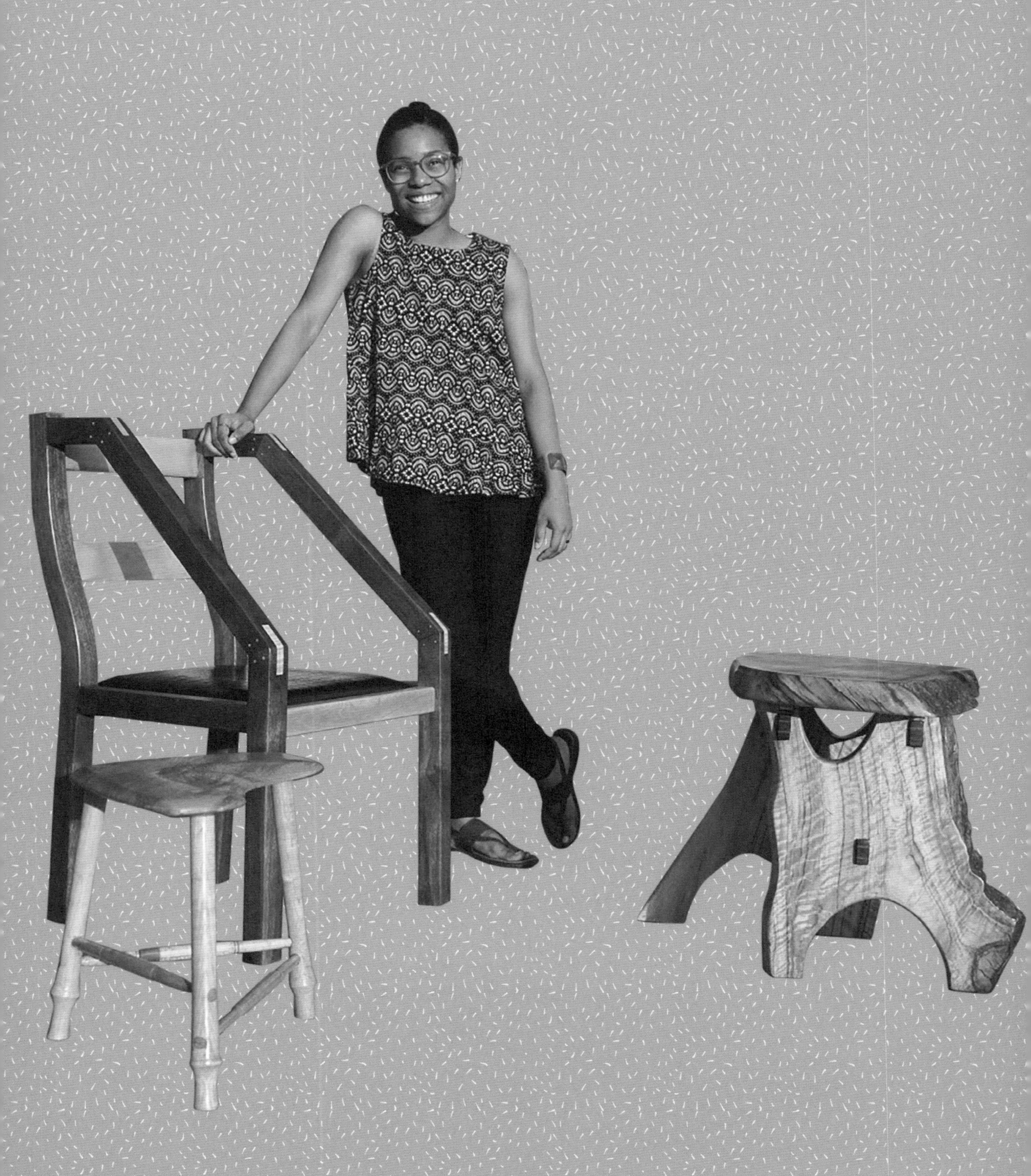

said, ‘Tiarra, that’s simply not the case! You’ve created good work before!’ I pushed myself to become a better designer and shook off those negative feelings. And I’m pushing forward every day.

“I want to design and build products to help improve life for everyone—from household items that fit a person’s unique needs to a community project to improve a whole neighborhood. I want to bring my ideas into reality using a variety of tools—from heavy machinery and modern technology to the simplest of hand tools (I adore the Japanese handsaw and spokeshave!).

“To the young girls who want to learn how to build, I say throw caution to the wind and do it, despite your fear or what you think people might think. Do it despite what YOU might think about yourself. There will be people who will discourage you or, as in my case, you may discourage yourself, but after the first time you create something you’re really proud of, you will never stop. Designing and building has changed my life. From the time I started building up until now, it has been an amazing journey that made me the girl I am today. Who would have known that my first building project would help me narrow down what I wanted to do for the rest of my life or give me the opportunity to attend one of the best design schools in the world?”

RIGHT-ANGLE BIRDHOUSE

I did this project with our first-ever group of Girls Garage summer campers. I happened to be looking at the art of Piet Mondrian, who is famous for compositions made up of black lines all at right angles and fields of red, yellow, white, and blue. His art inspired the geometry of this birdhouse! Using inexpensive cedar fence boards assembled on their sides at right angles, you can make so many fun shapes for many kinds of birds.

Materials

- 5½-inch-wide cedar fence board, approximately 10 feet total
- Wood glue (exterior, waterproof, such as Titebond III)
- 1¼-inch finish nails
- Strong rope or twine
- Quarter-sheet (2 feet by 4 feet) of ½-inch plywood
- ¼-inch wooden dowel, about 6 inches long (you can cut to a specific perch size later)
- Spray paint or other paint and paintbrushes, if desired
- Exterior polyurethane
- Disposable nitrile gloves to protect your hands when applying polyurethane

Tools

- Speed square
- Tape measure
- Pencil
- Miter saw, hand crosscut saw, or backsaw
- Sanding block and medium-grit sandpaper
- Clamps
- Hammer (or brad nailer and brads, if preferred)
- Drill
- ¼-inch drill bit
- Scissors
- Jigsaw
- 2-inch or 3-inch hole saw with a pilot bit

SAFETY CHECK!

- **An adult builder buddy**
- **Safety glasses (wear at all times!)**
- **Ear protection, if using a power saw**
- **Dust mask, if using a power saw**
- **Hair tied back**
- **No dangly jewelry or hoodie strings**
- **Sleeves pushed up to your elbows**
- **Closed-toe shoes**

[continued]

SAMPLE DIMENSIONS

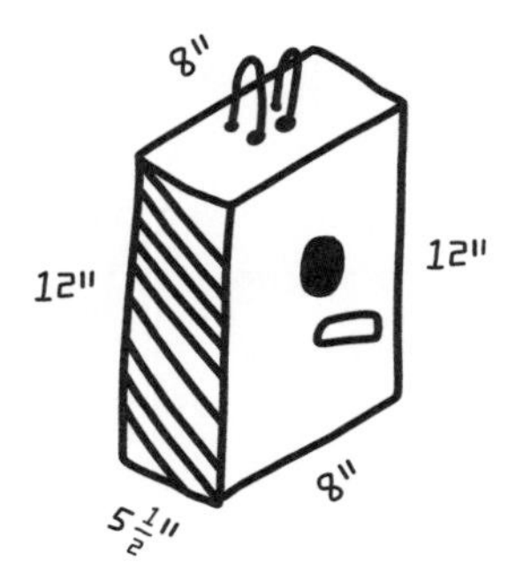

SO MANY SHAPE OPTIONS!

1. DRAW AND DIMENSION YOUR BIRDHOUSE SHAPE USING ONLY RIGHT ANGLES.

2. CUT YOUR FRAME PIECES OUT OF FENCE BOARD USING YOUR DIMENSIONS.

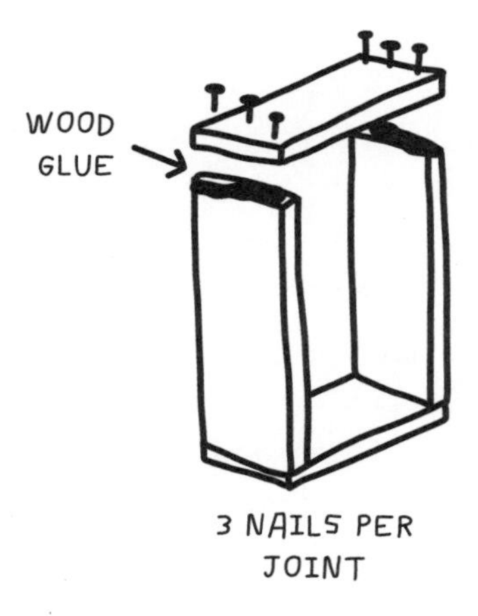

3. GLUE AND NAIL TOGETHER YOUR FRAME.

Step 1. Draw and dimension your birdhouse shape using only right angles.

Be the bird! Consider what kinds of birds are in your neighborhood. What size are they? How big of a house might they want or need? Start sketching shapes for the exterior profile of the house. The only rule is that your lines must all meet at right angles.

The first right-angle birdhouse I made was a squared-off Z-shape, like a Tetris piece, with the perch on the upper left side of the Z. Note where you will put your door hole and your perch (likely just under the door hole).

The fence board pieces will define the depth of the birdhouse (5½ inches). Do your best to dimension this drawing so you know how long to cut each of your fence board pieces (remember the thickness of the fence board is ½ inch, so you will need to account for that dimension as part of the overall length of your birdhouse sides, where two pieces meet).

Step 2. Cut your frame pieces out of fence board using your dimensions.

Using the dimensions you just figured out, measure and cut your frame pieces.

Use your speed square and tape measure to draw your cut lines on the cedar fence boards.

Using a miter saw, if you have one, or a handsaw, like a crosscut saw or backsaw, cut your pieces. Measure and cut one piece, and then measure your next piece (because of the kerf—remember from page 124). You might cut and lay out your pieces, and then realize some are a bit short or long. Trim or recut as necessary.

Give the ends a quick sanding so there aren't unnecessary splinters.

Step 3. Glue and nail together your frame.

Once you've cut all your pieces, "dry fit" them by laying them out on a flat surface with the pieces butting into each other at right angles, ready to be attached.

Start with just one joint and apply a line of wood glue. Make sure you use clamps to hold this joint together so you can get started.

Hammer in 3 nails across the width of the joint, evenly spaced between each other.

Go around the frame and do this for all the joints until your frame is complete. If you have a brad nailer, you could use that for this step instead.

Step 4. Choose your hang point and install your hanging rope.

Once you put the front and back faces on, you won't be able to access the inside, so now's your chance to install your hang rope.

Figure out which way is up and find a balanced spot to hang the birdhouse from, where it will hang evenly and not lean to one side.

With your drill and ¼-inch drill bit, drill 2 holes near this spot, about 2 inches apart.

Cut a 1-foot length of rope or twine and loop it down through one hole and out the birdhouse through the other. Tie the ends together.

Step 5. Trace and cut out the front and back faces.

Now you've got your outside shape of the birdhouse, but no front or back walls. Make these front and back faces:

- Lay the frame on top of your plywood and trace it. Write "back" on this piece.
- Flip your frame over and trace it on the plywood. Write "front" on this piece.
- Secure the plywood to sawhorses or a table surface and use a jigsaw to cut out the shapes.

[continued]

4. CHOOSE YOUR HANG POINT AND INSTALL YOUR HANGING ROPE.

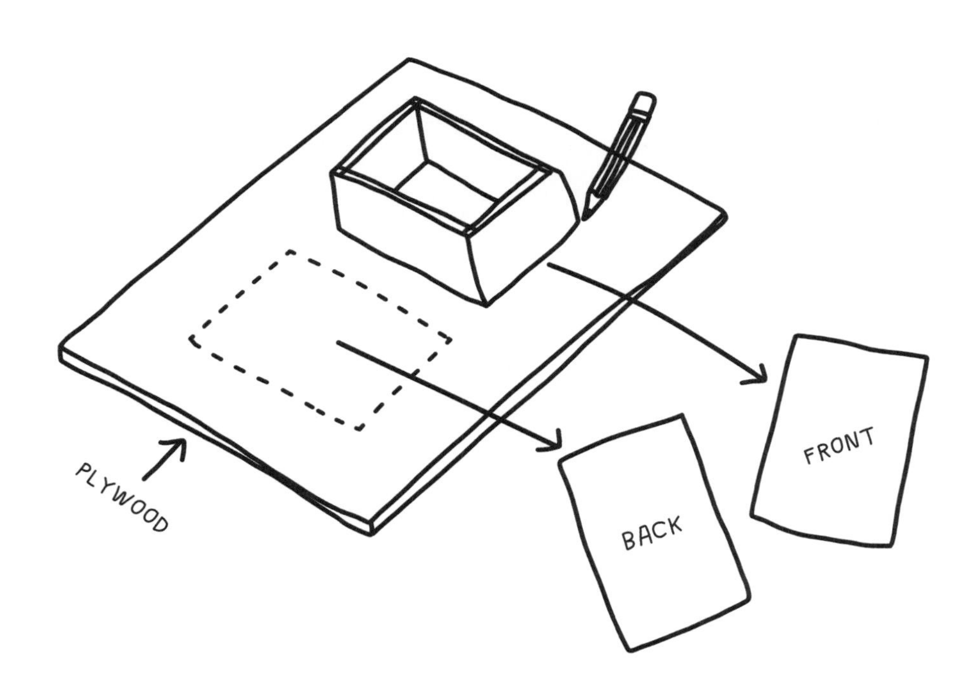

5. TRACE AND CUT OUT THE FRONT AND BACK FACES.

Step 6. Cut out your door hole.

You now have front and back walls, but no way for your little birdie to enter its house!

Go back and reference your drawing to locate the door hole and perch.

Mark its location on the front face you just cut. You can just put an X at the center of the bird hole.

Put your 2- or 3-inch hole saw bit into your drill or drill press, and align the built-in pilot bit with the X you just drew. Make sure to clamp your plywood securely, then drill out your hole!

Step 7. Install your perch.

Grab your wooden dowel and cut a 4-inch length using a handsaw or miter saw.

Use the ¼-inch drill bit to drill a hole in the front face of your birdhouse, just below your door hole.

Put a little bit of wood glue on the end of the dowel and in the hole you just drilled.

Stick the dowel in the hole while the glue is still wet. You might have to twist it and shimmy it in there!

Step 8. Sand and attach the front and back faces to your frame.

Now your front and back faces are ready to be attached!

Before you attach the front and back faces to the frame, give the edges a quick sanding.

Start with your back face (the one without the perch sticking out of it). Set your birdhouse frame upside down (so the front of the frame is facedown on the table), and drizzle a line of wood glue along the edge of the back of your cedar fence board frame.

Now clamp your back face to the frame so all the edges match up.

Hammer in your nails, about 3 inches apart, all the way around the frame. Make sure your nails are about ⅜ inch away from the edge so they go through the face plywood and into the middle of the fence board.

Once your back face is attached, flip your birdhouse over and attach the front face the same way.

Step 9. Paint and hang.

Go to town with some colorful paint that will attract your bird friends and make them feel at home! You can stain your wood or paint it with patterns. You will also want to give the birdhouse a coat of polyurethane, especially the plywood faces. Now go find a tree branch and hang your new bird home!

+ + BEYOND + +

An alternative to hanging your birdhouse would be to mount it on a post. You can find tons of ways to create this base, but a simple solution is to pour concrete into a bucket, and set a 4×4 metal post base into it (get a post base with two fins that sink into the wet concrete, and two fins that stick up to attach to your post). When the concrete dries, set a 4×4 lumber post into the base, screw it in, and use one additional metal bracket to attach the top of the post to your birdhouse. You can now bury this bucket in the ground (or just leave it exposed) for a mounted standing birdhouse.

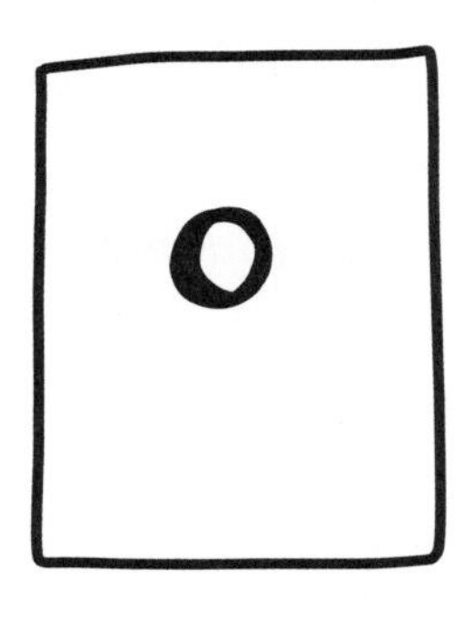

6. CUT OUT YOUR DOOR HOLE.

7. INSTALL YOUR PERCH.

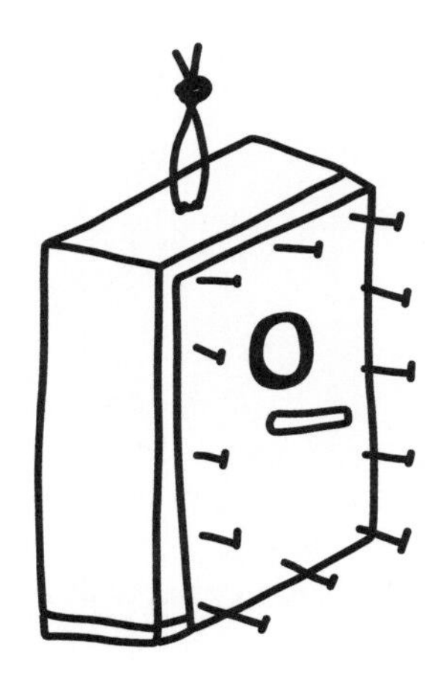

8. SAND AND ATTACH THE FRONT AND BACK FACES TO YOUR FRAME.

9. PAINT AND HANG.

WALL-MOUNTED BIKE RACK

Bikes are just awesome, but such a pain to store safely and securely (especially once you've graduated beyond the kickstand). My bike is always precariously tilted against a wall, only to come crashing down with the slightest bump. This bike rack is so easy to install on any wall that has wooden studs in it and gets your bike up off the floor for safe storage. The materials list below is for a two-bike rack, but you could also just do it for one (you'll only need two bike hooks in that case).

Materials

- Two 8-foot-long 2×4s
- One 6-foot-long 1×4
- Interior house paint and paintbrushes, if desired
- 3-inch wood screws
- 4 bike hooks (shaped like a question mark, with a threaded end that comes to a point)
- 2-inch wood screws
- 5-inch wood screws (the superstrong GRK brand is great)

Tools

- Hammer
- Stud finder, if desired
- Pencil
- Level or plumb bob
- Tape measure
- Speed square
- Miter saw or circular saw (or crosscut handsaw if you don't have a power saw)
- Sanding block and sandpaper
- Drill and drill bit set
- Driver and bit to match your screws

SAFETY CHECK!

- An adult builder buddy
- Safety glasses (wear at all times!)
- Ear protection, if using a power saw
- Dust mask, if using a power saw
- Hair tied back
- No dangly jewelry or hoodie strings
- Sleeves pushed up to your elbows
- Closed-toe shoes

Step 1. **Find the stud in your wall.**

This is without question the most important step! Locating a stud gives you a strong structural place to attach your bike rack to the wall. If you just slap it on there and only attach it to drywall, chances are it will come tumbling down—and take a chunk of your drywall with it. Reference Essential Skills (page 214) if you need a refresher on how to do this by hand using a hammer, or you can use an electronic stud finder.

[continued]

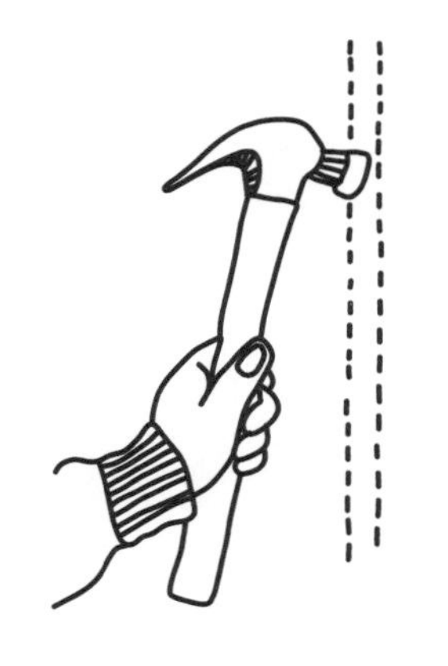

1. FIND THE STUD IN YOUR WALL.

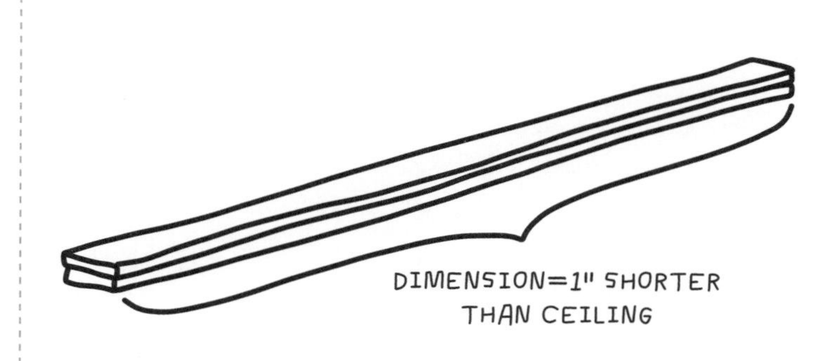

2. CUT YOUR 2X4S TO LENGTH.

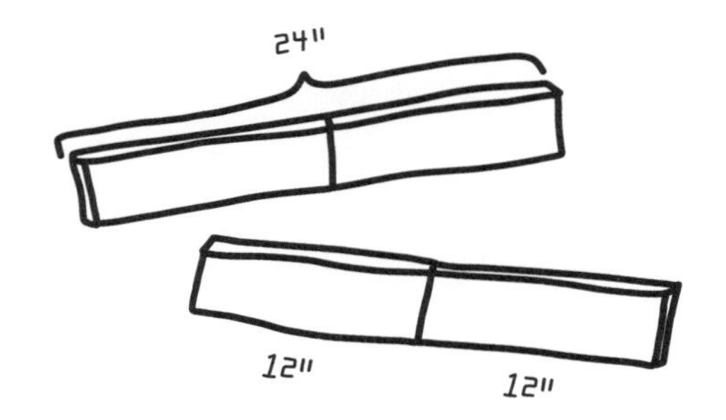

3. CUT AND MARK YOUR 1X4 CROSSBARS.

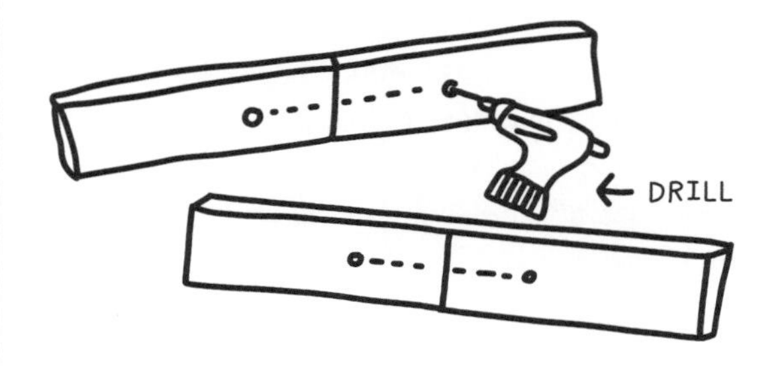

4. DRILL HOLES IN THE CROSSBARS FOR THE BIKE HOOKS.

5. PAINT ALL THE PIECES, IF DESIRED.

Once you've located a stud, mark its location with an X using a pencil. Make a few Xs at different vertical points. Connect these Xs with a straight pencil line using your level or plumb bob so you have a clear marking of where the center of your stud is. This is where you'll attach the vertical 2×4 "spine" of your bike rack.

Step 2. **Cut your 2×4s to length.**

You'll use two 2×4s sandwiched together and screwed to the wall to make up the vertical spine of your bike rack. If you only use one, there isn't enough distance between the wall and the bike for the handlebars to hang square.

First, cut those two 2×4s to match the height of your room. Measure the height of your wall from floor to ceiling (use the extended measuring tape technique on page 74). If it's more than 8 feet, congratulations! Your work is done and you don't need to make any cuts—just use the full 8-foot-long pieces. If your wall height is less than 8 feet, cut your 2×4s to match the height of your wall, minus an inch so you don't end up having to jam it between the floor and ceiling and risk scratching up your walls.

If you have to make cuts, use your tape measure, speed square, and pencil to mark your cut, then cut your pieces with a miter saw or handsaw.

Sand the edges if they're splintering.

Step 3. **Cut and mark your 1×4 crossbars.**

Your 1×4 boards will form the crossbar to hold your bike.

For 2 bikes, cut 2 pieces, each 24 inches.

Mark the center point of each (12 inches from either end) and draw a line across the face using your speed square. This will help you in the next steps.

Give the cut edges a quick sanding, if necessary.

Step 4. **Drill holes in the crossbars for the bike hooks.**

Find a pilot-hole drill bit that matches the shank size of your bike hooks (the threaded end that screws into the wood). This bit should be slightly smaller than the diameter of the bike-hook screw only (not including the threads). My best guess is that you'll use a 3⁄16-inch or 1⁄4-inch pilot-hole drill bit.

On each crossbar, drill 2 holes, each 6 inches from the center point. You should now have 2 holes, 12 inches apart from each other, centered on the length of your board.

Step 5. **Paint all the pieces, if desired.**

If you want to paint your wood pieces, now's your chance! Use an interior house paint; you can match the color of your wall to make the bike rack "disappear," or go with a contrast color, or just leave it as unfinished wood.

Step 6. **Attach the first 2×4 to the wall.**

Place one of your 2×4s against the wall vertically, so the center of it lines up with the location of the stud you marked previously. You'll have to take your best guess here, because you won't be able to see the line you drew once the 2×4 is in place. If you want to get precise, you could measure and mark 1¾ inches to each side of the line you drew on your wall (because a 2×4 is actually 3½ inches wide, remember?), and then place the 2×4 on the wall between the lines.

Use a level or a plumb bob to make sure the 2×4 is perfectly vertical, and that the entire length of it is in line with the stud.

Use your 3-inch screws and your driver to secure it in place, placing 1 screw every 8 inches or so, along the center of the 2×4. These screws will go through your 2×4, drywall, and into the stud.

[continued]

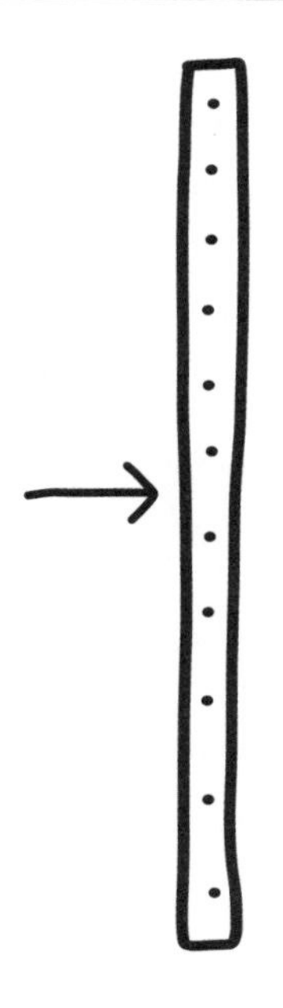

6. ATTACH FIRST 2X4 TO THE WALL.

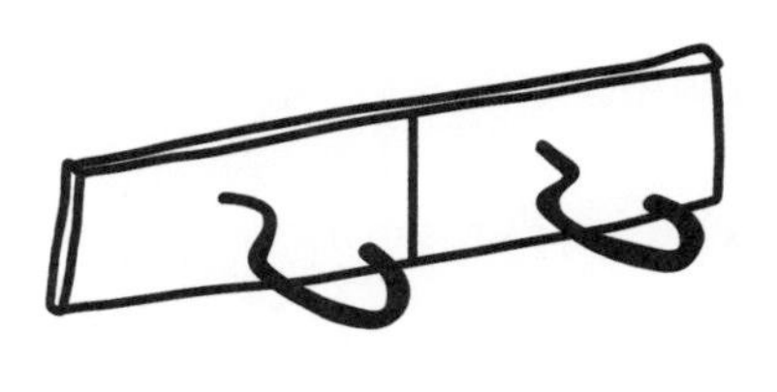

7. ATTACH THE BIKE HOOKS TO THE CROSSBARS.

8. ATTACH THE CROSSBARS TO YOUR SECOND 2X4.

9. ATTACH THE SECOND 2X4 TO THE FIRST 2X4 ON THE WALL AND HANG YOUR BIKES!

Step 7. Attach the bike hooks to the crossbars.

Now that your paint is dry, you can install your bike hooks in your predrilled holes. You should be able to screw them in by hand. Your 1×4 is only ¾ inch thick, but screw the bike hooks in all the way, up to the rubber coating on the hook. The screw part might be sticking out the back of your 1×4 board just a bit, and that's okay.

Step 8. Attach the crossbars to your second 2×4.

First, you'll want to grab the bikes you plan to hang and figure out the best spots to place your crossbars based on the bikes' heights. If you hang your crosspieces too low or too close together, your bikes won't have enough room to hang freely.

Lay the second 2×4 vertical piece flat on the floor.

Lay the bikes on top to estimate the approximate location to place your crossbars (the crossbars will support the top tube [crossbar] of the bike). Make an approximate mark on your 2×4 to note these locations.

Attach your crossbar, perpendicular to your 2×4 board, with the center point of your crossbars matching the centerline of your 2×4. Your 2×4 is 3½ inches wide, so you can find the center of the 2×4 (1¾ inches from each edge) and mark it, then line it up with the center of your 1×4. Matching the center points is really important, because it creates balance and stability when you hang your bike.

Once you have everything in line, attach the crossbars to your 2×4 using your 2-inch screws. Use 3 screws in a triangular pattern for a stable connection.

Step 9. Attach the second 2×4 to the first 2×4 on the wall and hang your bikes!

Now just line up your second 2×4 with your first and connect them using extra-long 5-inch screws. These screws will go through both 2×4s, your drywall, and into the stud, creating a super-strong connection. Place these screws just off the center line of the 2×4s to avoid hitting the first set of screws you drove through the first 2×4. Now grab your bikes and hang 'em up!

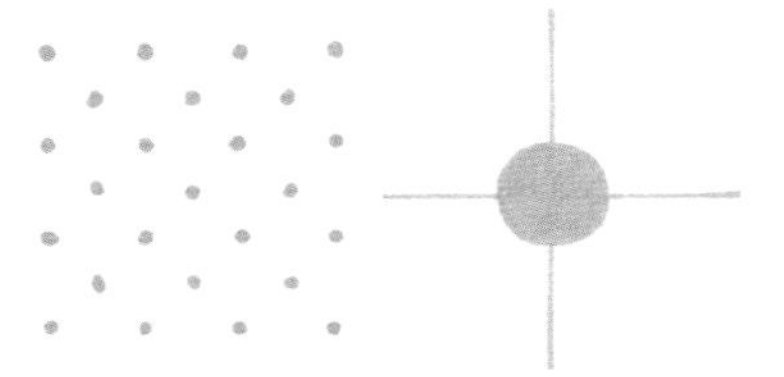

MODULAR BOOKSHELF

During every week of our summer camp at Girls Garage, twenty-four girls work together to build a project for a community partner. One summer, our local women's shelter asked for a bookshelf that would fill one entire wall of their front office. We built this modular bookshelf for them, which allowed each girl to make one cube, and then we assembled all twenty-four cubes together into a giant shelf symbolizing that "our whole is greater than the sum of our parts." You can also modify this project to make cubes of different sizes, rectangles, and more, then piece them together like a mosaic. I like to use 1×12 wood boards (usually common pine or poplar) because the width (11¼ inches, not actually 12 inches) is perfect for the depth of the bookshelf.

Materials

- 1×12 wood planks (just over 4 feet per cube for a 12-inch-by-12-inch-by-12-inch cube). You can also use ¾-inch plywood instead, but you'll need to cut it down into 11¼-inch-wide strips.
- 1⅝-inch screws
- Interior house paint and paintbrushes, if desired
- 1¼-inch screws
- Brush-on polyurethane (clear or colored stain)
- Disposable nitrile gloves to protect your hands when applying polyurethane
- L-brackets to secure the shelf to the wall

Tools

- Circular saw or table saw (if you use plywood and need to cut it into strips)
- Speed square
- Tape measure
- Pencil
- Miter saw (ideal) or circular saw or handsaw
- Sanding block and medium-grit sandpaper
- Drill and 3/32-inch drill bit for pilot holes
- Driver and bit to match screws
- Clamps

SAFETY CHECK!

- An adult builder buddy
- Safety glasses (wear at all times!)
- Ear protection, if using a power saw
- Dust mask, if using a power saw
- Hair tied back
- No dangly jewelry or hoodie strings
- Sleeves pushed up to your elbows
- Closed-toe shoes

[continued]

Step 1. Decide how big you want your bookshelf to be (and how many cubes to make).

This bookshelf is made up of multiple cubes that are attached together like a set of building blocks. Each cube measures approximately 1 cubic foot. You might want to sketch this out and measure where it's going, to make sure it will fit. One of my favorite configurations is 3 rows of 4 cubes, for a total of 12 cubes, with the middle row staggered to the side a bit, like bricks. But have some fun with it and design your own arrangement!

Step 2. If using plywood, rip cut your 11¼-inch strips.

If using plywood, whether you have a full or half sheet of plywood or something in between, it's almost certainly wider than 11¼ inches. Cut your 11¼-inch strips—remember, our 1×12 wood boards are not actually 12 inches, but rather 11¼ inches, so we're matching that dimension for good measure (pun intended). Skip this step if you're using 1×12 boards.

Use a circular saw or a table saw, if you have access to one, to cut 11¼-inch-wide strips out of your plywood. You'll need just over 4 feet of length for each cube that you intend to build, so calculate accordingly.

If you have a full sheet of plywood (4 feet by 8 feet), you should be able to get four 8-foot-long strips, each measuring 11¼ inches wide.

Step 3. Cut the 12-inch pieces for all your cube sides.

Your ideal tool for this is a miter saw, but you'll need a large 12-inch miter saw that has a table surface big enough to accommodate the width of your 1×12 board. If you don't have one, use a circular saw or handsaw (and some muscle!).

Grab your speed square and tape measure.

Measure 12-inch lengths out of your 1×12 board. *Measure, then cut. Measure, then cut.* You'll need four 12-inch-long pieces for each cube, so do your multiplication and get cutting! Your finished cut pieces should measure 11¼ inches wide by 12 inches long. The 11¼-inch dimension will correspond to the depth of the bookshelf.

Step 4. Sand your edges.

Once your cubes are assembled, it is harder to sand all the crevices where your pieces meet, so now's a good time to run a sanding block with a medium-grit sandpaper over the edges of all your cut pieces.

Step 5. Assemble each cube in a pinwheel pattern.

To assemble, take 4 pieces and stand them up on their 12-inch sides (remember that the 11¼-inch dimension is the depth of the cube). Arrange them into a square, in a pinwheel pattern, so that each piece butts into the next (this is tricky to explain: see the illustration!). If you've done this correctly, the total length of an assembled side of a cube should be 12¾ inches (12 inches for one piece's length and ¾ inch for the width of the plywood it connects to).

You will need 3 of your 1⅝-inch screws per corner joint. You might want to mark your screw locations with a pencil first. Starting with one corner joint, drill a pilot hole for your first screw, and then drive the screw. Repeat until you have installed all 3 screws. Then repeat for each corner joint, and for each cube. Whew, that's a lot of screws!

[continued]

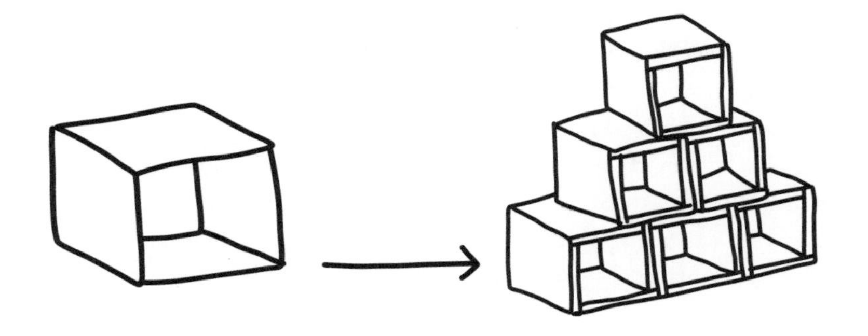

1. DECIDE HOW BIG YOU WANT YOUR BOOKSHELF TO BE (AND HOW MANY CUBES TO MAKE).

2. IF USING PLYWOOD, RIP CUT YOUR $11\frac{1}{4}$" STRIPS.

3. CUT THE 12" PIECES FOR ALL YOUR CUBE SIDES.

4. SAND YOUR EDGES.

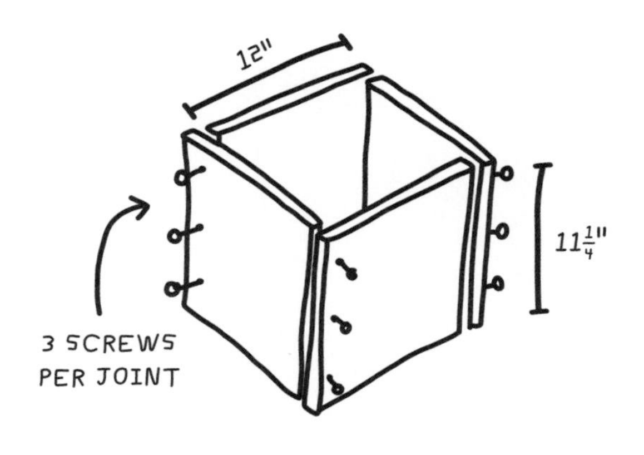

5. ASSEMBLE EACH CUBE IN A PINWHEEL PATTERN.

Step 6. Paint and arrange.

It is much easier to paint your cubes now than after they're assembled, so tap into your inner toddler and stack 'em up like blocks until you figure out an awesome configuration and paint scheme. Paint any of your cubes in any colors, letting them dry before the next step.

Step 7. Attach the cubes together.

Now that you have a pile of beautifully constructed and possibly painted cubes, it's time to put them all together!

Beginning with your lowest row of cubes, place 2 cubes next to each other. Use a clamp or two to hold them together snugly.

Drive four 1¼-inch screws in a square pattern to connect their adjacent faces to each other. There's no need for pilot holes this time because you're far enough from the edge of your piece of wood to avoid any splintering.

Connect each cube with its neighbor on the bottom row, then move up to the next row.

If you've designed your shelf with cubes that purposely meet in a staggered configuration, use your best judgment for placing your connecting screws.

Step 8. Clear-coat and fill with books!

If you didn't paint your shelf, you'll want to give it a quick clear coat to protect it from stains and discoloration. Dust off your shelf, then use a brush-on polyurethane (do this outside!), and let it dry. Then gather up your personal library and fill 'er up! You could also use the shelf as a display case for all your other hand-built projects (wink, wink). If you're placing your shelf against a wall, use 4 to 6 small L-brackets to secure the shelf to the wall, with one leg of the L connecting to the shelf and the other to the wall.

Step 9. (Optional) Add a back piece.

If you want a solid back for your shelf, lay your shelf down flat on the floor on a piece of plywood. Trace its profile on the plywood, then cut it out using a circular saw or jigsaw. Attach the plywood to the back of the shelf using 1¼-inch screws, around the edge, and in a few key spots in the middle. If you do add a back, you can secure the shelf to the wall directly through the back piece and into the studs in the wall.

+ + BEYOND + +

Even though it's "hip to be square," there's no requirement that your individual cubes have to be 12 inches by 12 inches by 12 inches! While it's easy to make all your cuts and cubes consistent, you could certainly experiment with cubes and rectangles of different sizes and assemble them in asymmetrical ways. Have fun with it and try out different shapes and sizes!

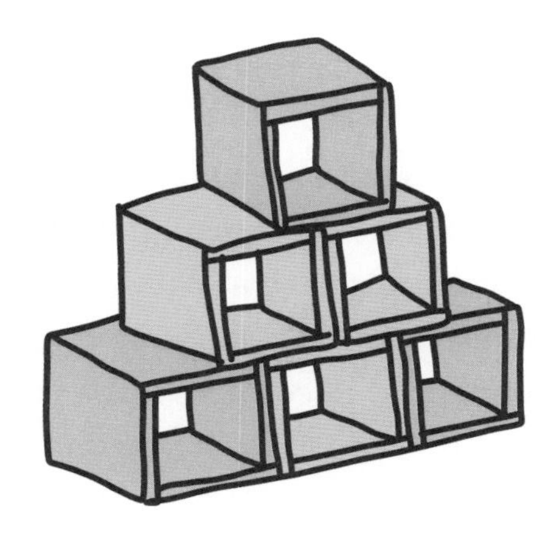

6. PAINT AND ARRANGE.

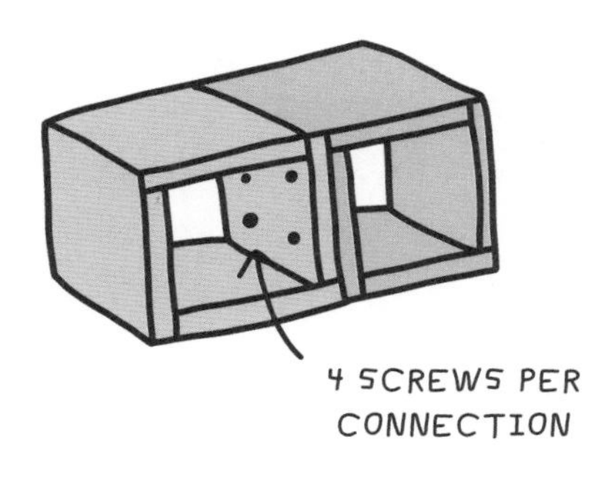

7. ATTACH THE CUBES TOGETHER.

8. CLEAR-COAT AND FILL WITH BOOKS!

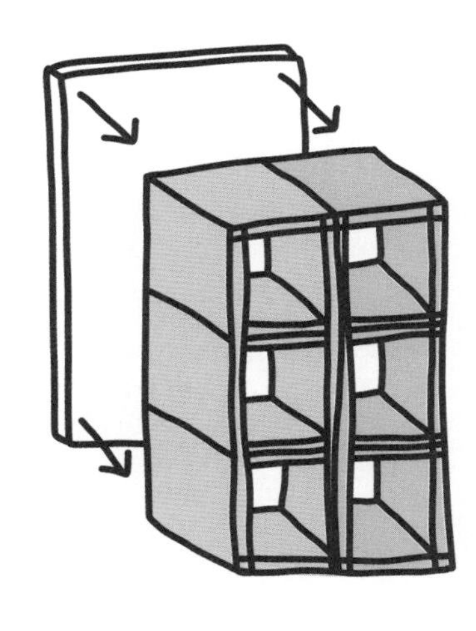

9. (OPTIONAL) ADD A BACK PIECE.

MILK CARTON CONCRETE PLANTER

Mixing and pouring concrete brings me back to my sandbox days, when I used to mix sand and water to create the perfect sandcastle consistency. It turns out that concrete works similarly, and you can pour concrete into any mold ("formwork"), as long as you can remove the formwork afterward to get your concrete piece out. The principles of this project can be adapted in so many other ways, too, as you can experiment with other molds, like plastic bottles, or you can build your own wooden formwork (basically a box that you pour concrete into, then remove it after the concrete has set).

Materials

- ½-gallon paper milk carton
- Bag of concrete (any quick-setting mix, or a finer countertop mix if you want a smoother pour and finished product)
- Water
- Small disposable plastic cup (with a diameter about half of the milk carton's)
- Coins or gravel or miscellaneous hardware (to weigh-down the cup)
- Paint and paintbrushes, if desired (spray or any art/craft paint)
- Spray-on water-based polyurethane
- Potting soil
- Succulent or other flora

Tools

- Scissors
- Mixing bowl
- Durable mixing spoon
- Wooden chopstick or stir stick
- 2 clamps
- Box cutter
- Pliers
- Fine-grit sandpaper (220 or higher)

SAFETY CHECK!

- An adult builder buddy
- Safety glasses (wear at all times!)
- Dust mask

Step 1. **Cut your carton.**

Use a pair of scissors to cut your milk carton down to your desired height. A cube shape tends to work well and is a nice proportion for planting a succulent, so I usually cut my carton down to about 6 inches tall (though I only fill it with about 4 inches of concrete).

Step 2. **Mix your concrete (not too wet!).**

When you pour or scoop your concrete out of the bag, it can be super dusty, so now is a good time to put on your dust mask.

[continued]

1. CUT YOUR CARTON.

2. MIX YOUR CONCRETE (NOT TOO WET!).

3. POUR CONCRETE, INSERT CUP, POUR MORE CONCRETE.

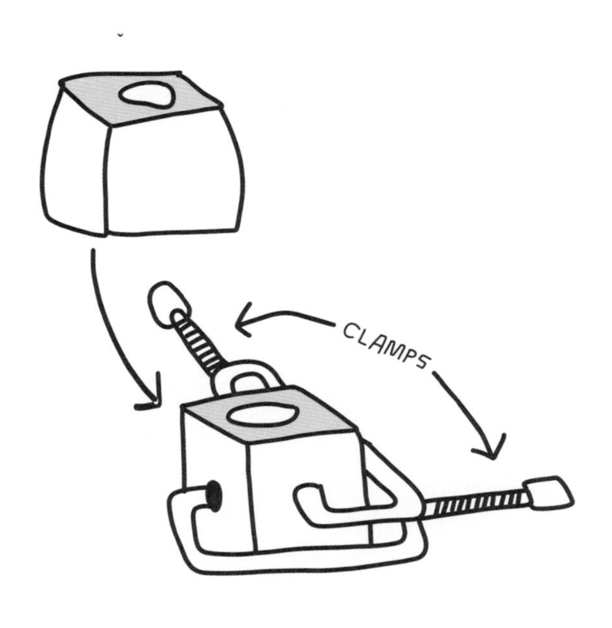

4. REMOVE AIR BUBBLES AND "SQUARE" YOUR MOLD.

Follow the mixing guidelines on whatever concrete mixture you have. Use a mixing bowl and a durable mixing spoon (you'll need to put some muscle into it!).

A good general rule is to pour water a little bit at a time into your dry mix. Your concrete should be the consistency of wet sand, not runny, and should be able to hold its own form. Using too-wet concrete can make your finished product brittle and prone to cracking!

Step 3. Pour concrete, insert cup, pour more concrete.

Use your spoon to scoop your mixed concrete into the carton until it's about halfway full.

Fill your cup with some gravel or coins or miscellaneous hardware to give it a bit of weight. Then set the cup in the carton on top of the concrete you just poured.

Now fill in the rest of the carton around the cup with the concrete to your desired planter height. It's helpful to stop filling the carton about an inch below the top, to make it easier to rip off and remove the carton later. The weights in your cup should prevent it from rising up to the top.

Step 4. Remove air bubbles and "square" your mold.

You're almost there! Because of the concrete's consistency, you'll likely have pockets of air that you'll definitely need to get out.

Use a chopstick to poke down into the concrete to remove air bubbles, especially along the sides of the carton, and also tap the sides lightly with your hands. When the top of the pour looks smooth and consistent, you should be good to go.

You'll also notice that the sides of your carton are probably bulging out, so it isn't quite square anymore, but more like a blobby square. Use 2 clamps, one spanning the square in each direction, to square up the sides. Don't squeeze too tightly—clamp just enough to get your sides square. Now you're done with the concrete!

Step 5. Let it cure for 24 hours, remove, and sand.

Walk away and let the concrete set for the specified amount of time for your mix, probably about 24 hours. It's important to allow the entire curing process to happen, otherwise your planter will be prone to cracks. Keep your piece in a cool area and away from direct sunlight to ensure a good curing process.

Cleanup time! If you've used reusable bowls (or "borrowed" them from the kitchen), clean them thoroughly and put them through the wash before using them again.

Once your concrete is cured and set, tear off the carton (which should come off easily) or cut it with a box cutter.

The cup might not come out so easily. You can use pliers to pull it out, and don't worry about cracking or tearing it. While removing the cup, be careful not to crack the edges of your planter, which might be a little on the delicate side at this point.

Use a fine-grit sandpaper (220-grit or higher will work great) to smooth any jagged edges and get your planter buttery smooth!

[continued]

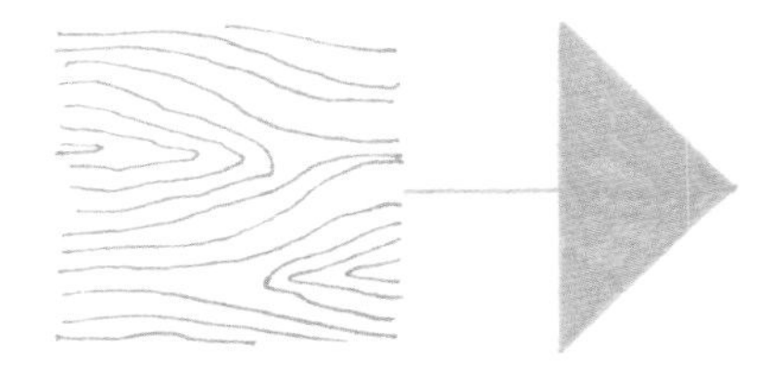

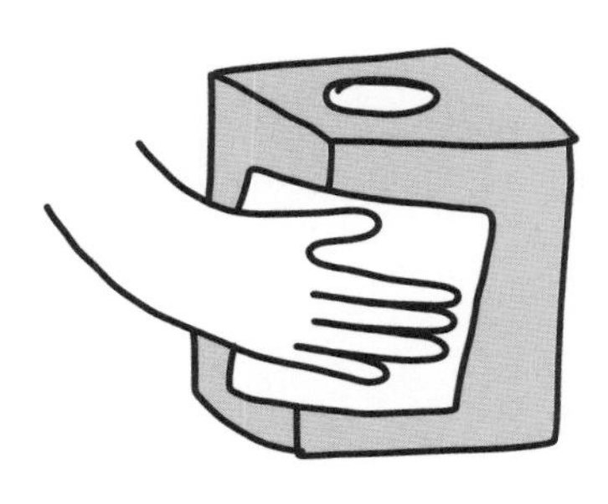

5. LET IT CURE FOR 24 HOURS, REMOVE, AND SAND.

6. PAINT, SEAL, AND PLANT!

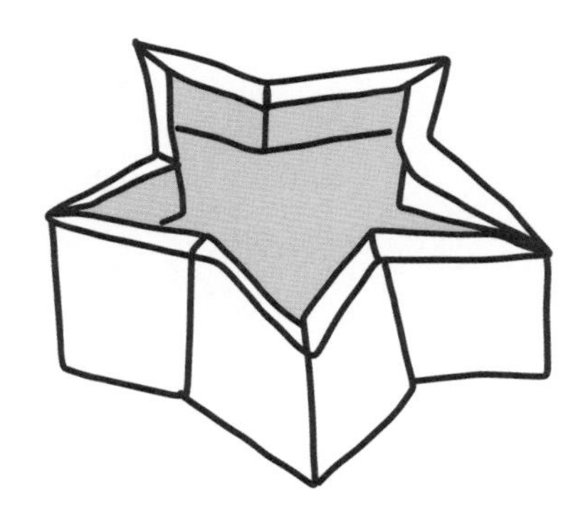

BEYOND. POUR CONCRETE INTO OTHER VESSELS OF ANY SHAPE, JUST MAKE SURE YOU HAVE A WAY TO REMOVE THE VESSEL.

Step 6. Paint, seal, and plant!

Now's your chance to paint your planter.

You can tape off certain parts to make patterns, spray-paint it, or do small painted details with paint pens. My favorite is taping off big geometric patterns and using metallic spray paint. Have fun with it!

Then give it a quick spray with a water-based polyurethane to seal it from moisture or staining.

Add a bit of potting soil and a small succulent, and you've got yourself a planter!

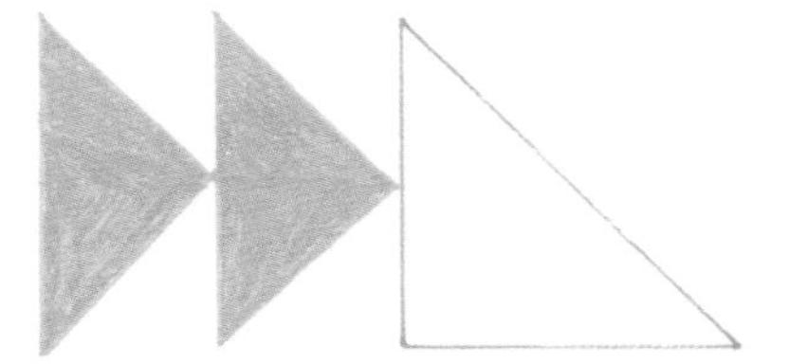

+ + BEYOND + +

There are infinite ways to modify this project! The paper milk carton is a great place to start to get comfortable with concrete as a material because it can be removed so easily. But you can also play with plastic water bottles that can be poured into and then cut off after the concrete cures. You can make taller vessels as flower vases, solid blocks as paperweights, and more.

For an advanced mold-making technique, use wood as your formwork. Basically, any box you can build out of plywood can be poured into (just use petroleum jelly or a spray-on release agent to make it easier to remove). Screw your plywood pieces together using short screws that can be easily removed when the concrete sets. You can pour large slabs for a coffee tabletop, small shapes as pavers and stepping-stones, and more. My favorite material to use as formwork is melamine board, which is particleboard with a front and back layer of a plastic laminate that makes it really easy to pull off once the concrete sets.

STANDING PLANTER BOX

I first saw a version of this planter box at the Ecology Center in Southern California, where an inspiring group of environmental activists promote simple solutions for a more sustainable life. Even for homes, offices, or schools that don't have much space, you can build this planter box and put it anywhere. And because it has legs, it isn't backbreaking to plant into and water. We've built probably fifty of them with girls, students, and teachers, and it's always a fun team-building project! The cedar fence boards used here are inexpensive and meant for outdoor use, so they won't rot in the rain. The redwood balusters used for the legs are also great for outdoor use, if you can find them. If not, use a 2×2 and seal it with a varnish or polyurethane.

Materials

- x Five 5-foot-long cedar fence boards
- x 4 balusters (balusters are precut for making stair banisters; look for some made of redwood or another weather-friendly wood, and that are square in profile without any decorative shapes)
- x 1⅝-inch wood screws
- x 1-inch screws (just a couple)

Tools

- x Miter saw or handsaw
- x Tape measure
- x Speed square
- x Pencil
- x Rubber sanding block and medium-grit sandpaper
- x Drill and 3⁄32-inch bit for pilot holes, plus a ½-inch bit for drainage holes
- x Driver and bit(s) to match your screws
- x Clamps
- x Coping saw or backsaw

SAFETY CHECK!

- ♦ An adult builder buddy
- ♦ Safety glasses (wear at all times!)
- ♦ Ear protection, if using a power saw
- ♦ Dust mask, if using a power saw
- ♦ Hair tied back
- ♦ No dangly jewelry or hoodie strings
- ♦ Sleeves pushed up to your elbows
- ♦ Closed-toe shoes

Step 1. **Cut your side, bottom, and leg pieces out of fence board.**

If you have a miter saw, use it; otherwise, you can use a crosscut saw or a handsaw (but make sure your cuts are straight!).

Set up your cut lines with your tape measure and speed square to make sure you have a perfectly straight perpendicular line to follow. Remember to measure, and then cut, then measure, then cut (don't mark all your cuts at once,

[continued]

because you'll lose length with the kerf of the saw blade). Here's your cut list:

- 4 balusters at 24 inches (legs)
- 4 fence boards at 16 inches (short side pieces)
- 4 fence boards at 24 inches (long side pieces)
- 3 fence boards at 24½ inches (bottom pieces)
- 1 fence board at 15½ inches (bottom crossbeam)

Now's a good time to give all your edges a quick sanding, too. Use a medium-grit sandpaper and a rubber sanding block.

Step 2. Attach the side pieces to the legs in a pinwheel pattern.

Just like with the modular bookshelf project (see page 285), we'll use a pinwheeled rectangle arrangement here to make our planter box a little more stable. I recommend building your planter box upside down, so you can stabilize your side pieces against the tabletop.

Grab 2 of your 16-inch fence boards and 2 of your 24-inch fence boards. Arrange them in a rectangle, with the pieces standing on their edges.

At the corners of the rectangles where they meet, pinwheel the joints, so every corner has one piece butting into the other (see illustration page 299).

Place the legs vertically in each corner and use your drill and ³⁄₃₂-inch bit to drill 2 pilot holes through the side pieces and into the legs, then use your driver and two 1⅝-inch screws to attach everything. Drive your screws from the outside in, going through the side piece and into the leg. You might need a friend, some clamps, and a speed square to help hold everything steady and square. Use 2 screws per attachment.

Move on to the second row of side pieces, which should be much easier now that the first row is installed!

Step 3. Cut two corner notches in two of your bottom pieces.

You should now have a planter box with sides but no bottom. Keep it upside down, so you can install the bottom pieces. Here's the trick: your 3 bottom pieces will span across the whole box and attach to the sides, but you'll need to cut some square notches out of the 2 outer pieces so they fit around the legs.

Grab your three 24½-inch bottom pieces. In two of them, cut 2 square (2¼-inch-by-2¼-inch) notches out of 2 corners. These notches can be easily cut with a coping saw or backsaw,

[continued]

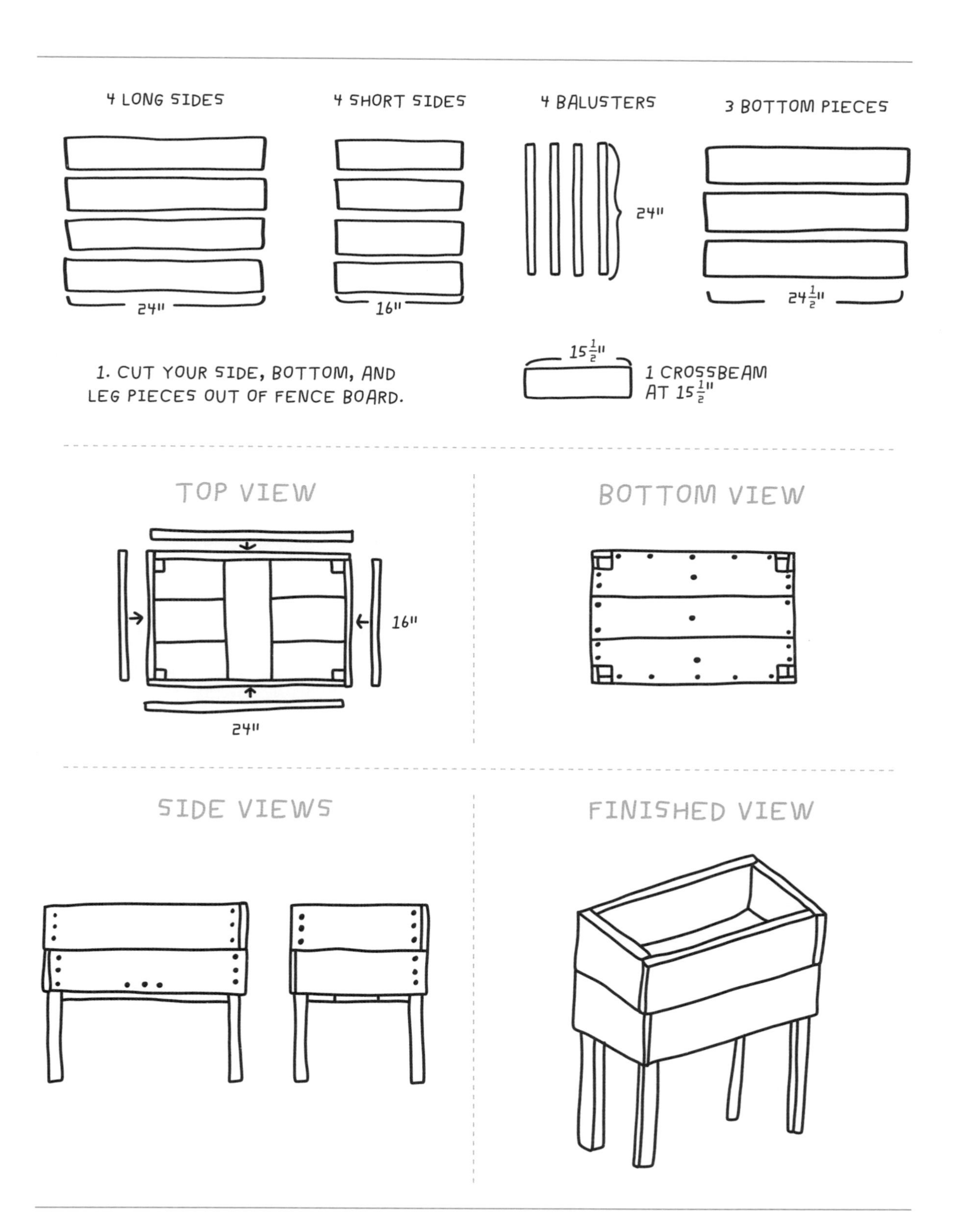

4 LONG SIDES
24"
4 SHORT SIDES
16"
4 BALUSTERS
24"
3 BOTTOM PIECES
24 1/2"
1. CUT YOUR SIDE, BOTTOM, AND LEG PIECES OUT OF FENCE BOARD.
15 1/2"
1 CROSSBEAM AT 15 1/2"
TOP VIEW
16"
24"
BOTTOM VIEW
SIDE VIEWS
FINISHED VIEW

no power saws necessary.

You should now have 2 bottom pieces with 2 corners cut out of each, and 1 bottom piece still intact.

Step 4. Attach the bottom.

Lay your bottom pieces across the bottom of your planter box, with the 2 notched pieces on the sides. Hopefully, your notches fit perfectly around your legs!

Drill a pilot hole, and then use your 1⅝-inch screws to attach the bottom pieces, installing screws along the edge where the bottom pieces meet the siding. Try to place the screws so they go through the bottom piece and into the center of the edge of your side pieces (if they're too close to the edge of your side pieces, the wood will split).

Step 5. Cut and install the bottom crossbeam inside the planter box.

You only have one piece left! The 15½-inch piece is a crossbeam, meant to give some more structure to the bottom of your planter box, so it doesn't bend or sag when filled with dirt.

Flip your planter box over now so it sits right-side up.

Place the crossbeam inside the planter box so it runs across the width, across all 3 bottom boards.

Use your short 1-inch screws to attach the crossbeam to the bottom pieces (no pilot holes necessary here; just keep your screws in the middle of your crossbeam and not too close to the edge). Then use a few 1⅝-inch screws (with a pilot hole) to attach the crossbeam to the sides (install these screws from the outside of the box, through the side boards and into the end of the crossbeam).

Step 6. Drill drainage holes.

There might be a few small gaps between your bottom boards (which is fine and good for drainage!), but it's a good idea to drill a few extra holes for water to drain out of your box. You can place them wherever you'd like on the bottom—shoot for 6 to 8 holes. A large-ish drill bit, like a ½-inch one, is perfect for this.

Step 7. Plant and enjoy!

Depending on what you want to plant, use potting soil and/or gravel. If you're planting small plants that don't need too much soil depth, fill the planter box with rocks or gravel, and then place potting soil on top. Plant away and watch your garden grow!

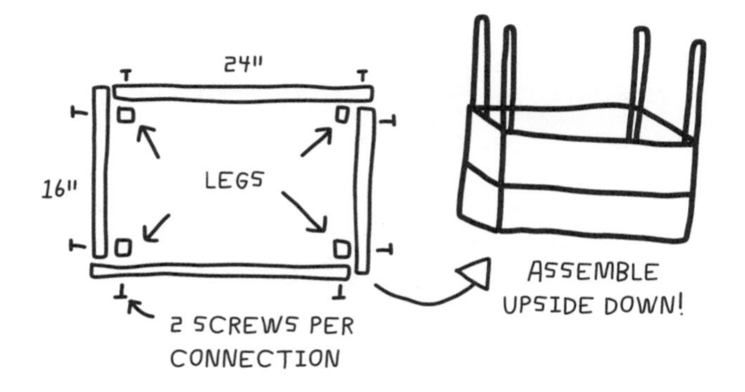

2. ATTACH THE SIDE PIECES TO THE LEGS IN A PINWHEEL PATTERN.

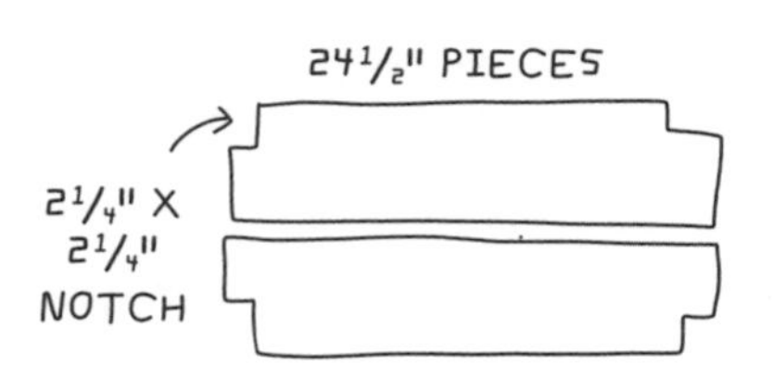

3. CUT TWO CORNER NOTCHES IN TWO OF YOUR BOTTOM PIECES.

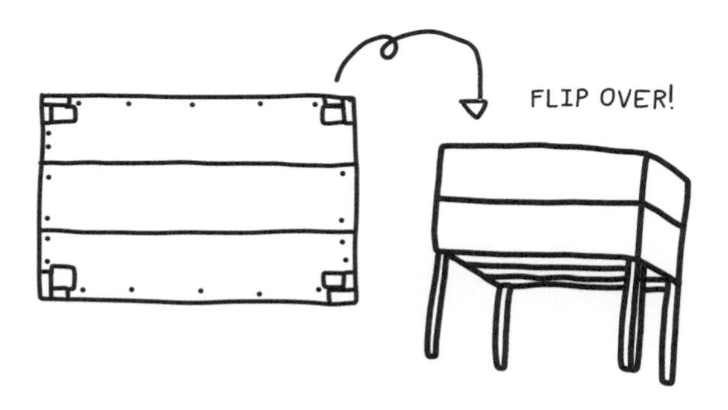

4. ATTACH THE BOTTOM.

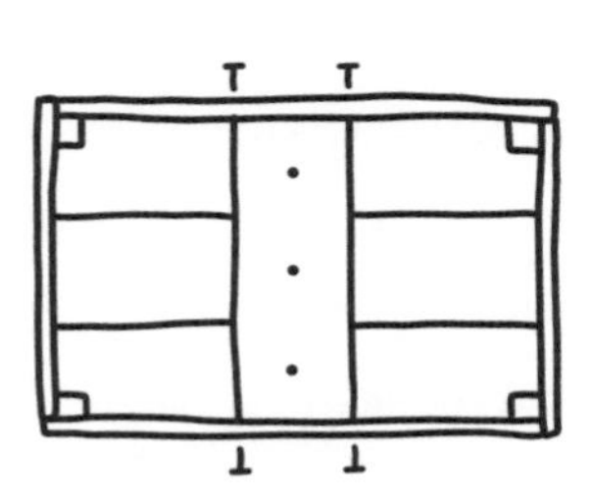

5. CUT AND INSTALL THE BOTTOM CROSSBEAM INSIDE THE PLANTER BOX.

6. DRILL DRAINAGE HOLES.

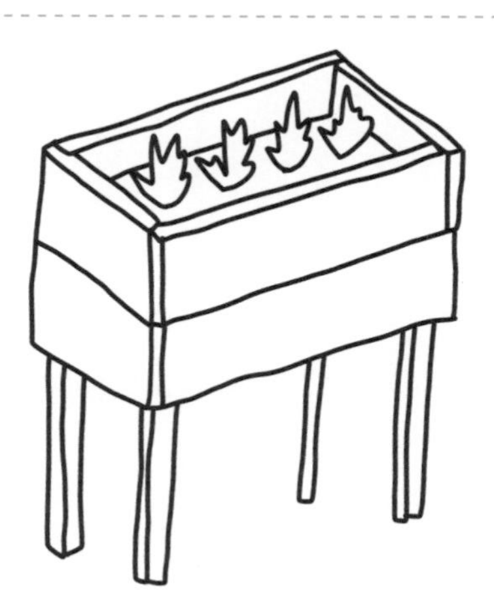

7. PLANT AND ENJOY!

STUD-FRAMED DOGHOUSE

I've always wanted to build my own house someday. There are so many things to learn about how houses are framed, sheathed, wired, insulated, and finished! This doghouse project is a great (miniature) place to start, and it will teach you some of the basics of wood framing, using many of the same materials used in building a house, like 2×4s. Until I can afford to build my own house, a lovely abode for my pup, Junebug, will have to do. The dimensions of this doghouse should work for most breeds (you may need to figure out your own extra-small or extra-large dimensions for a teacup poodle or Great Dane!). This project is really the culminating opus of all your building skills, requiring some precise miter cuts, multiple assemblies, and patience. You can do it!

Materials

- 2×4s, approximately 16 total feet
- 2×2s, approximately 64 total feet
- 2½-inch construction screws
- ¼-inch plywood: 2 full sheets (or 4 half-sheets or 8 quarter-sheets)
- 1⅝-inch wood screws
- Exterior paint and paintbrush, if desired (low- or no-VOC)

Tools

- Tape measure
- Speed square
- Pencil
- Miter saw
- Sanding block and sandpaper
- Driver and bit(s) to match your screws
- Rubber mallet or hammer
- Circular saw
- Chalk line
- 4 clamps
- Drill and a 3⁄32-inch bit for pilot holes
- Jigsaw

SAFETY CHECK!

- **An adult builder buddy**
- **Safety glasses (wear at all times!)**
- **Ear protection, if using a power saw**
- **Dust mask, if using a power saw**
- **Hair tied back**
- **No dangly jewelry or hoodie strings**
- **Sleeves pushed up to your elbows**
- **Closed-toe shoes**

[continued]

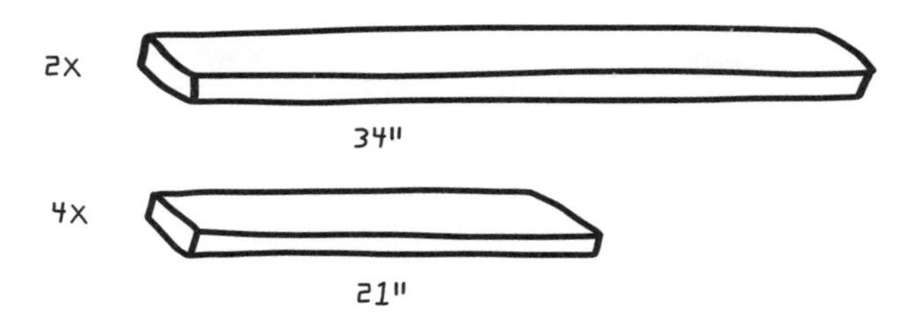

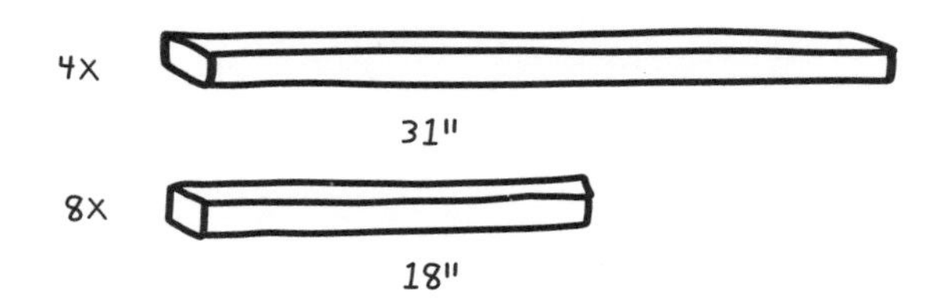

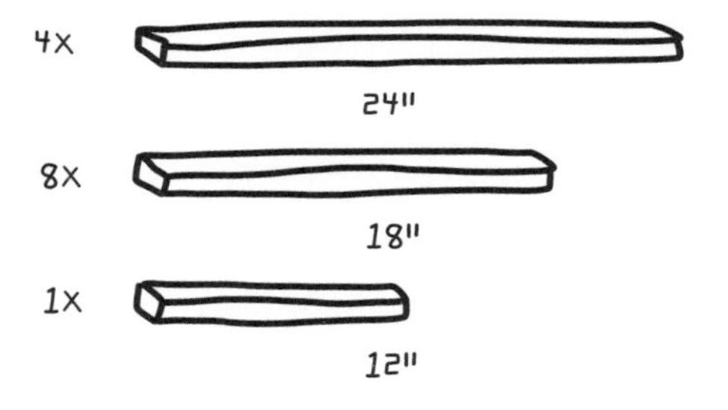

1. CUT YOUR 2X4S AND 2X2S.

Step 1. **Cut your 2×4s and 2×2s.**

This is great miter saw practice! Measure and cut all your pieces precisely, with the exception of your 45-degree-angle miter cuts for your roof, which you will do later. You can also run some sandpaper over your cut edges at this time, to make sure your pooch doesn't end up with splinters. Here is your cut list:

Base

- **2×4:** 2 at 34 inches
- **2×4:** 4 at 21 inches

Sides

- **2×2:** 4 at 31 inches
- **2×2:** 8 at 18 inches

Front and Back

- **2×2:** 4 at 24 inches
- **2×2:** 8 at 18 inches
- **2×2:** 1 at 12 inches (for doorway header)

You'll also need to cut your roof rafter and plywood siding pieces later. Don't worry about them until the later steps, but here is the list for reference:

Roof Rafters

- **2×2s:** 8 at 17½ inches at a 45-degree miter cut (17½ inches on the long edge)

Plywood Siding (¼ inch)

- 1 at 34 inches by 24 inches (base), which you'll cut and attach next during step 2
- 2 at 34 inches by 24½ inches (sides)
- 2 at 24 inches by 24½ inches (front and back)
- 2 at 36 inches by 19 inches (roof sides)

Step 2. **Build the base platform.**

Your 2×4s and some plywood will make up your base platform, which will give your doghouse a little bit of lift off the ground to keep the bottom from rotting. You'll frame the base just like you'd frame a house, with your 2×4s on their skinny edges.

Use the 34-inch pieces as the top and bottom (header and footer).

Use the first two of your 21-inch pieces as the short sides to complete the rectangular frame, with the ends of your 21-inch pieces butting into the 34-inch pieces. Attach these four pieces using a driver and two 2½-inch screws in each corner joint, going through the 34-inch piece into the end of the 21-inch piece (no need to drill pilot holes).

Place your 2 remaining 21-inch pieces in the middle of the rectangle, running parallel to the other 21-inch pieces like railroad tracks, evenly spaced (it doesn't need to be perfect, as you'll cover it up later). Use a rubber mallet to bump them into place if necessary. Use a speed square to make sure the pieces are perpendicular to the 34-inch pieces, and install them with two 2½-inch screws in each joint. You should now have a frame for your base measuring 24 inches by 34 inches. Note that the height is now 24 inches, which includes the height of your 21-inch sides plus the height of the top and bottom 2×4s, which are 1½ inches each.

Complete the base: Use your circular saw to cut a piece of plywood that measures 24 inches by 34 inches. You might need a chalk line (see page 86) so you have a full line across the plywood where you will cut.

Give your edges a quick sanding.

Place this plywood over the top of your base and attach it using 1⅝-inch screws, 6 inches or so apart. I like to make a small mark on the side

[continued]

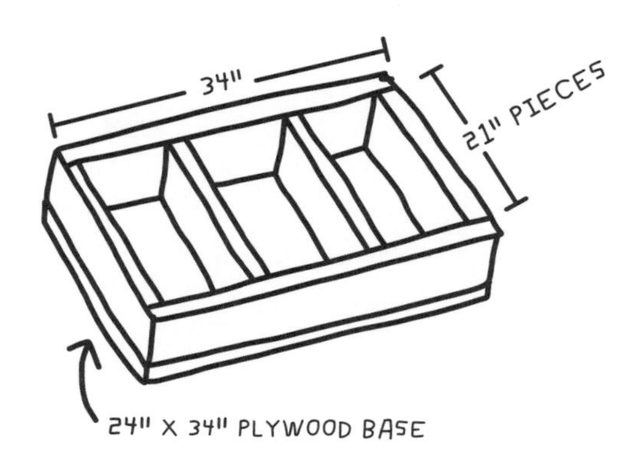

2. BUILD THE BASE PLATFORM.

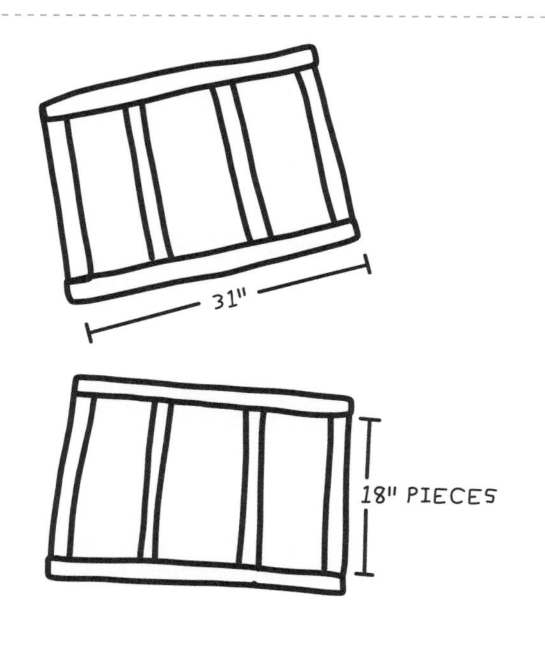

3. FRAME YOUR SIDE WALLS.

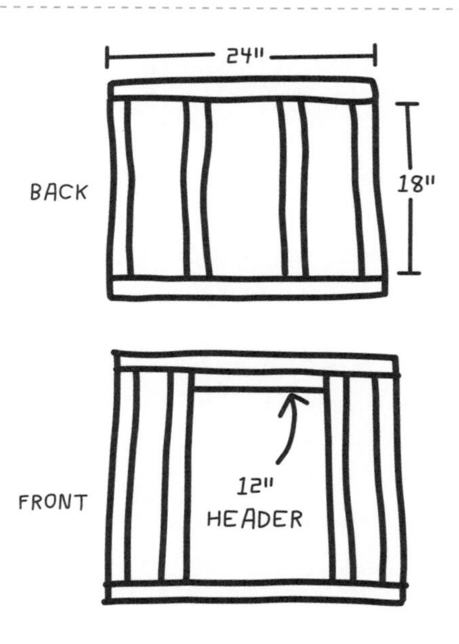

4. FRAME YOUR FRONT AND BACK WALLS.

of my base where I put screws in, so I know where they are and can avoid them when attaching my walls!

Step 3. Frame your side walls.

Your side walls will measure 31 inches wide by 21 inches tall and will sit on top of the base. They are only 31 inches long, instead of the 34-inch length of the base, because they will butt into the front and back walls, which are each 1½ inches thick.

Lay out two 31-inch-long 2×2 pieces parallel to each other as the top and bottom (header and footer) of your side walls.

Place four 18-inch 2×2 pieces vertically between them like a railroad track.

Starting with the two end pieces, drill pilot holes and attach the pieces using a driver and your 2½-inch screws, 2 in each joint.

The two middle 18-inch pieces should be placed 8 inches from the side of the frame. Measure and mark this point on your top and bottom pieces so you can line up the 18-inch pieces. Use a speed square to make sure they're perpendicular to the top and bottom pieces. You might need a rubber mallet to bump these middle pieces a bit before putting the screws in, because they might be squeezed in there between the top and bottom pieces.

Repeat this process to make the second side wall the exact same way.

Step 4. Frame your front and back walls.

Your front and back walls will be almost identical, but the front wall will have a slight variation to the framing pattern to accommodate the door opening.

Grab two 24-inch-long 2×2s. These are your top and bottom pieces (header and footer) that match the width of your base. Lay them out parallel to each other.

Place your four 18-inch-long pieces in between and perpendicular, like a railroad track. Attach the outside two first, making a complete frame. For the back wall, space the remaining two 18-inch pieces evenly inside the rectangle.

Repeat this process for the front wall, **with one exception!** When installing the two inside 18-inch pieces, place the 12-inch-long 2×2 piece between them, parallel to the top and bottom. This is your door header, or the top height of the door your dog will walk through. You can place it at any level you think will accommodate your dog's size!

Now attach that door header to the two 18-inch pieces, and then the two inside 18-inch pieces to the top and bottom of the frame.

Step 5. Attach the walls to the base and to each other.

Now you've got four walls and a base! Time for a barn (doghouse!) raising. Grab 4 clamps to help hold everything in place.

Tilt up your walls and place them on the edges of the base. Make sure the side walls are sandwiched between the front and back walls, as if the front and back walls are the bread and the side walls are the meat. Make sure the walls line up with each other and the walls' bases all line up with the edge of the base. Use a clamp to hold each wall to the one next to it.

[continued]

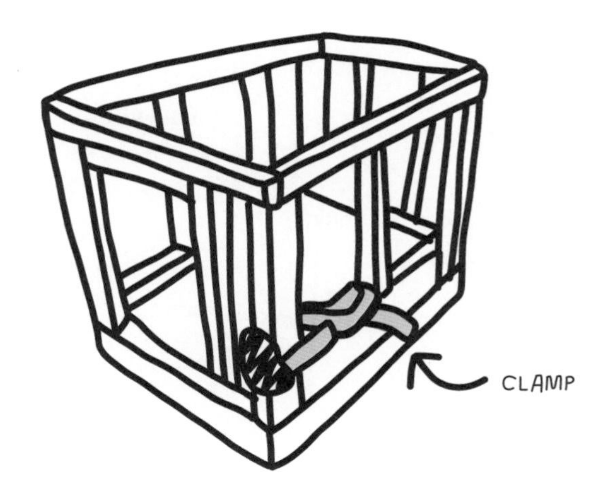

5. ATTACH THE WALLS TO THE BASE AND TO EACH OTHER.

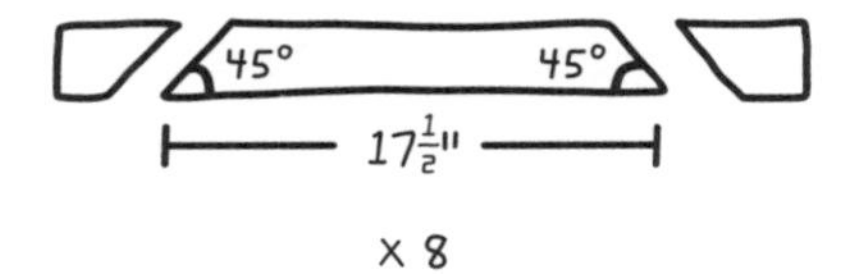

x 8

6. CUT YOUR ROOF PIECES (RAFTERS).

7. ATTACH THE ROOF RAFTERS TO THE TOP OF THE SIDE WALLS.

Use your driver and 2½-inch screws to attach the footer of each wall to the base (maybe 3 or 4 screws per wall). Be careful where you place these screws—you don't want to hit the screws you already put in when you attached the plywood to your base!

Attach the side of each wall to the wall next to it, with three 2½-inch screws per connection (one up high, one in the middle, and one down low). Remove the clamps once everything is attached!

Step 6. **Cut your roof pieces (rafters).**

This is probably the trickiest part, but also the most satisfying. Your roof will be made up of four rafters that slope up at a 45-degree angle and meet at the top. Set your miter saw to make the miter cuts. You'll need 8 pieces, with each end cut at a 45-degree angle (the profile will look like a trapezoid). One end of each piece will sit on top of the side wall, and the other will meet its twin at the top of the roof.

To get accurate cuts and angles, start by cutting off an end of a 2×2 piece at a 45-degree angle. No need to measure; you just need one angled end to start.

Measure from the long side of that angled cut (the pointy side), and mark 17½ inches.

Now use a speed square to draw a 45-degree line from that point across your 2×2. Make the miter cut using your saw. Your cut piece should be trapezoidal in shape.

Repeat the measuring, marking, and cutting 7 more times until you have 8 pieces total.

Step 7. **Attach the roof rafters to the top of the side walls.**

Take 2 of the pieces you've just miter cut and lay them out in a corner shape, like the corner of a picture frame.

Drill a pilot hole first so you don't split your wood, and then use two 1⅝-inch screws to attach that corner joint—one screw from one side, and one from the other.

Do this for each pair, so you have 4 total corner shapes as your rafters.

Now you'll attach each of these rafters to the top of your side walls, one at the front, one at the back, and two in the middle, lined up with the vertical posts in your side walls. Set the first rafter on top of your walls, spanning from one side wall to the other. Drill a pilot hole vertically where the rafter meets the side wall, and attach them with a 1⅝-inch screw. Do this for all 4 rafters.

Step 8. **Cut the plywood siding (sides, front, back, roof).**

You have already cut your plywood for your base and attached it. Now cut the plywood pieces for your siding, which will cover all four walls and the roof.

[continued]

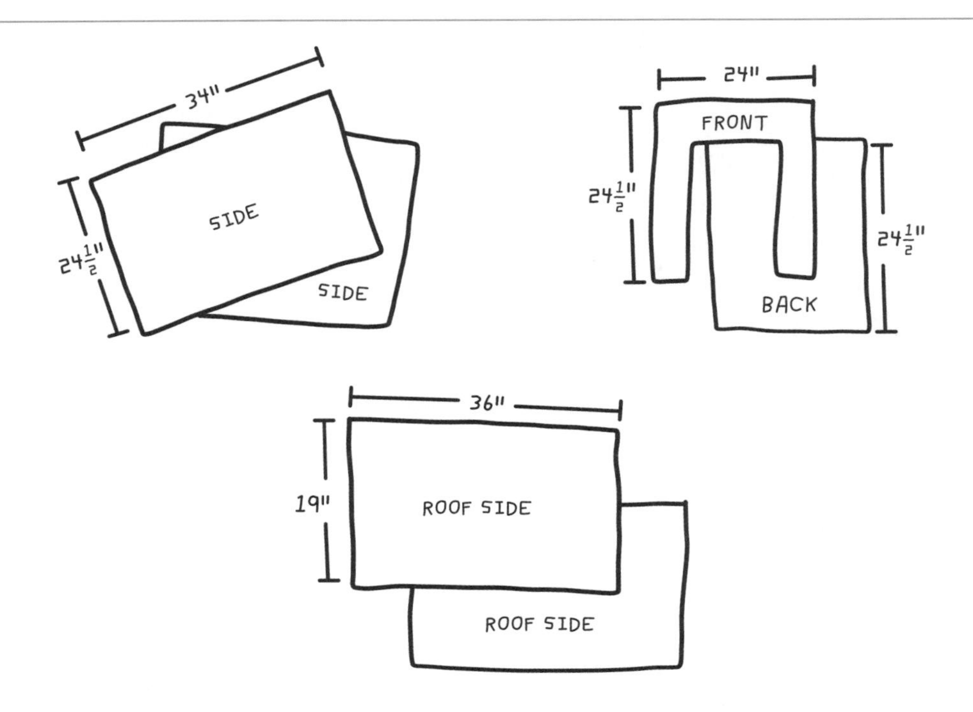

8. CUT THE PLYWOOD SIDING (SIDES, FRONT, BACK, ROOF).

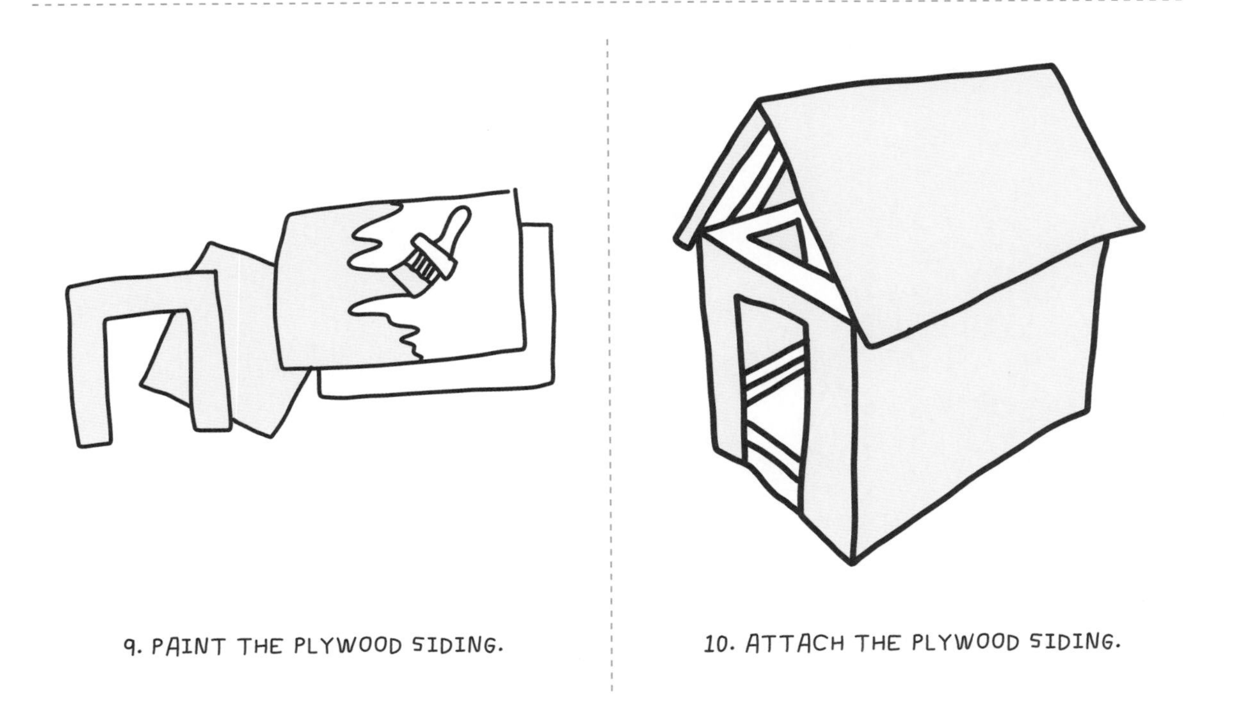

9. PAINT THE PLYWOOD SIDING.

10. ATTACH THE PLYWOOD SIDING.

Measure and mark your lines precisely before using a circular saw to make your cuts. You might want to use a chalk line for the larger pieces.

For your front piece, you need to cut out the front door, too. Place the plywood piece over the front wall of the doghouse, and, from the inside, trace the shape of the doorway with a pencil. Use a jigsaw to cut out the door.

Step 9. **Paint the plywood siding.**

If you want to paint your walls, now is your chance! Make sure to dust off all your wood pieces before applying paint. Since your beloved canine will be living in this doghouse, I recommend using an exterior house paint (so it's durable) that is low- or no-VOC (volatile organic compounds, which are toxic). Let your paint dry fully before the next step.

Step 10. **Attach the plywood siding.**

You're in the home stretch!

Match each plywood piece to the corresponding wall and the sloped sides of the roof. Note that the 24½-inch height of your side, front, and back walls will cover the wall (21 inches tall) *and* the height of your base frame (3½ inches). The plywood should cover the entire doghouse side, all the way down to the ground.

Use your 1⅝-inch screws (no pilot holes necessary) to attach them. Make sure your screws go through the plywood and into the frame pieces of the walls and roof, both around the outside of the frame of each wall and along the interior studs. Note that the triangular space above the front and back walls, under the roof, are purposely left open for ventilation. The roof should have a slight overhang, too.

Give your doghouse a last once-over, looking for splinters or sharp edges that need sanding. Pick out an extra-soft dog bed, a couple of toys, and make your doggie feel at home!

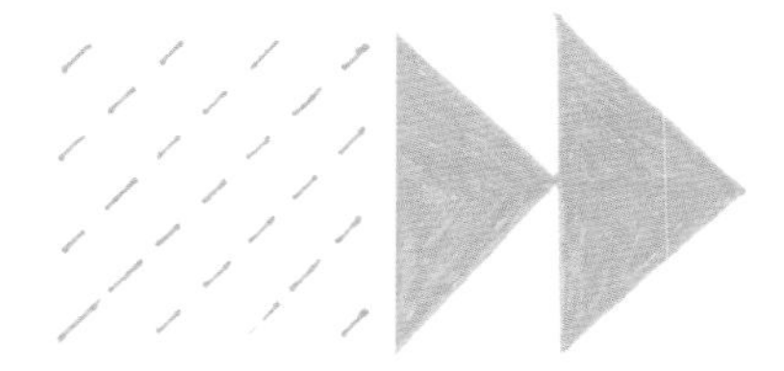

ACKNOWLEDGMENTS

In 2015, I sat at a table with Erica, Teah, and Eliza, three young girls from Girls Garage, sketching our favorite tools for "a book for girls about tools and building" that we might self-publish someday. Those first sketches and conversations, along with the ambitions of those girls and hundreds more, were the birth of this entire project. To the Girls Garage family: You are the reason I wrote this book. Thank you.

This project would not have been possible without the support, expertise, and patience of my dearest friends, family, and colleagues:

Kate Bingaman-Burt: You are more than a brilliant illustrator; you are my creative co-conspirator, unwavering cheerleader, and a ray of light who made this book a constant adventure.

Christina Jenkins, Maya Vilaplana, Allison Oropallo, Augusta Sitney, Meghann Harris, Hallie Chen, Kristy Higares, HyeYoon Song, Sarah Rich, Stephanie Mechura, Veleta Allen: You are the women I've been waiting for. Thank you for making Girls Garage the place we all want to work all day every day. A special fist bump goes to Allison Oropallo, for being the ultimate builder colleague, steadfast friend, and answerer of late-night texts about the use of the word "countersunk."

Semar Prom and Amanda N. Simons: Thank you for your fine-tooth-combed reading of these building "recipes" and your brilliant suggestions for how to make projects more achievable for all! And Liisa Pine: You are a welding genius! Infinite thanks for helping me wordsmith concepts like metallurgical phase change into accessible sentences.

The Firebelly Design team led by the indomitable Dawn Hancock: You are creative ninjas. I will forever be indebted to you for giving us the tools to tell the Girls Garage story in words, pictures, textures, and a custom typeface with secret fillet welds.

Thank you to the fifteen women and girls profiled in this book. Every one of you said "YES" without hesitation. Your faces and your stories will open the universe for girls everywhere.

Mom, Dad, Molly, and Maggie: If it weren't for our crazy family, I'd never have the chutzpah to do what I do. Thank you for always being my unconditional foundation of love and support.

Laila and Neruda: Thank you for sharing a life and home with me and for being the young people who ground my perspective and my hope for the future.

Victor Díaz: Thank you for talking through every aspect of this book with me, from subtitle to sentence structure. And for bringing me approximately five hundred cups of coffee over the course of the editing process. You are the sky.

Junebug: You are the best dog in the world. Thank you for giving me your best "you got this" look from your doggie bed across the room when I most need it.

Ariel Richardson and Jennifer Tolo Pierce, and the whole Chronicle Books team: You are the magic makers who brought a twinkle of an idea to life on paper, in binding, on shelves. Thank you for your endless well of patience and for being true collaborators and fearless builders in your own right.

INDEX

C

D

T

ABOUT THE AUTHOR AND ILLUSTRATOR

EMILY PILLOTON is the Founder and Executive Director of the nonprofit Girls Garage. A designer, builder, and educator, she has taught thousands of young girls how to use power tools, weld, and build projects for their communities. She has presented her work and ideas on the TED stage, *The Colbert Report*, and in the documentary film *If You Build It*. She is currently a lecturer in the College of Environmental Design at the University of California, Berkeley, and lives in the San Francisco Bay Area. Learn more at www.girlsgarage.org and @_GirlsGarage.

KATE BINGAMAN-BURT is an illustrator and hand letterer. As a critical component of her creative practice, Kate is a full-time educator and makes illustrations for all sorts of clients all around the world. She is an Associate Professor of Graphic Design at Portland State University. Kate owns Outlet, which hosts workshops, pop-up events, and a Risograph print studio. She also sits on the board of Design Week Portland and is a founding member of the organization.

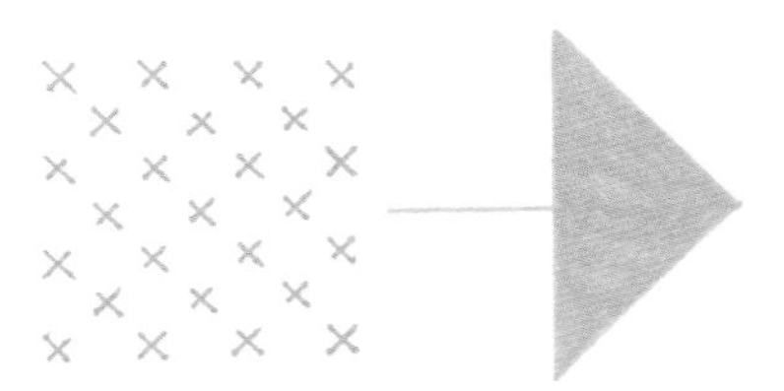